418, Time Series

OXFORD PAPERBACK REFERENCE

A Dictionary of

Statistics

18 auto correlation

Graham Upton is Professor of Statistics in the
Department of Mathematical Sciences at the
University of Essex. He is a graduate of Leicester and
Birmingham Universities and lectured at the
University of Newcastle upon Tyne for five years before
moving to Essex. He has given lecture courses at the
Universities of Dokkyo (Tokyo), Grenoble, Michigan,
and ITAM (Mexico). Dr Upton is the author of over 80
papers and five books, including *Understanding Statistics*
with Ian Cook (Oxford, 1996), and *Introducing Statistics*
(Oxford, 1998), also with Ian Cook.

Ian Cook is a graduate of Cambridge and London
Universities, and also of Hull, where he was a lecturer
in mathematics. He moved to the University of Essex as
a Senior Lecturer in 1964, taking early retirement in
1985. Dr Cook has been a Chief Examiner in
Mathematics at A Level for 30 years, setting questions
in all branches of mathematics and statistics. He is co-
author with Graham Upton of *Understanding Statistics*
and *Introducing Statistics*.

Oxford Paperback Reference

The most authoritative and up-to-date reference books for both students and the general reader.

ABC of Music
Accounting
Allusions
Archaeology
Architecture
Art and Artists
Art Terms
Astronomy
Better Wordpower
Bible
Biology
British History
British Place-Names
Buddhism*
Business
Card Games
Catchphrases
Celtic Mythology
Chemistry
Christian Art
Christian Church
Classical Literature
Classical Myth and Religion*
Classical World*
Computing
Contemporary World History
Countries of the World*
Dance
Dates
Dynasties of the World
Earth Sciences
Ecology
Economics
Encyclopedia*
Engineering*
English Etymology
English Folklore
English Grammar
English Literature
Euphemisms
Everyday Grammar
Finance and Banking
First Names
Food and Drink
Food and Nutrition
Foreign Words and Phrases
Geography
Humorous Quotations*
Idioms
Internet
Islam
Kings and Queens of Britain
Language Toolkit

Law
Linguistics
Literary Quotations
Literary Terms
Local and Family History
London Place-Names
Mathematics
Medical
Medicinal Drugs
Modern Design*
Modern Quotations*
Modern Slang
Music
Musical Terms*
Musical Works*
Nicknames*
Nursing
Ologies and Isms
Philosophy
Phrase and Fable
Physics
Plant Sciences
Pocket Fowler's Modern
    English Usage
Political Quotations
Politics
Popes
Proverbs
Psychology
Quotations
Quotations by Subject
Reverse Dictionary
Rhyming Slang
Saints
Science
Shakespeare
Slang
Sociology
Statistics
Synonyms and Antonyms
Twentieth-Century Art
Weather
Weights, Measures, and Units
Word Games*
Word Histories*
World History
World Mythology
World Place-Names*
World Religions
Zoology

*forthcoming

A Dictionary of
Statistics

GRAHAM UPTON AND IAN COOK

OXFORD
UNIVERSITY PRESS

OXFORD
UNIVERSITY PRESS

Great Clarendon Street, Oxford OX2 6DP

Oxford University Press is a department of the University of Oxford.
It furthers the University's objective of excellence in research, scholarship,
and education by publishing worldwide in

Oxford New York

Auckland Bangkok Buenos Aires Cape Town Chennai
Dar es Salaam Delhi Hong Kong Istanbul Karachi Kolkata
Kuala Lumpur Madrid Melbourne Mexico City Mumbai Nairobi
São Paulo Shanghai Singapore Taipei Tokyo Toronto

Oxford is a registered trade mark of Oxford University Press
in the UK and in certain other countries

Published in the United States
by Oxford University Press Inc., New York

British Library Cataloguing in Publication Data
Data available

Library of Congress Cataloging in Publication Data
Data available

ISBN 019-860950-7

1

Typeset by Kolam Information Services Pvt. Ltd, Pondicherry, India
Printed in Great Britain by Clays Ltd., St Ives plc

Contents

Preface

This dictionary has two aims: (1) to provide a satisfactory amount of accurate information about subjects of interest to the user, and (2) to induce the user to read about topics other than those of immediate concern. The achievement of the second aim has (we hope) been effected by a deliberate breadth in the dictionary's scope. As well as entries on statistical topics, there are entries on related topics in mathematics, operational research, and probability.

In deciding on the topics for inclusion, we have had to think of the probable users. Many of the people using this dictionary as an *aide-mémoire* in Statistics will be those who are meeting the subject for the first time, as students at school or university. Another large group of readers will be specialists in other subjects who have found the need to analyse their own data and have then encountered the gobbledegook associated with computer packages. Our selection of topics has been made with all of these people in mind.

Dictionaries vary widely in style. This became very apparent to us once we started on this project. We have taken the view that if a reader needs to look up an 'elementary' topic then that reader may well need a rather long explanation, possibly with an example. Conversely, a reader looking up an 'advanced' topic will be a reader who already has a deep statistical knowledge and needs rather less help.

We have included approximately 150 short biographies. The criterion for inclusion has been that the individual concerned has made an important contribution to the development of the subject of Statistics, or that the individual's name forms part of the title of a topic (or both). We have not restricted ourselves to the dead, and would be pleased to hear from any (of the living) who feel that our entry is a misrepresentation of their career. There will be surprise omissions as well as surprise inclusions, and we will be pleased to receive nominations for future inclusion. Certainly, no omission should be regarded as a comment on the achievements of that individual.

The dictionary concludes with a glossary, tables, a brief overview of the history of statistics and suggestions for further reading. We hope these will prove useful.

We have tried to be consistent in our presentation and accurate in what we present. However, we have feet of clay and we therefore welcome any correspondence that may improve future editions.

GRAHAM UPTON

IAN COOK

COLCHESTER
DECEMBER 2001

Abbe, Ernst Carl (1840–1905; b. Eisenach, Germany; d. Jena, Germany) A German mathematician and physicist. His father was a book printer and factory worker and his childhood was one of privation. Abbe studied at the Universities of Jena and Göttingen, receiving his PhD in 1861. In 1863 he was appointed to a lectureship at Jena on the basis of a dissertation that, in effect, derived the *chi-squared distribution. Following an approach from Carl Zeiss, most of his subsequent work was concerned with optics and astronomy. A lunar crater is named after him, also a minor planet, and several schools in Germany.

abscissa *See* CARTESIAN COORDINATES.

absolute difference The *absolute value of the difference between two numbers. *See also* MEAN DEVIATION.

absolute value (modulus) The value of a number disregarding its sign. Denoted by a pair of '|' signs: thus the modulus of −2.5 is $|-2.5| = 2.5$.

absorbing barrier (absorbing state) *See* MARKOV PROCESS.

acceptable quality level *See* ACCEPTANCE SAMPLING.

acceptance region The set of values of the *statistic, in a *hypothesis test, which lead to acceptance of the *null hypothesis.

acceptance–rejection algorithm A method for generating values of a *continuous random variable for use in a *simulation. Suppose that the random variable X, which takes values in the interval (a, b), has *probability density function f. Denote the maximum value of $f(x)$ by M. Let u and v be two random numbers *uniformly distributed in the interval (0, 1). Write $r = a + (b − a)u$ and $s = Mv$, so that r and s are uniformly distributed on (a, b) and $(0, M)$, respectively. Calculate $f(r)$. If $f(r) > s$ then r is accepted as a value of X. Otherwise, it is rejected and a new pair of values is taken for u and v. *See following diagram.*

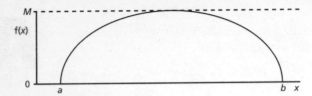

Acceptance–rejection algorithm. Uniform random numbers are generated in the intervals(a, b) and ($0, M$). If the point generated lies between the graph of f(x) and the x-axis, then the value of X is accepted.

acceptance sampling A method of *quality control. A random *sample is taken from a *batch of output and the decision to accept or reject the batch is based on either the number of *defectives in the sample **(inspection by attributes)** or on some summary *statistic such as the sample mean **(inspection by variables)**.

In the case of inspection by attributes, the *probability of accepting a batch is a function of the proportion of defectives in the batch. The maximum proportion of defectives that is regarded as desirable by the consumer is called the **acceptable quality level (AQL)**. The graph relating these quantities is called the **operating characteristic curve (OC-curve)**.

Any acceptance sampling scheme that does not sample 100% of a batch will lead to the occasional rejection of batches with very low proportions of defectives (the **producer's risk**), and to the occasional acceptance of batches with very high proportions of defectives (the **consumer's risk**).

Suppose that the proportion of defectives remains constant from batch to batch. Eventually a batch will be rejected. The **average run length (ARL)** is the average number of batches inspected up to and including the one that is rejected.

accessible *See* MARKOV PROCESS.

ACF Abbreviation for the *autocorrelation function.

action line *See* QUALITY CONTROL.

addition law for probabilities **(total probability law)** Law stating that if two events (*see* SAMPLE SPACE) A and B are *mutually exclusive then

$$P(A \cup B) = P(A) + P(B).$$

For example, the *probability that, when a normal six-sided die is rolled, it shows a multiple of 3 is

$$\frac{1}{3} = \frac{1}{6} + \frac{1}{6} = P(\text{shows } 3) + P(\text{shows } 6).$$

The generalization to n mutually exclusive events is the **law of total probability**:

$$P(A_1 \cup A_2 \cup \cdots \cup A_n) = P(A_1) + P(A_2) + \cdots + P(A_n).$$

See also UNION.

additive model A *model in which the combined effect of the explanatory variables (and their *interaction) is equal to the sum of their separate effects.

adjusted R^2 *See* ANOVA.

admissible decision *See* DECISION THEORY.

age-specific rate *See* MORTALITY RATE; INCIDENCE RATE; BIRTH RATE.

agglomerative clustering methods Methods for grouping *multivariate data into *clusters. Suppose there are n data items. The agglomerative clustering methods start by regarding these as n separate clusters of size 1. The two clusters judged closest together (on some criterion) are then merged to reduce the number of clusters to $(n - 1)$. This procedure continues until all the items are collected into a single cluster.

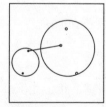 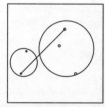 

Single linkage distance   Complete linkage distance   Group distances to be averaged

Agglomerative clustering methods. Examples of the distance definitions used in clustering.

The three simplest criteria are as follows. In **single linkage clustering** the distance between two clusters is defined as the least distance between an item in one cluster and an item in the other cluster. In **complete linkage clustering**, by contrast, the distance between two clusters is defined as the greatest distance between an item in one cluster and an item in the other cluster. As a compromise, **group-average clustering** uses the average of the distances between every member of one cluster and every member of the other cluster. The process of agglomeration is often represented using a *dendrogram. *See also* DISTANCE MEASURES; WARD'S METHOD.

AH *See* ALTERNATIVE HYPOTHESIS.

AI *See* ARTIFICIAL INTELLIGENCE.

AIC (Akaike's information criterion) Criterion, introduced by *Akaike in 1969, for choosing between competing statistical *models. For *categorical data this amounts to choosing the model that minimizes $G^2 - 2v$, where G^2 is the *likelihood-ratio goodness-of-fit statistic and v is the number of *degrees of freedom associated with the model. An alternative, that usually results in the selection of a simpler model, is the **Bayesian information criterion (BIC)**, for which the quantity minimized is $G^2 - v \ln n$, where ln is the *natural logarithm and n is the *sample size. The latter criterion is also called the *Schwarz criterion*. *See also* MALLOWS'S C_p; STEPWISE PROCEDURES.

AID Abbreviation for automatic interaction detector.

Ajne's test A test of the uniformity of *cyclic data. For this test, suggested by Bjorn Ajne in 1968, the *alternative hypothesis is that the data have come from a *unimodal distribution. For Ajne's test, the test statistic, M, is the number of observations contained within the semicircle for which this number is least. With n *observations,

$$P(M \leq k) = (n - 2k)\binom{n}{k} 2^{-(n-1)}, \qquad k < \frac{1}{2}n.$$

Akaike, Hirotugu (1927–) A Japanese statistician and mathematician. On graduating from the University of Tokyo in 1952, Akaike joined the staff of the *Institute of Statistical Mathematics. When he retired from the Institute, in 1994, he was its Director General.

Akaike's information criterion *See* AIC.

algorithm A completely defined set of operations that will produce a desired outcome. Examples are a sequence of instructions in a computer program, and a mathematical formula. *Compare* HEURISTIC.

alias *See* FACTORIAL EXPERIMENT.

allocation problem *See* ASSIGNMENT PROBLEM.

allometry The study of the interdependence of size and shape in living organisms.

Almon model *See* DISTRIBUTED LAGS MODEL.

alpha (α) The *probability, in a *hypothesis test, of rejecting the *null hypothesis when it is, in fact, true. Usually called the significance level.

alternating renewal process A *renewal process in which the process alternates between states A and B. Let the average length of a period in state A be μ_A and the average length of a period in state B be μ_B. The long-run proportion of the time that the system is in state A is

$$\frac{\mu_A}{\mu_A + \mu_B}.$$

An alternating renewal process is a special case of a *semi-Markov process. When the states are 'working' and 'under repair', the *probability that the system is working at time t is called the **availability**.

alternative hypothesis The hypothesis, in a *hypothesis test, that will be accepted if the *null hypothesis is rejected. The term was introduced by *Neyman and Egon *Pearson in 1933.

American Statistical Association (ASA) A scientific and educational society founded in 1839 in Boston, Massachusetts. It is the second oldest professional society in the United States. Its current headquarters are in Alexandria, Virginia. The Association has nearly 18 000 members. It publishes nine journals, including the *Journal of the American Statistical Association* and the *American Statistician*. Since 1965 the *Wilks Medal has been presented annually for distinguished contributions to Statistics.

American Statistician A journal published by the *American

Statistical Association, concentrating on statistical methodology. It was first published in 1947.

analysis of covariance *See* ANOCOVA.

analysis of variance *See* ANOVA.

ancillary statistic In the context of *estimation of an unknown *parameter, a statistic whose value provides information incidental to the estimation process. The most usual case is where the *sample size is not fixed. For example, we might wish to know the proportion of a sweet pea mixture that has red flowers. We plant N seeds, but our estimate will be based on the ancillary statistic n, the number of seeds that germinate and produce flowers. The term was introduced by Sir Ronald *Fisher in 1925.

ANCOVA *See* ANOCOVA.

Anderson, Theodore Wilbur (1918–) An American mathematical statistician. Anderson was a graduate of Northwestern University (1939) and Princeton University (PhD in 1945). In 1946 he joined the staff at Columbia University, moving in 1967 to Stanford University, where he is now Professor Emeritus. He was Editor of the *Annals of Mathematical Statistics* during 1950–2, and President of the *Institute of Mathematical Statistics in 1962. He was the *American Statistical Association's *Wilks Medal winner in 1988.

Anderson–Darling test A general test, published in 1952, that compares the fit of the observed *cumulative distribution function with that expected. It was derived by *Anderson and David A. Darling as a modification of the *Cramér-von Mises test. The test statistic A^2 is given by

$$A^2 = -\frac{1}{n}\sum_{j=1}^{n}(2j-1)[\ln\{F(x_{(j)})\} + \ln\{1 - F(x_{(j)})\}] - n,$$

where F is the hypothesized cumulative distribution function, n is the *sample size, and $x_{(j)}$ is the jth ordered *observation ($x_{(1)} \leq x_{(2)} \leq \cdots \leq x_{(n)}$). The statistic can also be used to test for *normal and *exponential distributions with unknown *parameters estimated by their sample equivalents. In some cases, as shown in the following table, an adjusted test statistic is required.

	TEST STATISTIC	UPPER TAIL PROBABILITY			
		0.10	0.05	0.025	0.01
Specified distribution	A^2	1.933	2.492	3.070	3.857
Normal, estimated mean ($n > 20$)	A^2	0.894	1.087	1.285	1.551
Normal, estimated variance ($n > 20$)	A^2	1.743	2.308	2.898	3.702
Normal, estimated mean and variance	$A^2(1 + \frac{3}{4n} + \frac{9}{4n^2})$	0.631	0.752	0.873	1.035
Exponential, estimated mean	$A^2(1 + \frac{3}{10n})$	1.062	1.321	1.591	1.959

Andrews plot A plot suggested in 1972 by David Andrews as an alternative method to *Chernoff faces for representing *multivariate data in two dimensions. The plot can help to identify *outliers and to establish similarities within groups of data items. For an item with values given by the vector ($a\ b\ c\ d\ e\ \dots$) the function $x(t)$ is defined as

$$x(t) = \frac{a}{\sqrt{2}} + b\sin(t) + c\cos(t) + d\sin(2t) + e\cos(2t) + \cdots.$$

The resulting graph is drawn for ($-\pi \le t \le \pi$), for each data item. The form of the graph is dependent upon the ordering of the characteristics $a,\ b, \dots$; the usual advice is to order the characteristics in declining order of their (supposed) importance. *See following diagram.*

angular histogram *See* CIRCULAR HISTOGRAM.

angular uniform distribution *See* CIRCULAR UNIFORM DISTRIBUTION.

anisotropy *See* ISOTROPY.

Annals of Applied Probability A publication of the *Institute of Mathematical Statistics.

Annals of Mathematical Statistics The single journal of the *Institute of Mathematical Statistics from 1930 to 1973, before its subdivision into three parts.

Annals of Probability A publication of the *Institute of Mathematical Statistics.

Annals of Statistics A publication of the *Institute of Mathematical Statistics.

Annals of the Institute of Statistical Mathematics An English language publication of this Japanese institute.

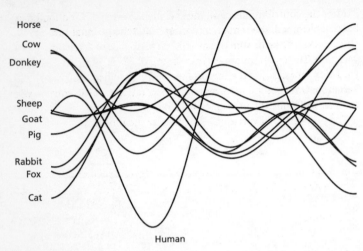

Horse
Cow
Donkey

Sheep
Goat
Pig

Rabbit
Fox

Cat

Human

Andrews plot. This plot compares humans with nine familiar animal species, using five characteristics (body weight, brain weight, hours of sleep, lifespan, and gestation). It appears that humankind is a race apart and that it is difficult to tell the sheep from the goats.

ANOCOVA (ANCOVA) (analysis of covariance) *ANOVA with a mixture of *continuous random variables and *qualitative variables. ANOCOVA *models can also be thought of as *multiple regression with some *dummy variables.

ANOVA (analysis of variance) The attribution of variation in a *variable to variations in one or more explanatory variables. The term was introduced by Sir Ronald *Fisher in 1918.

A measure of the total variability in a set of *data is given by the sum of squared differences of the *observations from their overall *mean. This is the **total sum of squares** (*TSS*). It is often possible to subdivide this quantity into components that are identified with different causes of variation. The full subdivision is usually set out in an **analysis of variance** table, as suggested by Sir Ronald *Fisher in his 1925 book *Statistical Methods for Research Workers*. Each row of the table is concerned with one or more of the components of the observed variation. The entries on a row usually include the **sum of squares** (*SS*), the corresponding number of *degrees of freedom (*v*), and their ratio, the **mean square** (= *SS*/*v*).

After the contributions of all the specified sources of variation have been determined, the remainder, often called the **residual sum of squares** (*RSS*) or **error sum of squares**, is attributed to *random variation. The mean square corresponding to *RSS* is often used as the yardstick for assessing the importance of the specified sources of variation. One method involves comparing ratios of mean squares with the *critical values of an *F-distribution.

The proportion of variation explained by the model is

$$R^2 = 1 - \frac{RSS}{TSS},$$

which is sometimes called the **squared multiple correlation**.

In an ANOVA analysis each explanatory variable takes one of a small number of values. If, instead, some explanatory variables are *continuous in nature, then the resulting models are called *ANOCOVA models. ANOVA can also be thought of as *multiple regression using only *dummy variables.

As an example, suppose that four varieties of tomatoes are grown in three grow-bags giving the yields (in g) shown below. The explanatory variables are the grow-bags and the varieties.

T_1	T_3	T_4	T_2
1890	1740	1620	1970

Bag 1

T_2	T_1	T_3	T_4
1850	1760	1800	1890

Bag 2

T_3	T_1	T_2	T_4
1810	1910	2100	1710

Bag 3

The following ANOVA table results.

SOURCE OF VARIATION	DEGREES OF FREEDOM	SUM OF SQUARES	MEAN SQUARE
Differences between grow-bags	2	12 950	6 475
Differences between varieties	3	93 425	31 142
Residual	6	72 250	12 042
Total	11	178 625	

Since the mean square for varieties is much greater than that for grow-bags we can conclude that differences between varieties are more important. However, the residual sum of squares amounts to nearly half the total sum of squares, indicating that there are major unexplained sources of variation.

Often, particular comparisons between the treatments are of interest. These are referred to as **contrasts**. The set of contrasts consisting of, say, (i) a comparison of treatment 1 with the average effect of treatments 2 to t, (ii) a comparison of treatment 2 with the average effect of treatments 3 to t, etc. are called **Helmert contrasts**. Both these contrasts and those corresponding to *orthogonal polynomials lead to an orthogonal *variance–covariance matrix.

Anscombe, Francis John (1918–2001; b. Hove, England; d. New Haven, Connecticut) An English statistician. Graduating from Cambridge University in 1939, Anscombe worked during the Second World War for the Ministry of Supply. His tasks included a mathematical solution for firing rockets during D-Day. In 1948 he joined the staff at Cambridge University, where, on behalf of the Fitzwilliam Museum he purchased a work by the then unknown painter Francis Bacon. When the Museum decided it was too modern, Anscombe retained it and subsequently used it to pay for his four children's education. Anscombe was keen that his PhD students should be involved with real problems. Both *Deming and *Tukey were fond of quoting Anscombe's maxim that it is better to 'realize what the problem really is, and solve that problem as well as we can, instead of inventing a substitute problem that can be solved exactly, but is irrelevant'. In 1956 Anscombe was recruited to a chair at Princeton University by Tukey, who 'wanted someone to talk to, not at'. In 1963 he moved to Yale University, where he founded its Department of Statistics, becoming Professor Emeritus on retirement.

Anscombe residuals Alternatives to the usual residuals (*see* REGRESSION DIAGNOSTICS) which *Anscombe proposed in 1953 for cases where the *random errors in a general *regression model do not have a *normal distribution. The idea is to produce quantities that do have near-normal distributions, and the form for the residual depends upon the error distribution assumed.

In the case of a *Poisson distribution, the Anscombe residual is given by

$$r = \frac{3\left(y^{\frac{2}{3}} - \hat{y}^{\frac{2}{3}}\right)}{2\hat{y}^{\frac{1}{6}}},$$

where y and $\hat{y}$ are, respectively, the *observed and *fitted values. In the case of a *gamma distribution the formula becomes

$$r = \frac{3\left(y^{\frac{1}{3}} - \hat{y}^{\frac{1}{3}}\right)}{\hat{y}^{\frac{1}{3}}},$$

and in the case of an *inverse normal distribution the formula is

$$r = \frac{\ln y - \ln \hat{y}}{\sqrt{\hat{y}}}.$$

Anscombe's regression data Artificial *data created by *Anscombe to illustrate the necessity for studying residuals. Each of the four data sets has the same fitted *regression line, $y = 3 + 0.5x$, and the same summary *ANOVA table (with the same regression sum of squares, total sum of squares, and value for R^2).

x	y	x	y	x	y	x	y
10	8.04	10	9.14	10	7.46	8	6.58
8	6.95	8	8.14	8	6.77	8	5.76
13	7.58	13	8.74	13	12.74	8	7.71
9	8.81	9	8.77	9	7.11	8	8.84
11	8.33	11	9.26	11	7.81	8	8.47
14	9.96	14	8.10	14	8.84	8	7.04
6	7.24	6	6.13	6	6.08	8	5.25
4	4.26	4	3.10	4	5.39	19	12.50
12	10.84	12	9.13	12	8.15	8	5.56
7	4.82	7	7.26	7	6.42	8	7.91
5	5.68	5	4.74	5	5.73	8	6.89

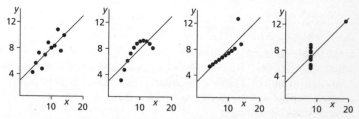

Anscombe's regression data. Each data set has the same mean and variance for x, the same mean and variance for y, the same fitted regression line, and the same residual sum of squares—Anscombe created these data sets to emphasize the need for the statistician to look carefully at data.

ant colony optimization An *optimization procedure that seeks to mimic an ant's apparent ability to find the shortest distance between two points. The ant's choices are based on the quantities of pheromones left by previous ants. These build up faster on shorter routes. The computer version similarly leaves markers behind to guide subsequent choices.

antecedent variable An explanatory variable, referring to an earlier time point, used as part of an explanation of the variation in a **consequent** (dependent) **variable**. *See* REGRESSION.

antedependence The dependence of the value of a quantity at time t on its value at one or more previous times. *See also* AUTOCORRELATION.

antilogarithm *See* LOGARITHM.

antithesis *See* ANTITHETIC VARIABLE.

antithetic variable A variable used in a *variance reduction technique for *simulation. It refers to the use of simulated values that are not *independent but are negatively correlated.

Suppose, for example, that we wish to estimate the expectation of the *random variable X, and that we have a simulation procedure using the function g of the *pseudo-random numbers $u_1, u_2, \ldots, u_n$ ($0 \le u_j \le 1$, for all j):

$$x = g(u_1, \ u_2, \ldots, u_n).$$

The antithetic variable procedure makes use of the **antitheses** of the random numbers, namely $(1 - u_1), \ (1 - u_2), \ldots, (1 - u_n)$, to form x' given by

$$x' = g(1 - u_1, \ 1 - u_2, \ldots, 1 - u_n).$$

It can be shown that the *variance of $\frac{1}{2}(x + x')$ is never greater than half the variance of x. There is therefore a gain in precision (since the mean of the two x-values is equal to the expectation of X) as well as an increase in efficiency, since each pseudo-random number is used twice.

A-optimal *See* OPTIMAL DESIGN.

Applied Statistics *See* JOURNAL OF THE ROYAL STATISTICAL SOCIETY.

AQL *See* ACCEPTANCE SAMPLING.

Aranda-Ordaz, Francisco Javier (1951–91; b. México City; d. México City) A Mexican statistician. Aranda-Ordaz obtained his BS and MS degrees at Universidad Nacional Autónoma de México (UNAM) before going to Imperial College, London, where the supervisor of his PhD (awarded in 1981) was Sir David *Cox. He returned to Mexico, joining the staff at UNAM, where he was Professor of Statistics at the time of his early death.

Aranda-Ordaz transformations Transformations, for a *proportion, p, suggested in 1981 by *Aranda-Ordaz. The transformations have the form

$$y = \ln\left[\frac{1}{\alpha}\{(1-p)^{-\alpha} - 1\}\right].$$

The case $\alpha = 1$ corresponds to the *logistic transformation.

arbitrage A sure-win betting scheme. If there are h horses in a race, with the odds quoted for horse j being o_j to 1 against that horse winning, then an arbitrage is possible if $\delta > 0$, where

$$\delta = 1 - \sum_{j=1}^{h} \frac{1}{1 + o_j}.$$

In this case a win of N (ignoring betting costs) is guaranteed by backing each horse, with x_j, the bet on horse j, being given by

$$x_j = \frac{N}{\delta(1 + o_j)}.$$

Arbuthnot, John (1667–1735; b. Inverbervie, Scotland; d. London, England) Scottish mathematician and royal physician. In 1692 he published the first English work on *probability, his translation of *De Ratiociniis in Ludo Aleae* by the Dutchman, Christiaan *Huygens. After graduating in medicine at St Andrews University in 1696, Arbuthnot moved to London. In 1700 his *Essay on the Usefulness of Mathematical Learning* was published. In 1704 he was elected a Fellow of the Royal Society. In the following year he was appointed physician to Queen Anne. His 1710 paper on the imbalance between male and female births may be regarded as the first application of probability to *social statistics.

arc *See* NETWORK FLOW PROBLEMS.

arc-sine law *See* RANDOM WALK.

arc-sine transformation For a *random variable X, having a *binomial distribution with *parameters n and p, a transformation that

*stabilizes the variance and produces a variable with an approximate
*normal distribution whose variance is almost independent of p is

$$\sin^{-1}\left(\sqrt{\frac{X}{n}}\right).$$

An improvement suggested by *Anscombe in 1948 is

$$\sin^{-1}\left\{\sqrt{\left(X+\frac{3}{8}\right)\bigg/\left(n+\frac{3}{4}\right)}\right\}.$$

Both transformed variables have expectation and variance given
approximately by $\sin^{-1}(\sqrt{p})$ and $1/(4n)$, respectively. *See also* FREEMAN–
TUKEY TRANSFORMATION.

arc-sinh transformation *See* NEGATIVE BINOMIAL DISTRIBUTION.

ARE *See* ASYMPTOTIC RELATIVE EFFICIENCY.

areal coordinates Relative to a given triangle of reference A_1 A_2 A_3
the areal coordinates of a point P inside, or on the boundary of, the
triangle are $(x_1,\ x_2,\ x_3)$, where $x_1,\ x_2,\ x_3$ are the areas of triangles
$PA_2A_3, PA_3A_1, PA_1A_2$, respectively. In the usual case, where the triangle of
reference is equilateral, the coordinates are called *barycentric
coordinates. The idea can be generalized to a tetrahedron of reference,
using volumes, and to a *simplex of reference.

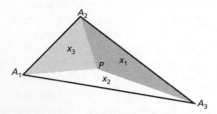

Areal coordinates. The location of a point is specified by the areas of the component
regions.

area sampling Sampling with primary sampling units (*see* CLUSTER
SAMPLING) that are non-overlapping regions of the earth's surface.
Examples of such regions are parishes, counties, fields, political
constituencies. Often the sampling procedure involves the use of aerial
photographs.

Arfwedson distribution A *distribution, presented in 1951 by the Swede, Gerhard Arfwedson, concerned with the case where each *observation takes one of k equally likely values. Let M be the *random variable denoting the number of values that do not occur in a *sample of size n. The distribution of M is given by

$$P(M = m) = \binom{k}{m} \sum_{j=0}^{k-m} (-1)^j \binom{k-m}{j} \left(\frac{k-m-j}{k}\right)^n, \quad m = 0, 1, \ldots, (k-1).$$

The distribution has also been called the **coupon-collecting distribution**, since one application is to find the probability that a person having n randomly selected coupons (which might be cigarette cards, plastic toys from cereal packets, etc.) will have at least one of each of the k equally likely varieties.

ARIMA models *Models for *time series which resemble *ARMA models except in that it is presumed the time series has a steady underlying *trend. The models therefore work with the differences between the successive observed values, instead of the values themselves. To retrieve the original data from the differences requires a form of integration and the models are therefore called **autoregressive integrated moving average** models. Models incorporating additional space-time variation are called **STARIMA models**.

arithmetic mean See MEAN (DATA).

ARL (average run length) See ACCEPTANCE SAMPLING.

ARMA models *Models for a *time series with no *trend (the constant *mean is taken as 0). They incorporate the terms in both an autoregressive model and a *moving average model. Models incorporating additional space-time variation are called **STARMA models**. For equivalent models for time series with a trend, see ARIMA MODELS.

AR models See AUTOREGRESSIVE MODELS.

artificial intelligence Research in which the aim is to construct an intelligent computing machine, including both hardware and software, that can tackle problems that usually require human intelligence. Such problems include *expert systems, the playing of games, and language translation. A major success is the chess-playing machine Deep Blue, although this machine does not emulate human analysis of chess, but uses vast computing power at each stage to consider millions of possible subsequent positions. In 1950, *Turing proposed a test for intelligence in a computing machine: an observer, posing questions via a keyboard and obtaining answers on a monitor, must be convinced that a human, rather than a machine, is responding to the questions.

artificial neural network (artificial neural net) A computing system built from a large number of simple processing elements dealing individually with parts of a large problem. The net may have several layers of processing elements and the processing elements are massively interconnected. Adaptively adjusted weights are applied to the inputs and to the connections between the processing elements. There are many applications, including, for example, speech and pattern recognition, oil and gas exploration, financial forecasting, and health care cost reduction.

ASA *See* AMERICAN STATISTICAL ASSOCIATION.

ASN (average sample number) *See* QUALITY CONTROL.

assignment problem (allocation problem) A *linear programming problem in which the aim is to allocate individuals to tasks in a manner that optimizes their overall effectiveness.

association Two *variables are associated if they are not *independent, i.e. if the value of one variable affects the value, or the *distribution of the values, of the other. Thus, for a human population, height and weight are associated, and so also are actual skin-colour and ethnicity. Association is usually measured in the case of numerical variables using a *correlation coefficient. The term was used by *Yule in 1900. *See also* ORDINAL VARIABLE.

asymmetric matrix *See* MATRIX.

asymptote If part of a graph is unbounded and there is a fixed straight line l such that the distance from a point P on the graph to l tends to 0 as $OP \rightarrow \infty$, where O is the origin, then l is an asymptote to the curve. Alternatively, it is the limiting position of the tangent to the graph at P, as $OP \rightarrow \infty$. For example, the asymptotes of the graph of

$$y = 2x + 3 - \frac{1}{x - 2}$$

are $y = 2x + 3$ and $x = 2$. *See following diagram.*

asymptotically efficient estimator An *estimator whose efficiency tends to unity as the *sample size increases.

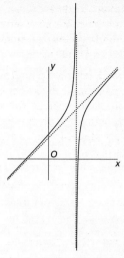

Asymptote. The dotted lines are the asymptotes (vertical, $x = 2$; oblique, $y = 2x + 3$) of the graph $y = 2x + 3 - \frac{1}{x-2}$.

asymptotically unbiased *See* ESTIMATOR.

asymptotic relative efficiency The limiting ratio of the *variances of two alternative *estimators of a *parameter as the *sample size increases.

asymptotic distribution The limiting *distribution of some *random variable as (usually) the *sample size is increased. For example, the asymptotic distribution of the sample *mean is a *normal distribution, because of the *central limit theorem.

attribute A basic property of an *experimental unit whose value cannot be changed by the experimenter. For example, the experimenter can vary the amount of fertilizer in an experiment, but cannot change the composition of the soil in a field. *See also* ACCEPTANCE SAMPLING.

attrition A problem that affects most *longitudinal studies. For example, in a *panel study, some of the original participants will not be participating at the end of the study—as a consequence of death, emigration, or failure to provide a forwarding address when moving house.

augmented design *See* ROTATABLE DESIGN.

Australia and New Zealand Journal of Statistics A journal, first published in 1998, formed by merging the *Australian Journal of Statistics* and the *New Zealand Statistician*. It is jointly published by the *Statistical Society of Australia, Inc. and the *New Zealand Statistical Association.

Australian Journal of Statistics The original journal published by the *Statistical Society of Australia, Inc. and superseded by the *Australia and New Zealand Journal of Statistics*.

autocorrelation A measure of the linear relationship between two separate instances of the same *random variable, as distinct from *correlation, which refers to the linear relationship between two distinct random variables. As with correlation the possible values lie between −1 and 1 inclusive, with unrelated instances having a theoretical autocorrelation of 0.

In the case of a *time series, autocorrelation measures the extent of the linear relation between values at time points that are a fixed interval (the **lag**) apart. Similarly, **spatial autocorrelation** quantifies the linear relationship between values at points in space that are a fixed distance apart (in any direction in the case of an *isotropic process). It is usually found that spatial autocorrelation is near 1 for points close together and decays to 0 as the distance increases—thus the daily rainfalls at the Lords and Oval cricket grounds in London will resemble each other closely, but will bear little or no resemblance to the rainfalls at the Kennington Oval in the West Indies.

For a *random variable X at time (or location) t, the *population **autocorrelation function (ACF)** for lag l, ρ_l, is given by

$$\rho_l = \frac{\text{Cov}(X_t, \ X_{t+l})}{\text{Var}(X_t)},$$

where the numerator is the **autocovariance function** for lag l and the denominator is the *variance of X_t (which, for a stationary process, is equal to that of X_{t+l}). At low lags autocorrelation is usually positive. It usually declines towards 0 as the lag increases. *See also* PARTIAL AUTOCORRELATION FUNCTION.

The **sample autocorrelation** for lag l, r_l, is given (for $l = 1, \ 2, \ldots, n - 1$) for the sequence of n values $x_1, \ x_2, \ldots, x_n$ (ordered in space or time) by

$$r_l = \frac{\sum_{t=1}^{n-l} (x_t - \bar{x})(x_{t+l} - \bar{x})}{\sum_{t=1}^{n} (x_t - \bar{x})^2},$$

where $\bar{x}$ is the sample mean.

A plot of the *variance of $\frac{1}{2}(x_t - x_{t+l})$ against l is called a **variogram** (or **semi-variogram**). The related plot of autocorrelation versus lag is called a **correlogram**, and a plot of the autocovariance against lag may be called a **covariogram**. *See also* PERIODOGRAM.

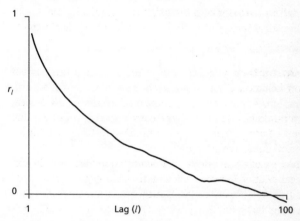

Autocorrelation. A correlogram showing a typical plot: the autocorrelation reduces as the lag increases.

autocovariance *See* AUTOCORRELATION.

autoregressive integrated moving average *See* ARIMA MODELS.

autoregressive models *Models for a *time series having no *trend (the constant *mean is taken as 0). Let $X_1, X_2, \ldots$ be successive instances of the *random variable X, measured at regular intervals of time. Let ε_j be the random variable denoting the *random error at time j. A pth-order **autoregressive model** (or **autoregressive process**) relates the value at time j to the preceding p values by

$$X_j = \alpha_1 X_{j-1} + \alpha_2 X_{j-2} + \cdots + \alpha_p X_{j-p} + \varepsilon_j,$$

where $\alpha_1, \alpha_2, \ldots, \alpha_p$ are constants. Such a model is written in brief as AR(p). The AR(1) process is a *Markov chain. Autoregressive models can also be expressed as *moving average models. Models combining both type of process include *ARMA models and *ARIMA models.

The **Yule–Walker equations**, introduced by *Yule in 1927 and Walker in 1931, relate $\alpha_1, \alpha_2, \ldots, \alpha_p$ to the population *autocorrelation values $\rho_1, \ \rho_2, \ldots, \rho_p$:

$$\rho_1 = \alpha_1 + \alpha_2 \rho_1 + \cdots + \alpha_p \rho_{p-1}$$
$$\rho_2 = \alpha_1 \rho_1 + \alpha_2 + \cdots + \alpha_p \rho_{p-2}$$
$$\vdots$$
$$\rho_p = \alpha_1 \rho_{p-1} + \alpha_2 \rho_{p-2} + \cdots + \alpha_p$$

availability *See* ALTERNATING RENEWAL PROCESS.

average For a set of data, a loosely used term for a *measure of location—either the *mode, or, in the case of numerical data, the *median or the *mean. Its meaning is often restricted to this last case. *See also* MOVING AVERAGE; WEIGHTED AVERAGE.

average run length (ARL) *See* ACCEPTANCE SAMPLING.

average sample number (ASN) *See* QUALITY CONTROL.

axes *See* CARTESIAN COORDINATES.

axial data *See* CIRCULAR DATA.

background variable An explanatory variable that can affect other (dependent) variables but cannot be affected by them. For example, one's schooling may affect one's subsequent career, but the reverse is unlikely to be true. *See* REGRESSION.

back-to-back stem and leaf diagram *See* STEM AND LEAF DIAGRAM.

backward elimination *See* STEPWISE PROCEDURES.

balanced design An *experimental design in which the same number of *observations is taken for each experimental condition.

balanced incomplete block design An *experimental design in which t *treatments are compared in b *blocks of size k, where $k < t$ and b and k are chosen so that bk is a multiple of t. The balanced nature of the design is reflected in the fact that each treatment appears the same number (r) of times, and every pair of treatments appear together in a block on the same number (λ) of occasions. To satisfy these requirements

$$bk = rt \qquad \text{and} \qquad \lambda(t - 1) = r(k - 1).$$

An example with four treatments (A–D) and blocks of size 3 is shown below.

B	A	C		D	A	B		D	A	C		C	B	D

Banach, Stefan (1892–1945; b. Kraków, Poland; d. Lvov, Ukraine) A Polish mathematician. He was given away by his parents and brought up by a laundress. At the age of fifteen he was earning a living by coaching in mathematics. In 1919 he was appointed as a lecturer at the University of Lvov, Ukraine, and was awarded his doctorate despite having no undergraduate qualification. He was imprisoned in Lvov during the Second World War and died shortly afterwards from his consequent ill-health.

Banach's matchbox problem A classic problem in *probability. A pipe smoker has a full box of n matches in each pocket. When he wants a match, he is equally likely to select either pocket. The question concerns the number of matches left in the second box on the first occasion that he chooses a box and finds it empty. The probability that there are exactly k matches left in the second box is

$$\binom{2n-k}{n}\left(\frac{1}{2}\right)^{2n-k}.$$

bandit problems Problems concerned with the determination of an optimal strategy. The bandit referred to is the 'one-armed bandit' otherwise known as a 'fruit machine'. For an actual machine in an amusement arcade the general advice would be not to play it, since it is the machine owner who will benefit in the long run. However, the term 'one-armed bandit' in statistics refers to the problem of deciding whether to 'play' when the expected pay-off may not be negative. Statisticians also consider k-armed bandits for which the question is 'Which of the k arms should be played?' One application of the resulting theory is to the medical problem of deciding which of a number of possible treatments should be given to a patient—here the pay-off is measured in terms of the patient's future health.

An optimal strategy is based on the **Gittins index**, which is defined as the maximum value, over all N, of the quantity

$$\frac{\sum_{t=1}^{N} \beta^{t-1} \mathrm{E}\{X(t)\}}{\sum_{t=1}^{N} \beta^{t-1}},$$

where $\mathrm{E}\{X(t)\}$ is the *expected value of the payout at the tth play of the bandit and β is the discount rate.

bandwidth *See* KERNEL METHODS.

bar chart A diagram (*see following*) for showing the *frequencies of a *variable that is *categorical or *discrete. The lengths of the bars are proportional to the frequencies. The widths of the bars should be equal. If the widths of the bars are negligible then the diagram may be called a **line graph**. Diagrams resembling bar charts were used in a theoretical context in the 14th century. *See also* MULTIPLE BAR CHART; COMPOUND BAR CHART.

Barnard, George Alfred (1915–2002; b. Walthamstow, England; d. Brightlingsea, England) An English logician, with a special interest in *statistical inference. Barnard graduated from St John's College, Cambridge in 1937. His interest in statistics developed after being recruited to the Ministry of Supply in 1942. In 1945 he joined the

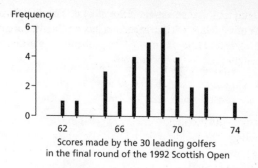

Scores made by the 30 leading golfers
in the final round of the 1992 Scottish Open

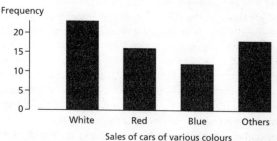

Sales of cars of various colours

Bar chart. The top diagram is a bar chart for a discrete variable (note that 0 need not appear); the bottom diagram illustrates data for a categorical variable. The choice of width for the bars is arbitrary, but the width should not be the same as that of the gap between successive bars.

---

faculty at Imperial College, London, moving in 1966 to the University of Essex as the founding Professor of Statistics. Barnard was Chairman of the *Institute of Statisticians in 1961, President of the Operational Research Society in 1963, President of the Institute of Mathematics and Its Applications in 1970, and President of the *Royal Statistical Society in 1972. He received the Royal Statistical Society's *Guy Medal in Silver in 1958 and in Gold in 1975. He was made an Honorary Fellow of the Society in 1993. Barnard was one of the first to advocate *Monte Carlo tests.

Bartlett, Maurice Stephenson (1910–2002; b. Scrooby, England; d. Exmouth, England) An English statistician. Bartlett graduated from Cambridge University in 1932. After four years at ICI, Bartlett joined the faculty at Cambridge University in 1938. He was appointed Professor of Mathematical Statistics at Manchester in 1947. In 1960 he moved to University College, London and in 1967 to Oxford University, where he was Professor of Biomathematics. He was elected a Fellow of the Royal

Society in 1961 and was President of the *Royal Statistical Society in 1966. He received the Royal Statistical Society's *Guy Medal in Silver in 1952 and in Gold in 1969.

Bartlett's identity A *matrix identity used in *multivariate analysis. If $\mathbf{M}$ is a non-singular $n \times n$ matrix, $\mathbf{v}$ is an $n \times 1$ vector, $\mathbf{v}'$ is the transpose of $\mathbf{v}$, and k is a constant such that $1 + k\mathbf{v}'\mathbf{M}^{-1}\mathbf{v} \neq 0$, then

$$(\mathbf{M} + k\mathbf{v}\mathbf{v}')^{-1} = \mathbf{M}^{-1} - \frac{k\mathbf{M}^{-1}\mathbf{v}\mathbf{v}'\mathbf{M}^{-1}}{1 + k\mathbf{v}'\mathbf{M}^{-1}\mathbf{v}}.$$

Bartlett's test for eigenvalues A *test used in *principal components analysis. The test is of the *null hypothesis that the smallest m eigenvalues of a $n \times n$ *variance–covariance matrix are equal to 0. Writing $k = n - m + 1$, the test statistic is

$$X^2 = v\left\{ m \ln\left(\frac{1}{m} \sum_{j=k}^{n} \lambda_j \right) - \sum_{j=k}^{n} \ln(\lambda_j) \right\},$$

where v is the number of *degrees of freedom associated with the variance–covariance matrix, and $\lambda_1, \lambda_2, \ldots, \lambda_{k+m}$ are the eigenvalues. Under the null hypothesis, X^2 has an approximate *chi-squared distribution with $\frac{1}{2}(m-1)(m-2)$ degrees of freedom.

Bartlett's test for homogeneity of variance A test, introduced by *Bartlett in 1937, for equality of *variance in k *populations having *normal distributions with unknown means. The *data consist of random *samples of sizes $n_1, n_2, \ldots, n_k$, with the total number of *observations being denoted by n and the *unbiased estimates of the population variances being $s_1^2, s_2^2, \ldots, s_k^2$. Denoting $(n_j - 1)$ by v_j, for all j, Bartlett's test statistic, L, is defined by

$$L = \frac{v \times \left\{ (s_1^2)^{v_1/v} \times (s_2^2)^{v_2/v} \times \cdots \times (s_k^2)^{v_k/v} \right\}}{v_1 s_1^2 + v_2 s_2^2 + \cdots + v_k s_k^2},$$

where $v = v_1 + v_2 + \cdots + v_k$. Unusually low values of L indicate unequal variances. The test is affected by departures from normality. *See also* COCHRAN'S C-TEST, LEVENE'S TEST.

barycentric coordinates Coordinates describing the position of a point in a triangle or *simplex. The position of any point P inside, or on the boundary of, a given triangle of reference $A_1 A_2 A_3$ can be specified by finding masses m_1, m_2, m_3 such that P is the centre of mass of particles of masses m_1, m_2, m_3 placed at A_1, A_2, A_3 respectively. In this case P is said to have barycentric coordinates (m_1, m_2, m_3). For an internal point of the triangle the masses are all positive. If $k > 0$ then (km_1, km_2, km_3) represents the same point, so barycentric coordinates are usually chosen so that $m_1 + m_2 + m_3 = 1$. With this convention, the barycentric coordinates of A_1 are $(1, 0, 0)$, and those of the midpoint of A_1A_2 are $(\frac{1}{2}, \frac{1}{2}, 0)$. Suppose a random trial can result in one of three possibilities with probabilities p_1, p_2, p_3, such that $p_1 + p_2 + p_3 = 1$, then allocation of these probabilities can be represented by a point inside, or on the boundary of, the triangle. The idea of barycentric coordinates can be generalized to the case of a tetrahedron in three dimensions or a simplex. *See* AREAL COORDINATES.

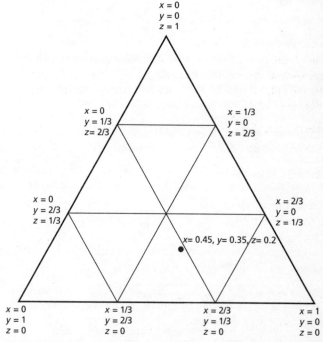

Barycentric coordinates. Illustrated for compositions of three classes having proportions x, y, and z. The case where the three classes are equally likely corresponds to the centroid of the equilateral triangle.

base-weighted index *See* INDEX.

BASIC (Beginner's All-purpose Symbolic Instruction Code) An elementary computer programming language.

basic feasible solution *See* LINEAR PROGRAMMING.

Basu, Debrabrata (1924–2001; b. Dhaka, Bangladesh; d. Calcutta, India) A Bangladeshi statistician whose career was divided between India and the United States. He was a faculty member at Florida State University and the *Indian Statistical Institute.

Basu theorems These theorems, due to *Basu, are concerned with the notions of sufficient statistics, *conditional independence, and *ancillary statistics.

batch A collection of items forming part of the output from some production process.

bathtub curve *See* HAZARD RATE.

Bayes, Reverend Thomas (1701–61; b. London, England; d. Tunbridge Wells, England) A Nonconformist minister in Tunbridge Wells, England. He was elected a Fellow of the Royal Society in 1742. The eponymous theorem has led to the development of an approach to *Statistics that runs parallel to the methods of *hypothesis testing. This approach is referred to as *Bayesian inference and its advocates are referred to as Bayesians. The theorem was contained in an essay not published until after Bayes's death and was largely ignored at the time.

Bayes factor (Bayes ratio) A measure of the evidence provided by the data, D, in favour of model M_1 as opposed to model M_2. The Bayes factor is B, given by

$$B = \frac{P(D|M_1)}{P(D|M_2)}.$$

The value of B may be assessed with the following table, derived from that suggested by *Jeffreys.

$2\ln(B)$	EVIDENCE IN FAVOUR OF M_1
< 0	Negative
0–2.2	Not worth more than a bare mention
2.2–6	Positive
6–10	Strong
> 10	Very strong

Bayesian A statistician who analyses data using the methods of *Bayesian inference. The term was used by Sir Ronald *Fisher in 1950.

Bayesian inference An approach concerned with the consequences of modifying our previous beliefs as a result of receiving new *data. By contrast with the 'classical' approach, which begins with a *hypothesis test that proposes a specific value for an unknown *parameter, θ, Bayesian inference proposes a **prior distribution**, $p(\theta)$, for this parameter. Data, $x_1, x_2, \ldots, x_n$, are collected and the *likelihood, $f(x_1, x_2, \ldots, x_n | \theta)$, is calculated. *Bayes's theorem is now used to calculate the **posterior distribution**, $g(\theta | x_1, x_2, \ldots, x_n)$. The change from the prior to the posterior distribution reflects the information provided by the data about the parameter value.

If nothing is known about the value of a parameter, then a **non-informative prior** is used—typically, this is a *rectangular distribution over the feasible set of values of the parameter. Another approach, the **empirical Bayes method**, utilizes the data to inform the prior distribution.

A useful choice for the form of a prior distribution is a member of a family of distributions which is such that the posterior distribution is another member of that family, so that the effect of the data can be interpreted in terms of changes in parameter values. Such a prior is called a **conjugate prior**.

Bayesian information criterion *See* AIC.

Bayes net; Bayes network *See* GRAPHICAL MODELS.

Bayes ratio *See* BAYES FACTOR.

Bayes's theorem A simple form of Bayes's theorem is

$$P(A|B) = \frac{P(A)P(B|A)}{P(A)P(B|A) + P(A')P(B|A')},$$

where A' denotes the complementary event (*see* SAMPLE SPACE) to the *event A. For example, suppose a man has two coins in his pocket. One is *unbiased, whereas the other is double headed. He takes one coin at random from his pocket and tosses it. Given that the coin falls heads, the *probability that it is the double-headed coin is

$$\frac{\frac{1}{2} \times 1}{\left(\frac{1}{2} \times 1\right) + \left(\frac{1}{2} \times \frac{1}{2}\right)} = \frac{2}{3}.$$

The general form is

$$P(A_j|B) = \frac{P(A_j)P(B|A_j)}{\sum_{k=1}^{n} P(A_k)P(B|A_k)},$$

where the events $A_1, A_2, \ldots, A_n$ are *mutually exclusive and *exhaustive.

Behrens–Fisher problem A problem concerned with the comparison of the *means of two *populations having *normal distributions with different *variances. The problem was first discussed by B. V. Behrens in 1929. Although Behrens's method of solution was unclear, his conclusions were confirmed by Sir Ronald *Fisher in 1935.

The *null hypothesis is that the populations have the same mean, and the suggested solution is an application of *Welch's statistic. In this case, the test statistic t is given by

$$t = (\bar{x}_1 - \bar{x}_2) \Big/ \sqrt{\frac{s_1^2}{n_1} + \frac{s_2^2}{n_2}},$$

where $\bar{x}_1$ and $\bar{x}_2$ are the sample means, s_1^2 and s_2^2 are the sample variances (using the divisors $n_1 - 1$ and $n_2 - 1$), and n_1 and n_2 are the *sample sizes. If the populations do have the same mean then t is an observation from an approximate *t-distribution with v *degrees of freedom, where v is taken as the nearest integer to

$$(a + b)^2 \Big/ \left(\frac{a^2}{n_1 - 1} + \frac{b^2}{n_2 - 1}\right),$$

where $a = s_1^2/n_1$ and $b = s_2^2/n_2$. If $n_1 \leq n_2$, then the above formula ensures that $(n_1 - 1) \leq v \leq (n_1 + n_2 - 2)$.

If the populations can be assumed to have the same variance then the problem is simple and the standard *t-test is appropriate.

Belgian Statistical Society A society, founded in 1937, to provide a focus for statisticians in Belgium. The Society has more than two hundred members and publishes a regular newsletter.

bell-curve A curve that resembles the axial cross-section of a bell. One example is the graph of the *probability density of a *normal distribution. It is often assumed that every bell-curve must arise from a normal distribution, but this is not the case. For example, all the *t-distributions (including the *Cauchy distribution) also give rise to bell-curves. The term can also be used to describe a *histogram that approximates a bell-curve.

Bell–Doksum tests Tests corresponding to *Kruskal–Wallis tests, in which the *ranks used by those tests are replaced by corresponding *normal scores.

Benders's decomposition A method of simplifying the maximization (or minimization) of a semi-linear objective function of the form $\mathbf{c}'\mathbf{x} + g(\mathbf{y})$, subject to constraints such as $\mathbf{A}\mathbf{x} + h(\mathbf{y}) \leq \mathbf{b}$, where g and h are known functions, $\mathbf{A}$ is a known *matrix, $\mathbf{b}$ and $\mathbf{c}$ are known *vectors, and $\mathbf{c}'$ is the transpose of $\mathbf{c}$.

Berkson, Joseph B. (1899–1982; b. New York City; d. Rochester, Minnesota) An American statistician and doctor. He obtained his BS from City College (now City University) in 1920. He obtained his AM from Columbia University in 1922 and his MD from Johns Hopkins University in 1927. From 1932 to 1964 he worked at the Mayo Clinic (a centre for medical education and research) and the University of Minnesota. Berkson coined the word *'logit'. He was elected to the National Academy of Sciences in 1979.

Bernoulli, Daniel (1700–82; b. Gröningen, Netherlands; d. Basel, Switzerland) A Swiss mathematical physicist. Daniel Bernoulli is best known to statisticians for his solution of the *St Petersburg paradox posed by his cousin Nicolaus *Bernoulli. Daniel, a famous prodigy, was the son of Johann Bernoulli and the nephew of Jacob *Bernoulli, both being Professors of Mathematics. He gained a succession of degrees from the University of Basel whilst still a teenager: BA in philosophy and logic in 1715, MA in 1716, Doctor of Medicine in 1720. In 1724, whilst practising medicine in Venice, he applied mathematics to the design of an hour-glass for use at sea. This won him a prize and a post as Professor of Mathematics at St Petersburg, where he worked on a number of *probability problems. He returned to Basel in 1734, initially as Professor of Botany, then as Professor of Physiology and finally, in 1750, as Professor of Physics.

Bernoulli, Jacob (Jacques) (1654–1705; b. Basel, Switzerland; d. Basel, Switzerland) A Swiss mathematician. Jacob Bernoulli is best known to statisticians for his *Ars Conjectandi* (*The Art of Conjecture*), a treatise on *probability, published posthumously in 1713, in which he derived the form of the *binomial distribution. Jacob was the uncle of Nicolaus *Bernoulli. At the University of Basel he studied philosophy and theology according to his parents' wishes, whilst studying mathematics and astronomy for his own satisfaction. After graduation Jacob travelled around Europe studying with fellow mathematicians. In 1683 he returned to the University of Basel to teach mechanics. He studied and published in many areas of mathematics, including, in 1689, a statement of the *law of large numbers.

Bernoulli, Nicolaus (1687–1759; b. Basel, Switzerland; d. Basel, Switzerland) A Swiss mathematician. Nicolaus Bernoulli was a prolific correspondent and poser of the *St Petersburg Paradox, solved by his cousin Daniel *Bernoulli. He was a nephew of Jacob *Bernoulli, who supervised his Master's degree in mathematics at the University of Basel. He was awarded his PhD at Basel in 1709 for a study of the application of probability theory to legal problems. In 1716 he was appointed to Galileo's chair at Padua. In 1722 he left Italy and returned to the University of Basel, as Professor of Logic and later Professor of Law.

Bernoulli distribution The *distribution of a *discrete random variable taking two values, usually 0 and 1. An experiment or trial that has exactly two possible results, often classified as 'success' or 'failure', is called a **Bernoulli trial**. If the *probability of a success is p and the number of successes in a single experiment is the *random variable X, then X is a **Bernoulli variable** and is said to have a Bernoulli distribution with *parameter p. The *mean of the distribution is p and the *variance is $p(1 - p)$. The *probability function is given by

$$P(X = 1) = p, \qquad P(X = 0) = 1 - p.$$

A *binomial variable with parameters n and p is the number of successes in n *independent Bernoulli trials and may be regarded as the sum of n independent *observations of a Bernoulli variable with parameter p. The phrase 'Bernoullian trial' was used in a 1937 book on probability.

Bernoulli Society Society founded in 1973 as an autonomous section of the *International Statistical Institute. Its object is the advancement, through international contacts, of the sciences of *probability (including the theory of *stochastic processes) and mathematical statistics and of their applications.

Bernoulli trials; Bernoulli variable *See* BERNOULLI DISTRIBUTION.

Bernstein, Sergi Natanovich (1880–1968; b. Odessa, Ukraine; d. Moscow, Russia) A Russian mathematician and probabilist. He obtained a doctorate from the Sorbonne University in Paris in 1904 and another from Kharkov University in 1913. From 1908 to 1933 he taught at Kharkov and subsequently at the Mathematical Institute in Moscow.

Bernstein's inequality A stronger version of the *Chebyshev inequality. Let $X_1, X_2, \ldots, X_n$ be *independent *random variables, such that, for $j = 1, 2, \ldots, n$, X_j has expectation 0 and $|X_j| \leq M$. *Bernstein showed, in a 1926 paper, that, for all positive ε,

$$\mathrm{P}\left(\left| \sum_{j=1}^{n} X_j \right| > \varepsilon \right) \leq 2 \exp\left\{ -\frac{1}{2} \varepsilon^2 \middle/ \left(\sum_{j=1}^{n} \sigma_j^2 + M\varepsilon \right) \right\},$$

where $\mathrm{Var}(X_j) = \sigma_j^2$.

best linear unbiased estimator *See* BLUE.

best-prize selection An optimization problem concerning the selection of one of n prizes. The prizes are presented one at a time and n is known. We either accept the prize presented or we decline it and ask to see the next prize. We cannot accept a prize once we have declined it. It may seem that it will be difficult to do well, but in fact there is a good strategy, i.e. Reject the first n/e prizes (*see* EXPONENTIAL) and then accept either the first subsequent prize that is better than all the previous ones, or the last prize. With this strategy the probability that the prize selected is the best of all the n prizes is about $1/e \approx 0.368$.

beta (β) The *probability, in a *hypothesis test concerning the value of a *parameter, of accepting the *null hypothesis when the *alternative hypothesis is, in fact, true. Also referred to as the probability of a Type II error. The value of β depends upon the true parameter value. The probability of rejecting the null hypothesis when the alternative hypothesis is true is the power of the test. Its value is $1 - \beta$.

beta-binomial distribution (Polya distribution) A *compound distribution that results from allowing the success *probability in a sequence of Bernoulli trials to have a *beta distribution. If the *parameters of that distribution are α and β, then the probability of obtaining r successes in n trials is given by the beta-binomial distribution as

$$\binom{n}{r} \frac{B(\alpha + r, \beta + n - r)}{B(\alpha, \beta)}, \qquad r = 0, 1, \ldots, n$$

where B is the *beta function.

The distribution has *mean np and *variance

$$np(1 - p) \frac{(n + \alpha + \beta)}{(1 + \alpha + \beta)},$$

where

$$p = \frac{\alpha}{\alpha + \beta}.$$

beta distribution A distribution often used as a prior distribution for a *proportion. The *probability density function, for a *random variable X having a beta distribution is

$$f(x) = \frac{1}{B(\alpha, \beta)} x^{\alpha-1}(1 - x)^{\beta-1}, \qquad 0 < x < 1,$$

where α and β are positive *parameters, and B is the *beta function. The name appears in a 1911 publication by *Gini. The distribution has *mean

$$\frac{\alpha}{\alpha + \beta}$$

and *variance

$$\frac{\alpha\beta}{(\alpha + \beta)^2(\alpha + \beta + 1)}.$$

If both $\alpha > 1$ and $\beta > 1$, the distribution has *mode at

$$\frac{(\alpha - 1)}{(\alpha + \beta - 2)}.$$

If both $\alpha < 1$ and $\beta < 1$ then the distribution is U-shaped, whereas, if just one of α and β is < 1, then the distribution is J-shaped.

If $Y_1, Y_2, \ldots, Y_k$ are *independent random variables, with Y_j having a *chi-squared distribution with v_j *degrees of freedom, then the ratio

$$\sum_{j=1}^{k-1} Y_j \Big/ \sum_{j=1}^{k} Y_j$$

has a beta distribution with $\alpha = \frac{1}{2}\sum_{j=1}^{k-1} v_j$ and $\beta = \frac{1}{2}v_k$.

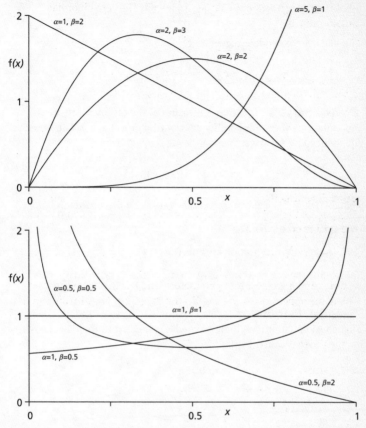

Beta distributions. The distribution has a variety of shapes, which depend on the values of α and β.

---

beta function The beta function B is given by

$$B(a, b) = \int_0^1 t^{a-1}(1 - t)^{b-1}\mathrm{d}t,$$

where $a > 0$, $b > 0$. The function is symmetric, i.e. $B(a, b) = B(b, a)$, for all a, b.

The beta function is related to the *gamma function Γ by

$$B(a, b) = \frac{\Gamma(a)\Gamma(b)}{\Gamma(a + b)}.$$

If a and b are positive integers then

$$B(a, b) = \frac{(a - 1)!(b - 1)!}{(a + b - 1)!}.$$

Bhattacharya's distance A measure of the distance between *populations with *probability density functions f and g. The measure, introduced by A. Bhattacharya in 1946, is given by

$$\cos^{-1}\left(\int_{-\infty}^{\infty} \{f(x)g(x)\}^{\frac{1}{2}} dx \right).$$

See also HELLINGER DISTANCE; KULLBACK–LEIBLER INFORMATION.

bias; biased estimator *See* ESTIMATOR.

BIC (Bayesian information criterion) *See* AIC.

bimodal Having two *modes or modal classes. The word appears in a 1903 natural history text.

bimodal distribution A *distribution having two *modes. A distribution with more than two modes is called multimodal; one with a single mode is called *unimodal.

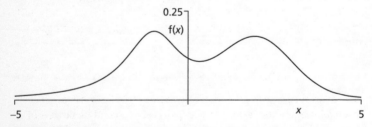

Bimodal distribution. The distribution illustrated is a mixture of two normal distributions.

binary Adjective describing a variable that can have only two possible values. The values are usually 0 and 1. The **binary system** is a system of counting using just these digits. In this system the decimal numbers 1, 2, 3, 4,..., 8,..., 16,..., 32,..., become 1, 10, 11, 100,..., 1000,..., 10 000,..., 100 000,....

binary system *See* BINARY.

binomial coefficient The coefficient of a power of x in the *binomial expansion. The general binomial coefficient is defined, for any value of n and for any positive integer r, by

$$\binom{n}{r} = \frac{n \times (n-1) \times (n-2) \times (n-3) \times \cdots \times (n-r+1)}{1 \times 2 \times 3 \times \cdots \times r}.$$

By convention, $\binom{n}{0} = 1$. Suppose now that n is a non-negative integer. In this case $\binom{n}{r}$ is an integer and $\binom{n}{n} = 1$. Furthermore,

$$\binom{n}{r} = 0 \qquad \text{for } r > n.$$

If $r \leq n$ then $\binom{n}{r}$ is also written as ^nC_r (*see* COMBINATION) and

$$\binom{n}{r} = \binom{n}{n-r} = \frac{n!}{r!(n-r)!}.$$

The phrase 'binomial coefficient' appears in English in a mathematics text of 1876.

binomial distribution The distribution associated with the *random variable, X, defined as the number of 'successes' in n *independent trials each having the same *probability, p, of success. The random variable X is said to be a **binomial variable** and to have a binomial distribution with *parameters n and p. This is written as $X \sim B(n, p)$. The *mean of this distribution is np and the *variance is $np(1 - p)$. The *probability function is given by

$$P(X = r) = \binom{n}{r} p^r (1 - p)^{n-r}, \qquad r = 0, 1, \ldots, n.$$

The distribution takes its name from the fact that successive probabilities are the terms in the expansion in ascending powers of p, by the binomial theorem, of $(q + p)^n$, where $q = 1 - p$. The first published derivation of the distribution was by Jacob *Bernoulli in 1713.

As an example, suppose that a computer generates fifteen random integers between 0 and 9 inclusive. The number of these integers that are odd has a B(15, 0.5) distribution. The number that are non-zero has a B(15, 0.9) distribution, and the number that are greater than 7 has a B(15, 0.2) distribution. The diagram shows the graphs of the probability functions for these distributions.

If we note that $P(X = 0) = q^n$, successive probabilities can be calculated using the recurrence relation

$$P(X = r) = \frac{n - r + 1}{r} \times \frac{p}{q} \times P(X = r - 1).$$

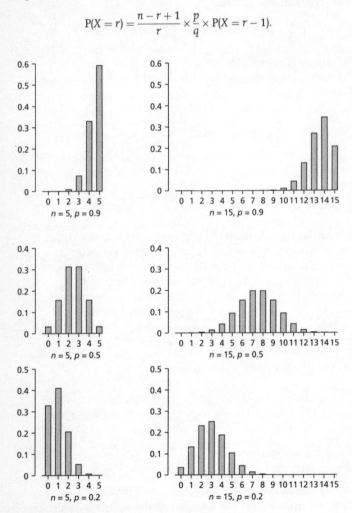

Binomial distribution. The distribution is skewed to the right if $p > 0.5$ and to the left if $p < 0.5$. It is symmetric if $p = 0.5$.

If $(n + 1)p$ is not an integer the graph is *unimodal, with *mode at the (integer) value of r such that

$$(n + 1)p - 1 < r < (n + 1)p.$$

If $(n + 1)p$ is an integer then

$$P[X = \{(n + 1)p - 1\}] = P[X = (n + 1)p],$$

and $(n + 1)p - 1$ and $(n + 1)p$ are both modal values (as in the B(15, 0.5) case illustrated).

A binomial random variable with parameters n and p may be regarded as the sum of n independent observations of a *Bernoulli variable with parameter p. The sum of two independent binomial variables with parameters n_1, p and n_2, p, respectively, is also a binomial variable with parameters $(n_1 + n_2), p$.

For large values of np and nq the **normal approximation to the binomial distribution** may be used:

$$P(X \le r) \approx \Phi(z), \qquad \text{where } z = \frac{r + \frac{1}{2} - np}{\sqrt{np(1 - p)}}$$

and Φ is the *cumulative distribution function for a *standard normal variable. The '$\frac{1}{2}$' is a *continuity correction. For large values of n and small values of p the **Poisson approximation to the binomial distribution** may be used:

$$P(X = r) \approx \frac{(np)^r e^{-np}}{r!}.$$

The word 'binomial' was used in its mathematical sense in a 1557 text entitled *The Whetstone of Witte* by Robert Recorde. The 'binomial distribution' was so named by *Yule in 1911. A tabulation of the distribution for small n is given in Appendix IV.

binomial expansion The application of the Maclaurin expansion to $(1 + x)^n$, for any value of n:

$$(1 + x)^n = 1 + nx + \frac{n(n - 1)}{2}x^2 + \cdots = \sum_{r=0}^{\infty} \binom{n}{r} x^r,$$

where $\binom{n}{r}$ is a *binomial coefficient. The series converges and the expansion is valid in the following cases:

(i) for any value of n, if $-1 < x < 1$,
(ii) for any value of x, if n is a non-negative integer, in which case the series is finite, since $\binom{n}{r} = 0$ for $r > n$, and the expansion is that given in the *binomial theorem.

binomial theorem The theorem that gives the expansion of the nth power, where n is a non-negative integer, of a binomial:

$$(a + b)^n = a^n + \binom{n}{1}a^{n-1}b + \cdots + \binom{n}{r}a^{n-r}b^r + \cdots + \binom{n}{n-1}ab^{n-1} + b^n,$$

where

$$\binom{n}{r} = \binom{n}{n-r}.$$

The *binomial coefficient $\binom{n}{r}$ is often written as ^nC_r (*see* COMBINATION). Setting $a = b = 1$ gives the relation

$$\binom{n}{0} + \binom{n}{1} + \cdots + \binom{n}{n} = 2^n.$$

The theorem appears in the 1742 *Treatise of Fluxions* by *Maclaurin.

binomial variable *See* BINOMIAL DISTRIBUTION.

bin-packing problem An *optimization problem. There is a supply of bins, all of the same size. These bins are to be filled with collections of different numbers of items (each of the same size). Each collection must go in a single bin. The problem is to minimize the number of bins required.

Biometrics The journal of the *International Biometric Society. It was first published under that name in 1947, with Gertrude *Cox as the first editor.

Biometrika The first *Statistics journal to specialize in *biometry. The first issue of *Biometrika* appeared in 1901, with Karl *Pearson as editor. Pearson remained as editor until 1936. Subsequent editors include Egon *Pearson (1936–65) and Sir David *Cox (1965–91).

biometry The measurement of quantities in the living world. The word is often used as a synonym for *biostatistics.

biostatistics *Statistics applied to the living world. It includes *demography, epidemiology, and clinical trials. Specialized measurement techniques include *capture–recapture methods and the analysis of *line transects.

biplot (Gabriel biplot) A diagram similar to a *scatter diagram that attempts to represent *observations having several coordinates on a diagram having (usually) just two coordinates. It was introduced in 1971 by Kuno Ruben Gabriel, currently Professor Emeritus at the University of Rochester, New York.

birth-and-death process A continuous-time *Markov process with states $\{0, 1, \ldots\}$ for which the only possible transitions from state n are to $n + 1$ (a birth) or to $n - 1$ (a death). If the *probability of a death is 0 then the process is a **pure birth process**.

birthday problem A well-known, but intriguing *probability problem. There are n people in a room. Assume that none is born on 29 February and that the remaining 365 days are all equally likely as birthdays. What is the smallest value of n for which the probability that at least two have the same birthday is greater than 0.5?

The answer is not 183, but 23. The *complementary event is that all n people have different birthdays. The probability of this, p_n, is

$$\frac{365}{365} \times \frac{364}{365} \times \cdots \times \frac{366 - n}{365},$$

which reduces surprisingly quickly as n increases.

n	3	5	9	13	16	19	22	23	26	30	34	40	46
p_n	0.99	0.97	0.91	0.81	0.72	0.62	0.52	0.49	0.40	0.29	0.20	0.11	0.05

birth rate The number of births occurring in a stated population during the stated period of time, usually a year, as a proportion of the number in the stated population. A total or crude birth rate utilizes all births, usually expressed as births per 1 000, whereas an **age-specific birth rate** includes births to one age group only, and is usually reported on the basis of the number of births per 100 000 persons in this age group. The birth rate may be standardized when comparing birth rates over time, or between countries, to take account of differences in the age structures of the populations.

biserial correlation A measure of the *association between a *binary variable, X, taking values 0 and 1, and a *continuous random variable, Y. If it is assumed that for each value of X the *distribution of Y is *normal, with different *means but the same *variance, then an appropriate measure is the **point biserial correlation coefficient**. This is estimated from a sample as r_{pb} $(-1 \le r_{pb} \le 1)$, given by

$$r_{pb} = \frac{(\bar{y}_1 - \bar{y}_0)\sqrt{p(1-p)}}{s_y},$$

where $\bar{y}_1$ and $\bar{y}_0$ are the mean Y-values corresponding to the two values of X, s_y^2 is the *sample variance (using the $n - 1$ divisor) of the combined set of n Y-values, and p is the proportion of X values equal to 1.

If it can be assumed that X is a *dichotomous representation of an underlying continuous random variable, W, with W and Y having a *bivariate normal distribution, then an appropriate measure is the **biserial correlation coefficient**. This is estimated as r_b, given by

$$r_b = \frac{\bar{y}_1 - \bar{y}_0}{s_y} \sqrt{\frac{p(1-p)}{u}} = \frac{r_{pb}}{\sqrt{u}},$$

where

$$u = \frac{1}{\sqrt{2\pi}} e^{-\frac{1}{2}h^2},$$

and h is the value defined by $P(Z \geq h) = p$, for a *standard normal variable Z.

bit A *binary digit. Computers store information in the form of bits. A sequence of (usually) eight bits is called a **byte**. The term 'bit' was coined by *Tukey; the term 'byte' by Dr Werner Buchholz of IBM.

bivariate data Data consisting of pairs of values (x_1, y_1), (x_2, y_2), ... taken from a *bivariate distribution. The term 'bivariate' dates from an article published in 1920.

bivariate distribution When, as a result of an experiment, values for two *random variables X and Y are obtained, it is said that there is a bivariate distribution. In the case of a sample, with n pairs of values (x_1, y_1), (x_2, y_2), ..., (x_n, y_n) being obtained, the methods of *correlation and *regression may be appropriate. Associated with the experiment is a **bivariate probability distribution**. In the case when X and Y are *discrete random variables, the distribution is specified by the **joint distribution** giving $P(X = x_j \ \& \ Y = y_k)$ for all values of j and k.

The **marginal distribution** of X is given by

$$P(X = x_j) = \sum_k P(X = x_j \ \& \ Y = y_k)$$

and the marginal distribution of Y is given by

$$P(Y = y_k) = \sum_j P(X = x_j \ \& \ Y = y_k).$$

The *expected values and *variances of X and Y are given in the usual way from these marginal distributions. For example,

$$E(X) = \sum_j x_j P(X = x_j).$$

The **conditional distribution** of X, given that $Y = y_k$, is given by

$$P(X = x_j | Y = y_k) = \frac{P(X = x_j \ \& \ Y = y_k)}{P(Y = y_k)},$$

and the **conditional expectation** of X, given that $Y = y_k$, written as $E(X | Y = y_k)$, is defined in the usual way as

$$E(X | Y = y_k) = \sum_j x_j P(X = x_j | Y = y_k).$$

The conditional distribution of Y and the conditional expectation of Y, given that $X = x_j$, are defined similarly.

In the case when X and Y are *continuous random variables, the distribution is specified by the **joint probability density function** $f(x, y)$ with the property that if R is any region of the (x, y) plane then

$$P[(X, Y) \in R] = \int \int_R f(x, y) dx dy.$$

The **marginal distribution** of X then has *probability density function $\int_{-\infty}^{\infty} f(x, y) dy$ and the marginal distribution of Y has probability density function $\int_{-\infty}^{\infty} f(x, y) dx$. The expected values and variances of X and Y are given in the usual way from these marginal distributions.

The probability density function for the **conditional distribution** of X, given that $Y = y_k$, is

$$\frac{f(x, y_k)}{\int_{-\infty}^{\infty} f(x, y_k) dx},$$

and

$$E(X | Y = y_k) = \frac{\int_{-\infty}^{\infty} x f(x, y_k) dx}{\int_{-\infty}^{\infty} f(x, y_k) dx}.$$

See MULTIVARIATE DISTRIBUTION.

bivariate normal distribution *See* MULTIVARIATE NORMAL DISTRIBUTION.

bivariate probability distribution *See* BIVARIATE DISTRIBUTION.

biweight function *See* M-ESTIMATES.

Blackwell, David (Harold) (1919–; b. Centralia, Illinois) An American mathematical statistician. He was educated at the University of Illinois, gaining his doctorate at the age of 22. In 1944 he joined the faculty of Howard University, where he published the work now called the *Rao–Blackwell theorem. In 1954 he moved to the University of California at Berkeley, where he is now Professor Emeritus. He was President of the *Institute of Mathematical Statistics in 1956. In 1965 he was elected to the National Academy of Sciences (the first African American to be so honoured). He was President of the *Bernoulli Society in 1975 and was made an Honorary Fellow of the *Royal Statistical Society in 1976.

blinding A method of avoiding bias in the context of treating a disease. In a medical experiment the comparison of treatments could be biased if either the patient, the doctor administering the treatment, or the data analyst knew which treatment was allocated to which patient. If the patient (or the doctor, or the data analyst) is unaware of which treatment is being given, then they are said to be 'blind to' the treatment allocation process. If neither the patient nor the doctor is aware of the treatment allocation then the process is said to be **double blind**.

In such an experiment, one treatment is often a dummy—for example, a pill of the same shape and flavour as the genuine pill, but with no active ingredients. This is called a **placebo**. In practice there is often a **placebo effect**, in which the treated person shows a benefit despite the treatment theoretically having no effect. The effect is a reflection of the psychological benefit of believing that one is being given an effective treatment.

block In the context of *experimental design, a homogeneous group of *experimental units.

BLUE (best linear unbiased estimator) An *estimator that is unbiased, is formed from a linear combination of the *observations, and has the smallest *variance of all such estimators.

BMDP (Biomedical Computer Programs) A computer package permitting many types of statistical analysis.

body mass index *See* QUETELET INDEX.

bomb packing A *probability fallacy. At a time when plane hijackers used the threat that they would explode a bomb in their luggage if the pilot did not do as they asked, it was suggested that one should pack a bomb in one's own luggage, since the probability of there being two bombs on the same aircraft was minimal. The logic fails, since the event of interest is not that there is a bomb on the plane, but that one of the other passengers has packed a bomb.

Bonferroni, Carlo Emilio (1892–1960; b. Bergamo, Italy; d. Florence, Italy) An Italian mathematician. Bonferroni was educated at the University of Turin. He specialized in the mathematics of finance, holding chairs successively in Bari and in Florence.

Bonferroni inequality An inequality concerning the joint *probabilities of occurrence of combinations of events. Let $E_1, E_2, \ldots, E_m$ be m events, with E'_j denoting the *complementary event to the event E_j, for $j = 1, 2, \ldots, m$. *Bonferroni developed various bounds and the inequality that is most often cited is

$$1 - \left\{ P(E'_1) + P(E'_2) + \cdots + P(E'_m) \right\} \leq P(E_1 \cap E_2 \cap \cdots \cap E_m).$$

The most usual context has the event E_j defined as 'hypothesis test j produces a non-significant result'.

Boole, George (1815–64; b. Lincoln, England; d. Ballintemple, Ireland) A self-taught English mathematician with a flair for languages. At school Boole excelled at Latin and by the age of 16 was an assistant school teacher, whilst studying mathematics for his own interest. To support his parents, Boole opened his own school, continuing with his work in mathematics. This work became so well known that, at the age of 35, Boole was appointed Professor of Mathematics at Queen's College, Cork. *An Investigation into the Laws of Thought, on Which are founded the Mathematical Theories of Logic and Probabilities*, in which he introduced what is now known as *Boolean algebra, was published in 1854. This algebra has found many applications, particularly in the design of computers. He was elected a Fellow of the Royal Society in 1857.

Boolean algebra The algebra, developed by *Boole, of events, *unions, *intersections and complementary events in a *sample space S. For any events A, B, C the following algebraic laws hold:

$$A \cup A = A, \qquad\qquad A \cap A = A,$$
$$A \cup A' = S, \qquad\qquad A \cap A' = \phi,$$
$$A \cup B = B \cup A, \qquad\qquad A \cap B = B \cap A,$$
$$A \cup S = S, \qquad\qquad A \cap S = A,$$
$$A \cup \phi = A, \qquad\qquad A \cap \phi = \phi,$$
$$A \cup (B \cup C) = (A \cup B) \cup C, \qquad A \cap (B \cap C) = (A \cap B) \cap C,$$
$$A \cup (B \cap C) = (A \cup B) \cap (A \cup C), \qquad A \cap (B \cup C) = (A \cap B) \cup (A \cap C),$$
$$(A \cup B)' = A' \cap B', \qquad\qquad (A \cap B)' = A' \cup B',$$

where A' and B' are the complementary events of A and B respectively, and ϕ is the empty set. The last line of the above comprises **de Morgan's laws**. The penultimate line comprises the distributive laws and the ante-penultimate line comprises the associative laws.

bootstrap A *computer-intensive *resampling method for estimating the properties of a *distribution while making minimal assumptions. In this respect it resembles the *jackknife. The idea is simple. Suppose we have n *observations $x_1, x_2, \ldots, x_n$ from an unknown distribution. We assume that the *population being sampled has $\frac{1}{n}$ of its observations equal to x_1, $\frac{1}{n}$ equal to x_2, and so on. Using *pseudo-random numbers we now select m sets of n observations from this hypothetical distribution. If we are interested in, for example, the *median of the original distribution, then we can use the m 'sample' medians to give an overall estimate and a *confidence interval for that estimate.

In practice bootstrap estimators are often slightly biased. However, this bias can be estimated—again using bootstrap methods. Estimation using this form of bias correction is referred to as the **double bootstrap**. The term 'bootstrap' was introduced by *Efron in 1979.

Bortkiewicz, Ladislaus Josephowitsch von (1868–1931; b. St Petersburg, Russia; d. Berlin, Germany) A Prussian economist and statistician. Bortkiewicz studied law as an undergraduate at the University of St Petersburg. His doctorate was obtained from the University of Göttingen in 1893. After a period in Strasbourg he returned to St Petersburg and there, in 1898, published a work entitled *The Law of Small Numbers* that dealt with properties of the *Poisson distribution. Bortkiewicz introduced the *Prussian horse-kick data that have become familiar to generations of students of *Statistics. In 1901 he settled at the University of Berlin, becoming Professor of Statistics and Political Economy in 1920.

Bowley, Sir Arthur Lyon (1869–1957; b. Bristol, England; d. Haslemere, England) An English social statistician. Bowley was a mathematics graduate at Cambridge University. In 1895 he joined the London School of Economics, being appointed as their first Professor of Statistics in 1919. He was the author of the influential textbook *Elements of Statistics*, first published in 1901, with a seventh edition in 1937. He was President of the *Royal Statistical Society in 1938 and was awarded the Society's Guy Medal in Silver in 1895 and in Gold in 1935.

Box, George E. P. (1919–; b. Gravesend, England) An English statistician initially trained as a chemist. During the Second World War, when determining the effects of poisonous gases, Box noted that the results varied and required statistical analysis. After the War, he studied mathematics and statistics at University College, London. He was appointed Professor of Statistics at the University of Wisconsin-Madison in 1960. He was President of the *American Statistical Association in 1978 and was awarded its *Wilks Medal in 1972. He was President of the *Institute of Mathematical Statistics in 1980 and was elected a Fellow of the Royal Society in 1985. He was awarded the *Royal Statistical Society's *Guy Medal in Gold in 1993 and was made an Honorary Fellow of the Society in 1993.

box-and-whisker diagram *See* BOXPLOT.

Box–Cox transformation A transformation to normality suggested by *Box and Sir David *Cox in 1964. They proposed a family of transformations that might be used to convert a general set of n *observations into a set of n *independent observations from a *normal distribution with constant *variance. The transformation involves a *parameter λ that can be estimated from the *data using the method of *maximum likelihood. The transformed observation, $y^{(\lambda)}$, is related to the original observation, y, by

$$y^{(\lambda)} = \begin{cases} \frac{1}{\lambda}\left(y^{\lambda} - 1\right) & \lambda \neq 0, \\ \ln y & \lambda = 0. \end{cases}$$

Box–Jenkins procedure A general strategy for the analysis of
*time series based on the use of *ARIMA models or, for seasonal
*data, *SARIMA models. The procedure was set out by *Box and
*Jenkins in their 1970 book *Time Series Analysis: Forecasting and
Control*. The first stage consists of removing *trends or *cycles
from the data. An appropriate type of *model must then be
identified and its *parameters estimated. The estimated model is
then compared with the original data and adjustments are made if
necessary.

Box–Ljung test A test of whether a proposed *ARMA model describes
a *time series. Suppose that the model has p *parameters and the time
series has n *observations. Let x_j and $\hat{x}_j$ be, respectively, the jth observed
and fitted values, define a_j by $a_j = x_j - \hat{x}_j$ and define r_k by

$$r_k = \sum_{t=1}^{n-k} a_t a_{t+k} \bigg/ \sum_{t=1}^{n} a_t^2.$$

The Box–Ljung statistic, Q, is given by

$$Q = n(n+2) \sum_{k=1}^{m} \frac{r_k^2}{(n-k)},$$

where m is arbitrary and very much smaller than n, but usually about 24.
If the model is correct then Q is an observation from a *chi-squared
distribution with $(m - p)$ *degrees of freedom.

Box–Muller transformation A procedure, suggested by *Box and
Muller in 1958, for the *simulation of *observations from a *normal
distribution. If u_1 and u_2 are two *independent observations from a
continuous *uniform distribution on the interval $(0, 1)$, then the
quantities x and y, given by

$$x = \cos(2\pi u_1)\sqrt{-2\ln(u_2)}, \qquad y = \sin(2\pi u_1)\sqrt{-2\ln(u_2)},$$

where $2\pi u_1$ is taken to be in radians, are independent observations from
a standard normal distribution.

Box–Pierce test A test to determine whether a *time series consists
simply of random values (white *noise). The test statistic is Q_m, given by

$$Q_m = n(n+2) \sum_{l=1}^{m} \frac{r_l^2}{n-l},$$

where r_l is the sample *autocorrelation at lag l, m is the maximum lag of interest, and n is the number of *observations. If n is much greater than m and if the *null hypothesis of white noise is correct, then Q_m has a *chi-squared distribution with m *degrees of freedom.

boxplot (box-whisker diagram) A graphical representation of numerical *data, introduced by *Tukey and based on the *five-number summary. The diagram has a scale in one direction only. A rectangular box is drawn, extending from the lower *quartile to the upper quartile, with the *median shown dividing the box. 'Whiskers' are then drawn extending from the end of the box to the greatest and least values. Multiple boxplots, arranged side by side, can be used for the comparison of several samples.

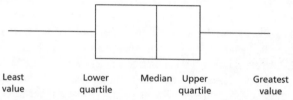

| Least
value | Lower
quartile | Median | Upper
quartile | Greatest
value |

Boxplot. The basic boxplot illustrates five key values: the minimum value, the maximum value, the quartiles, and the median.

In **refined boxplots** the whiskers have a length not exceeding 1.5 × the *interquartile range. Any values beyond the ends of the whiskers are shown individually as *outliers.

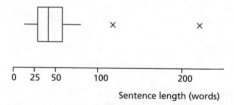

Sentence length (words)

Refined boxplot. In this diagram, individual outliers are indicated. The data illustrated are the lengths of the first eighteen sentences of *A Tale of Two Cities* by Charles Dickens. The first sentence—'It was the best of times, it was the worst of times, . . .'—stretches to 118 words, and the fourteenth sentence has 221 words.

Box–Tidwell transformation A general power transformation, in which the observation y is transformed to the value $y^{(p)}$, where

$$y^{(p)} = \begin{cases} y^p & \text{if } p \neq 0, \\ \ln y & \text{if } p = 0. \end{cases}$$

box–whisker diagram *See* BOXPLOT.

Bradley, Ralph Allen (1923–2001; b. Smith Falls, Ontario; d. Athens, Georgia) A Canadian statistician. Bradley gained his MA from Queen's University, Ontario in 1944, followed by a PhD from the University of North Carolina in 1949. From 1950 to 1959 he taught Statistics at Virginia Polytechnic Institute, before taking up a post at Florida State University (until 1982). For the next 10 years, he was Research Professor at the University of Georgia. He was Editor of *Biometrics* from 1957 to 1962, President of the *International Biometric Society in 1965, and President of the *American Statistical Association in 1981.

Bradley–Terry model A *model, proposed by *Bradley and Terry in 1952, that describes the *probability that one treatment is preferred to another. An experiment compares t treatments, two at a time. When two treatments (j and k, say) are compared, the outcome is a statement of which of the two is preferred. The model specifies that there are constants $\pi_1, \pi_2, \ldots, \pi_t$, with $0 \leq \pi_s \leq 1$, for all s, and $\sum_{s=1}^{t} \pi_s = 1$, which are such that the probability that treatment j is preferred to treatment k is

$$\frac{\pi_j}{\pi_j + \pi_k}.$$

branch and bound A procedure for solving an *integer programming problem. The problem is first solved ignoring the integer constraint. The solution obtained being noted, a variable is given an integer value either above or below the apparent maximum. Each resulting reduced problem is solved in its turn. The result is a branching tree reporting bounds on the optimum derived from problems related to the original.

branching process A *stochastic process with varying numbers of states. Consider a *population in which each individual has *probability p_k of providing k descendants in the next generation. Let X_j be the number of individuals in the jth generation. The sequence $X_1, X_2, \ldots$, is a *Markov process known as a branching process. The question of interest is the probability of ultimate **extinction**. Extinction is certain if and only if

$$\sum_{k=0}^{\infty} kp_k \leq 1$$

and otherwise the probability of extinction is equal to P, the solution (between 0 and 1) of

$$P = \sum_{k=0}^{\infty} P^k p_k.$$

Brown–Forsythe test *See* LEVENE'S TEST.

Brownian motion A continuous-time version of the *random walk, named after Robert Brown (1773–1858), a Scottish botanist. In 1827, Brown noticed the erratic movement of pollen grains under water. The first explanation of this motion (in terms

Brownian motion. This simulation, in two dimensions, illustrates the typical mixture of apparently random wanderings mixed with apparently purposive movement.

of the bombardment of the pollen by the surrounding water molecules) was given by Einstein in 1905. Norbert *Wiener provided a concise mathematical description in 1918 and the motion is also called a **Wiener process**.

Starting from the origin at time 0, the path of a particle is made up of independent increments (in d dimensions) which are such that its distance from the origin at time t is an *observation from a *normal distribution with *mean 0 and *variance proportional to t.

An alternative description is provided by assuming that it is the velocity rather than the position which is changing through time as a consequence of collisions and friction. This model is called the **Ornstein–Uhlenbeck process**.

brushing An interactive method for the *exploratory data analysis of *multivariate data. In a simultaneous display of *scatter diagrams of the values of pairs of *variables, the highlighting ('brushing') of data points in one diagram results in the same data items being highlighted in the other diagrams. This helps in understanding relations between variables and in identifying *outliers.

bubble plot A *scatter diagram using circles as plotting symbols in which the areas of the circles indicate the values of a third *variable.

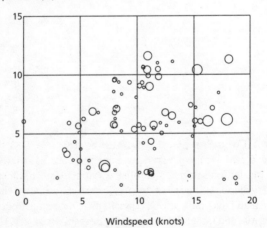

Bubble plot. The diagram displays the relation between windspeed and temperature near Bolton, England during March 2000. Each circle corresponds to an hour during which rain was recorded. The areas of the circles are proportional to the amount of rain.

bubble sort A simple but not very efficient *algorithm for arranging a set of n numbers in order of increasing magnitude. The method starts with the right-hand pair of numbers, swapping if necessary so that the smaller number is on the left of the pair, and proceeds towards the left, making a total of $(n - 1)$ comparisons. At the end of this pass the smallest number is at the left end of the line. The algorithm recommences with the right-hand pair of numbers and again proceeds towards the left. This time $(n - 2)$ comparisons are made and the pass ends with the second-smallest number in second position. The procedure repeats until the complete ordering is achieved: this requires a total of $\frac{1}{2}n(n - 1)$ comparisons.

16	8	13	4
16	8	4	13
16	4	8	13
4	16	8	13
4	16	8	13
4	8	16	13
4	8	13	16

Bubble sort. Illustration of the simple neighbour-swapping algorithm.

Buffon (Comte de), Georges Louis Leclerc (1707–88; b. Montbard, France; d. Paris, France) A French aristocrat educated in law and medicine. Buffon's principal interest was in nature: he was struck by the diversity of life. At the age of twenty, Buffon encountered the *binomial theorem and his first work, *Sur le jeu de franc-carreau*, introduced differential and integral calculus into *probability theory. He was appointed keeper of the French botanical gardens and his study of plants and their development led to his publication (in about forty volumes) of *Histoire Naturelle* (*Natural History*), which tried to show the continuity of nature. Buffon had no concept of evolution, believing that species are fixed. Buffon speculated that the Earth might have been created by the collision of a comet with the Sun. Based on the cooling rate of iron, he proclaimed that the age of the Earth was 75 000 years; as a result the Catholic Church (which, at that time, claimed 6 000 years as the Earth's age) ordered that his books should be burnt. A street in Paris and a lunar crater are named after him.

Buffon's needle problem A *probability problem posed by *Buffon in 1777. A needle of length l is to be randomly thrown on to a piece of paper that is covered by parallel lines that are a distance d apart; the problem is 'What is the probability that the needle crosses a line?' The answer is

$$\frac{2l}{\pi d}.$$

This provides an empirical method for the estimation of π (though many throws are needed if an accurate answer is required).

Bulletin of the International Statistical Institute A journal, which first appeared in 1886, that reports the proceedings of the biennial sessions of the *International Statistical Institute.

Burman, J. Peter (1924–98; b. London, England; d. Blackheath, England) An English statistician. Burman was educated at Cambridge University and then spent two years working for the Ministry of Supply. He was the co-author with *Plackett of the 1946 paper that introduced the influential Plackett and Burman designs (*See* FACTORIAL EXPERIMENT). In 1946 he commenced a long career at the Bank of England. His later work was concerned with the development of *ARMA models.

burn-in period *See* MARKOV CHAIN MONTE CARLO METHODS.

Burt, Sir Cyril Lodowicz (1883–1971; b. Stratford-on-Avon, England; d. London, England) An English pioneer of educational psychology. At Oxford University he studied philosophy. In 1908, on graduating, he joined the faculty at the University of Liverpool, where he worked with delinquent boys in the docklands area. In 1913 he was appointed as the first ever educational psychologist for London County Council. Burt joined University College, London in 1924 as Professor of Education, becoming Professor of Psychology in 1931. In 1942 he was President of the British Psychological Society and he was knighted in 1946. After his death, his work on identical twins was declared to be fraudulent, though this remains a subject of debate.

Burt table A symmetric table that is used in *correspondence analysis. It shows the *frequencies for all combinations of categories of pairs of *variables in a *data set. As an example, the Burt table shown displays simultaneous information on the occurrence of category combinations for the variables Age, Health, and Class.

		AGE		HEALTH		CLASS	
		YOUNG	OLD	ILL	WELL	MIDDLE	WORKING
AGE	YOUNG	64	0	9	55	47	17
	OLD	0	36	17	19	21	15
HEALTH	ILL	9	17	26	0	7	19
	WELL	55	19	0	74	61	13
CLASS	MIDDLE	47	21	7	61	68	0
	WORKING	17	15	19	13	0	32

byte See BIT.

C, C⁺⁺ The programming language C^{++} is an advanced form of the language C, which originated at Bell Laboratories in the USA.

Canadian Journal of Statistics Journal of the *Statistical Society of Canada. The first volume appeared in 1973. The journal includes articles written in either English or French.

canonical correlation analysis Method of assessing the relationship between two groups of *variables (for example, the relationship between three measures of a worker's ability and four measures of his or her performance).

capability analysis Method determining the extent to which the long-term performance of an industrial process complies with engineering requirements or managerial goals. Often it is required that output should lie between upper and lower **specification limits** (*USL*; *LSL*). One measure of the extent to which this is achieved is the **potential capability**, C_p, given by

$$C_p = \frac{USL - LSL}{6\sigma},$$

where σ is the *standard deviation of the output *distribution. An equivalent statistic is C_r, the **capability ratio**, which is $1/C_p$.

capability ratio *See* CAPABILITY ANALYSIS.

capture–recapture methods Methods for estimating the size of a *population. The method is generally applied to living organisms, such as fish in a lake.

    The simplest example proceeds as follows. A *sample of n_1 individuals is obtained from the population (for example, by catching fish in a net). Each of these individuals is now marked in some fashion (for example, by attaching a tag) and is returned to the population. Later, a second sample (of n_2 individuals) is taken from the population. Suppose m of these individuals are tagged. The sample proportion of tagged individuals is

m/n_2, and the population proportion is n_1/N, so that an estimate of the population size, N, is $\hat{N}$, given by

$$\hat{N} = \frac{n_1 n_2}{m}.$$

carry-over effect *See* CROSSOVER TRIAL.

CART (Classification and Regression Trees) A program for constructing a *classification tree.

Cartesian coordinates The usual system for identifying the location of a point in two, or more, dimensions. The position of a point P in a plane can be represented by a pair of numbers (x, y), relative to two **axes** Ox and Oy which are straight lines meeting at the **origin** O represented by $(0, 0)$. In the usual case of rectangular (or perpendicular, or orthogonal) axes, the **abscissa** x is the perpendicular distance of P from Oy and the **ordinate** y is the perpendicular distance of P from Ox. The value of x is positive if P lies in the half-plane to the right of O, and is negative in the left-hand half-plane. Similarly, the value of y is positive in the upper half-plane and negative in the lower half-plane. The plane is therefore divided into four **quadrants** corresponding to the four combinations of coordinate signs. The units of distance in the directions Ox and Oy may be different and are usually indicated by numbers on the axes.

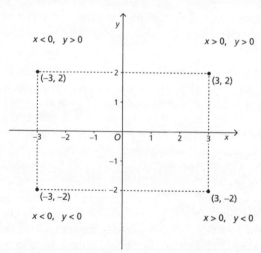

Cartesian coordinates. The axes meet at the origin O. The coordinates of a point are (x, y), where x and y are the corresponding signed distances along the two axes.

The term 'Cartesian' is derived from *Descartes who first introduced coordinates. The word 'coordinate' was introduced by *Leibniz in about 1693. The phrase 'Cartesian coordinates' was used in 1844.

The idea of coordinates can be generalized to three-dimensional space. The process can be reversed by considering any ordered set of three numbers (x, y, z) as a point in three-dimensional space. This can be further generalized by considering the row-vector $(x_1\ x_2 \ldots x_n)$ to be the coordinates of a point in an n-dimensional space. This space is usually denoted by $\mathbb{R}^n$.

cartogram A distorted map in which regions are drawn not to an areal scale but to some other scale such as population. Some cartograms attempt to retain the shape of the geographical region they represent; others, such as the **rectangular cartogram** illustrated as follows, use a single shape to represent all regions of equal importance.

case–control studies *See* RETROSPECTIVE STUDIES.

categorical variable (nominal variable) A *variable whose values are not numerical. Examples include gender (male, female), paint colour (red, white, blue), type of bird (duck, goose, owl). A variable with just two categories is said to be *dichotomous, whereas one with more than two categories is described as **polytomous**. The corresponding nouns are **dichotomy** and **polytomy**.

Cauchy, Baron Augustin-Louis (1789–1857; b. Paris, France; d. Sceaux, France) A French mathematician and engineer. After studying first mathematics and then engineering at the École Polytechnique, Cauchy was initially employed as an engineer. However, in 1815, he was appointed to the staff of the École Polytechnique to teach mathematical analysis. Despite his many publications, he was not promoted to full professor for 18 years because he refused to take the oath of allegiance to the republican government. He made major developments in the theories of determinants, partial differential equations, group theory, and complex variables. His last words were said to be (in translation) 'Men pass away—but their deeds abide.' A street in Paris and a lunar crater are named after him.

Cauchy distribution A *continuous random variable with *probability density function f given by

KEY: turnout
- 90%
- 85%
- 80%
- 75%
- 70%
- 65%
- 60%
- 55%

Rectangular cartogram. The diagram shows variation in turnout (the proportion of the electorate who actually cast their votes) in the British constituencies in the 1992 general election. Each constituency is represented by a rectangle whose area indicates the physical size of the constituency. The big conurbations (Greater London, Birmingham, Manchester, etc.) are indicated as outlined regions. Somewhat paradoxically, turnout is lower in city centres than in the rural constituencies.

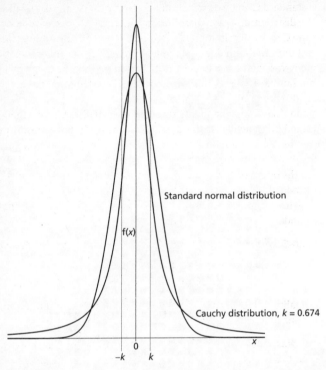

Cauchy distribution. The Cauchy distribution illustrated has $m = 0$ and $k = 0.674$. Also illustrated is the standard normal distribution. Both distributions have 25% of their area above 0.674 and 25% below -0.674. The fatter tails of the Cauchy distribution are apparent.

$$f(x) = \frac{k}{\pi\left\{k^2 + (x - m)^2\right\}}, \qquad -\infty < x < \infty,$$

where $k > 0$ and m are *parameters, is said to have a Cauchy distribution.

The graph of f is a *bell-curve centred on m. The *mode and the *median are both equal to m, and the *quartiles are $m \pm k$. A Cauchy distribution has no *mean or *variance, since, for example,

$$\int_{-\infty}^{\infty} \frac{kx}{\pi\left\{k^2 + (x - m)^2\right\}}\,dx$$

does not exist.

The standard Cauchy distribution is given by $k = 1, m = 0$, and in this case the distribution is a *t-distribution, with one *degree of freedom. Since the Cauchy distribution has neither a mean nor a variance, the *central limit theorem does not apply. Instead, any linear combination of Cauchy variables has a Cauchy distribution (so that the mean of a random sample of observations from a Cauchy distribution has a Cauchy distribution).

If X and Y have *independent *standard normal distributions then Y/X has a standard Cauchy distribution. Equivalently, if U has a *uniform continuous distribution on $-\frac{1}{2}\pi < u < \frac{1}{2}\pi$ then $\tan U$ has a standard Cauchy distribution. A geometrical representation of this is as follows. Let O be the origin of *Cartesian coordinates, and let A be the point $(0, 1)$. If the random point P, with coordinates $(X, 0)$, is such that the angle OAP ($= U$, say) has a uniform continuous distribution on $-\frac{1}{2}\pi < u < \frac{1}{2}\pi$, then X has a standard Cauchy distribution.

Cauchy–Schwarz inequality If X and Y are two *random variables then

$$\{E(XY)\}^2 \le E(X^2)E(Y^2),$$

with equality if and only if $P(aX = bY) = 1$ for some real constants a and b, at least one of which is non-zero.

causal diagram A graphical representation of the relationships between *variables with arrows from each explanatory variable to each of the dependent variables that it affects.

cdf *See* CUMULATIVE DISTRIBUTION FUNCTION.

cell; cell frequency *See* CONTINGENCY TABLE.

censored data Data items in which the true value is replaced by some other value. For example, suppose a set of components are being monitored to see how long they last before breaking. If the monitoring stops before all the components have broken, then the information concerning the lifetimes of the unbroken components has been **right-censored**. The score of a cricketer who is not out is an example of censored data, since it is not known what score would have been achieved if the cricketer's innings had been allowed to continue. In both cases the value used is the largest value so far achieved for that data item. To avoid bias, subsequent calculations should take account of the censoring.

censored regression models *Regression models in which the *data on the dependent variable may be regarded as being *censored data. The analysis of such models is also called **tobit analysis**.

census *Data for an entire *population—as opposed to a random *sample. In England the most famous early census is that of the 'Domesday Book'. The first modern census was the 1666 census of the 3,215 inhabitants of New France (now Canada!). In Europe the first complete demographic census was that in Sweden in 1749. The first federal census in the USA took place in 1790, and the first complete demographic censuses in both Britain and France occurred in 1801. In both the United Kingdom and the USA censuses now occur every ten years.

central limit theorem (clt) A theorem, proposed by *Laplace, explaining the importance of the *normal distribution in *Statistics. Let $X_1, X_2, \ldots, X_n$ be *independent *random variables each having the same *distribution, with *mean μ and *variance σ^2. Let $\bar{X}$, given by

$$\bar{X} = \frac{1}{n}(X_1 + X_2 + \cdots + X_n),$$

denote the sample mean. The central limit theorem states that, for large n, the distribution of $\bar{X}$ is approximately a *normal distribution with mean μ and variance $\frac{1}{n}\sigma^2$. Thus, for a large random *sample of *observations from a distribution with mean μ and variance σ^2, the distribution of the sample mean is approximately normal with mean μ and variance $\frac{1}{n}\sigma^2$ and the distribution of the sample total is approximately normal with mean $n\mu$ and variance $n\sigma^2$. The phrase 'central limit theorem' appears in a 1919 article by *von Mises.

central moment *See* MOMENT.

central tendency The tendency of quantitative *data to cluster around some central value. The central value is commonly estimated by the *mean, *median, or *mode, whereas the closeness with which the values surround the central value is commonly quantified using the *standard deviation or *variance. The phrase 'central tendency' was first used in the late 1920s.

centred moving average *See* MOVING AVERAGE.

cgf *See* CUMULANTS.

CHAID (Chi-Square Automatic Interaction Detection) A program for constructing a *classification tree.

chain graph *See* GRAPHICAL MODELS.

change-point problems Problems concerned with the identification of the point in a *time series at which some change occurs. Suppose that *data are collected on one or more *variables at a series of time points in order to examine the properties of the variables and possible relationships between them. The time of collection appears to be incidental. However, frequently this turns out to be not quite true. For example, in one case the explanation for an apparent change in the mean of a variable was found to be attributed to a tall meter reader who looked down at a dial being replaced by a short one who looked up at the same dial. Another example concerns changes in the responses to a question in an annual survey that turn out to be a consequence of a change in the survey design. In these examples, the cause of the change and hence the time that the change took place have been identified. Often, however, though it is evident that a change has taken place, it is not clear at what precise point this occurred.

Chapman, Sydney (1888–1970; b. Eccles, England; d. Boulder, Colorado) An English mathematician. In 1907, after graduating as an engineer at Manchester University, he studied mathematics at Cambridge University. In 1910 he became assistant to the Astronomer Royal, and this led to his research in geomagnetic theory. His subsequent academic career involved appointments at Manchester University (1919), Imperial College, London (1924), and Oxford University (1946). He was elected a Fellow of the Royal Society in 1919 and was awarded its Copley Medal in 1964. He was President of the London Mathematical Society in 1929.

Chapman–Kolmogorov equation *See* MARKOV PROCESS.

characteristic equation *See* MATRIX.

characteristic function The (complex-valued) function $E(e^{itX})$, where X is a random variable, t is a real variable and $i = \sqrt{-1}$. If X is continuous and has *probability density function f then the characteristic function is a multiple of the *Fourier transform of f. The characteristic function is closely related to the *moment-generating function and has the advantage of always existing. The first English use of the phrase 'characteristic function' was by *Kullback in 1934.

characteristic value; characteristic vector *See* MATRIX.

Chebyshev (Tchebycheff), Pafnuty Lvovich (1821–94; b. Okatovo, Russia; d. St Petersburg, Russia) A Russian mathematician. Chebyshev spent most of his career at St Petersburg University, where he developed a mathematical school with an international reputation (*Markov was one of his students). Chebyshev made major contributions in many branches of mathematics. To statisticians he is best remembered for his work on the *weak law of large numbers and the introduction of polynomials as a means of handling the fitting of curvilinear models. A lunar crater is named after him.

Chebyshev–Hermite polynomials Polynomials that play an important role in the *Edgeworth expansion and *Gram–Charlier expansion. The rth of these polynomials, $H_r(x)$, is the coefficient of $t^r/r!$ in the expansion of $\exp\left(tx - \frac{1}{2}t^2\right)$ and is given by

$$H_r(x) = x^r - \frac{r(r-1)}{2.1!}x^{r-2} + \frac{r(r-1)(r-2)(r-3)}{2^2.2!}x^{r-4} - \cdots.$$

Chebyshev's inequality For a *random variable X with expectation μ and *standard deviation σ,

$$P(|X - \mu| > k\sigma) \geq \frac{1}{k^2}$$

for all positive values of the constant k. *See also* BERNSTEIN'S INEQUALITY; HÖLDER'S INEQUALITY; MARKOV'S INEQUALITY; MINKOWSKI'S INEQUALITY.

Chernoff, Herman (1923–; b. New York City) An American applied mathematician and statistician. Chernoff was educated at City College (now City University), New York and Brown University, where he obtained his doctorate in 1948. After a period at the University of Illinois (Urbana) he moved to Stanford University in 1952. In 1974 he became Professor of Applied Mathematics at the Massachusetts Institute of Technology (MIT) and in 1985 moved to become Professor of Statistics at Harvard University. He now holds the title of Professor Emeritus at both MIT and Harvard University. His statistical interests include optimal *experimental design, pattern recognition, and *sequential sampling. He was the *American Statistical Association's *Wilks Medal winner in 1987. He was President of the *Institute of Mathematical Statistics in 1968.

Chernoff faces An alternative to *Andrews plots as a method of giving a two-dimensional representation of *multivariate data. The shape and size of the various characteristics of a face are varied according to the values of the *variables concerned.

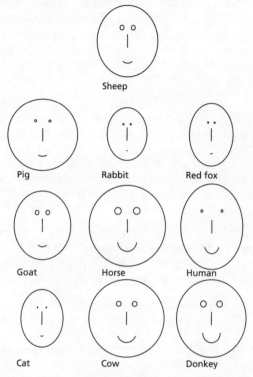

Chernoff faces. The horizontal scaling, vertical scaling, size of eyes, and shape of mouth are governed by the values of various attributes of the creatures being compared. Note the similarity between sheep and goat!

Chinese postman problem A *network problem that can be formulated as a *combinatorial optimization problem. A postman has to travel down every street in a town. The streets have different lengths. The problem is to devise a route that minimizes the distance the postman travels. The adjective 'Chinese' refers to the nationality of the original poser of the problem and not to any unusual behaviour by Chinese postmen.

chi-squared (χ^2) distribution If $Z_1, Z_2, \ldots, Z_v$ are v *independent *standard normal variables, and if Y is defined by

$$Y = \sum_{j=1}^{v} Z_j^2,$$

then Y has a chi-squared distribution with v *degrees of freedom (written as χ_v^2). The *probability density function f is given by

$$f(y) = \frac{e^{-\frac{1}{2}y} y^{(\frac{1}{2}v-1)}}{2^{\frac{1}{2}v} \Gamma(\frac{1}{2}v)}, \qquad y > 0,$$

where Γ is the *gamma function. The form of the distribution was first given by *Abbe in 1863 and was independently derived by *Helmert in 1875 and Karl *Pearson in 1900. It was Pearson who gave the distribution its current name.

The chi-squared distribution has *mean v and *variance $2v$. For $v \leq 2$ the *mode is at 0; otherwise it is at $(v - 2)$. A chi-squared distribution is a special case of a *gamma distribution. The case $v = 2$ corresponds to the *exponential distribution. Percentage points for chi-squared distributions are given in Appendix X.

chi-squared (χ^2) test (Pearson goodness-of-fit test) A *goodness-of-fit test, introduced by Karl *Pearson in 1900, that is popular because of its simplicity. Let O denote the observed *frequency of an outcome in a

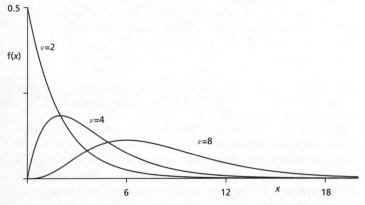

Chi-squared distribution. All chi-squared distributions have ranges from 0 to ∞. Their shape is determined by the value of v. If $v > 2$ then the distribution has a mode at $(v - 2)$; otherwise the mode is at 0.

*sample and let E denote the corresponding *expected frequency under some *model. The test statistic is X^2, defined by

$$X^2 = \Sigma r^2,$$

where the summation is over all the outcomes whose frequencies are being compared, and r, the **Pearson residual**, is defined by

$$r = \frac{O - E}{\sqrt{E}}.$$

It is important to note that the comparison involves frequencies and not proportions.

If there are m observed frequencies, and p *parameters have been estimated using these frequencies, then, if the model is correct, the observed value of X^2 will approximate an observation from a *chi-squared distribution with $(m - p - 1)$ *degrees of freedom. In the case of a *continuous random variable, the frequencies will refer to ranges of values. For other random variables it is usual to combine together neighbouring rare values into a single category, since the chi-squared approximation fails if there are too many small expected frequencies.

As an example, suppose it is hypothesized that a type of sweet pea occurs in shades of white, red, pink, and blue, with proportions $\frac{1}{4}, p, (\frac{3}{4} - 3p)$, and $2p$, respectively. A random sample of 120 seeds is sown. All germinate with 20 having white flowers, 10 having red flowers, 40 pink, and 50 blue. The question is whether these results are consistent with the hypothesis. In this case the *maximum likelihood estimate of p is 0.15, so the expected frequencies are 30, 18, 36, and 36. Thus

$$X^2 = \frac{(20 - 30)^2}{30} + \frac{(10 - 18)^2}{18} + \frac{(40 - 36)^2}{36} + \frac{(50 - 36)^2}{30} = 12.8.$$

There are $4 - 1 - 1 = 2$ degrees of freedom. Since 12.8 is a very large value (compared with the percentage points (Appendix X) of a χ^2_2-distribution), the hypothesis can confidently be rejected.

One situation in which the use of the chi-squared test is frequently encountered is as a test for *independence in a $J \times K$ *contingency table that cross-classifies the variables A and B. Let the observed frequency of data belonging to category j of variable A and to category k of variable B be f_{jk}. Write

$$f_{j0} = \sum_{k=1}^{K} f_{jk}, \qquad f_{0k} = \sum_{j=1}^{J} f_{jk}, \qquad f_{00} = \sum_{j=1}^{J} \sum_{k=1}^{K} f_{jk}.$$

Then, according to the *null hypothesis of *independence, the expected frequency e_{jk} is given by

$$e_{jk} = \frac{f_{j0} f_{0k}}{f_{00}},$$

and the test statistic X^2 is given by

$$X^2 = \sum_{j=1}^{J} \sum_{k=1}^{K} \frac{(f_{jk} - e_{jk})^2}{e_{jk}}.$$

If the null hypothesis of independence is correct, then the distribution of X^2 can be approximated by a chi-squared distribution with $(J-1)(K-1)$ degrees of freedom. In the special case where $J = K = 2$ (a two-by-two table), the chi-squared approximation is improved by using the *Yates-corrected chi-squared test. *See also* LOG-LINEAR MODELS.

chi-squared variable A *random variable having a *chi-squared distribution.

Cholesky, André-Louis (1875–1918; b. Montguyon, France) A French mathematician and geodesist. Cholesky was trained at L'École Polytechnique and entered the artillery branch of the French army. By 1905 he was attached to the Geodesic Section and spent the winter of 1907–8 mapping first Crete, then Algeria, Tunisia, and Morocco. He was killed in action in the First World War.

Cholesky decomposition If A is a non-singular symmetric *matrix, then it can be expressed as the product LL′, where L is a lower triangular matrix and L′ is the *transpose of L. The decomposition is useful in solving sets of linear equations and in matrix inversion.

circular data A type of *cyclic data in which measurements are directions. **Axial data** occur when a data item has known orientation but its direction is not known—for example, an iron filing will align itself either way around in a magnetic field. In this case the effective orientations lie in a 180° range. To utilize the standard methods for cyclic data it is usual to double all the angles (subtracting 360° if required).

circular distribution A distribution suitable for *circular data.

Examples include the *circular uniform distribution, the *von Mises distribution, and the *wrapped Cauchy distribution.

circular histogram The natural alternative to the *histogram for illustrating grouped *cyclic data. Apparently first used by Florence *Nightingale as a means of representing the deaths in the Crimean War. *See also* ROSE DIAGRAM.

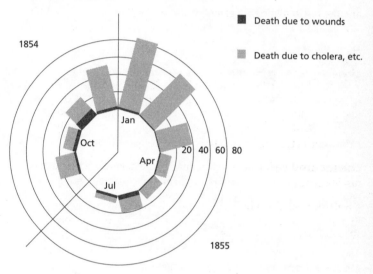

Circular histogram. Florence Nightingale invented this diagram to illustrate that the main cause of death in the Crimean War was disease rather than the enemy. The reduction in deaths after March 1855 illustrates the efficacy of Nightingale's measures to reduce disease.

circular mean *See* CYCLIC DATA.

circular normal distribution *See* VON MISES DISTRIBUTION.

circular uniform distribution The *uniform distribution for directions. The *probability density function f is constant for directions (in *radians):

$$f(\theta) = \frac{1}{2\pi}, \qquad -\pi < \theta < \pi.$$

city-block metric *See* DISTANCE MEASURES.

class boundary (class limit) When numerical *data are grouped into classes (e.g. $1 \leq x < 5, 5 \leq x < 10, \ldots$), the values marking the limits of these classes are called the lower and upper class boundaries. Some texts reserve the term 'class limits' for use with *continuous variables.

classification tree A rule for predicting the class of an object from the values of its predictor variables. Classification trees are much used in *data mining. They differ from *discriminant analysis in that judgements are reached by considering variables hierarchically rather than simultaneously. Classification tree programs include CART, CHAID, and QUEST.

class limit *See* CLASS BOUNDARY.

closed question (multiple choice question) A question in a *questionnaire for which the respondent must select one from a number of given answers. By contrast an **open question** allows the respondent to answer in any fashion.

clt *See* CENTRAL LIMIT THEOREM.

cluster analysis A method for identifying *data items that closely resemble one another, assembling them into **clusters**. A number of characteristics are measured for each of several items (which might be, for example, people, plants, machines, etc.). The process of formation of the clusters is often represented using a *dendrogram. The most commonly used methods are the *agglomerative clustering methods.

clustered points *See* INDEX OF DISPERSION.

clusters *See* CLUSTER ANALYSIS; CLUSTER SAMPLING.

cluster sampling An economical method for *sampling a scattered *population. When, as is usually the case, a geographical population is scattered, it would be uneconomic to visit the scattered individuals chosen by simple random sampling. Instead, the population is subdivided into a large number of geographically compact regions (the **clusters**) and a random sample of clusters is selected. These are referred to as the **primary sampling units (PSUs)**. In single-stage cluster sampling all members of the selected cluster are interviewed. In multi-stage cluster sampling, further subdivisions take place.

cobweb diagram A diagram that illustrates the *associations between the categories of two or more *categorical variables by means of lines whose widths indicate the strengths of the association and whose colour indicates whether the association is positive or negative.

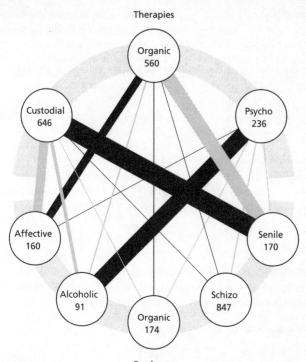

Therapies

Psychoses

Cobweb diagram. The diagram shows the relation between types of insanity and the prescribed treatment. The thickness of the lines indicates the strength of the association between the linked categories. Black lines indicate positive association and grey lines indicate negative association. The diagram demonstrates that senile patients are particularly likely to be given custodial therapy and particularly unlikely to be given organic therapy.

Cochran, William Gemmell (1909–80; b. Rutherglen, Scotland; d. Orleans, Massachusetts) A Scottish statistician who spent most of his career in the United States. Cochran studied mathematics first at Glasgow University and then at Cambridge University. During this time his first papers were published; the second of these (in 1934) introduced *Cochran's theorem. At the height of the depression Cochran abandoned his doctoral studies when offered a job as assistant to *Yates at Rothamsted (the agricultural research institute in Hertfordshire). In 1939 he moved to the United States, where, successively, he held chairs at Princeton, North Carolina, Johns Hopkins, and Harvard Universities. His co-authored books *Experimental Designs* (with Gertrude *Cox) and *Statistical Methods* (with George *Snedecor) were regarded as compulsory reading by the next generation of statisticians. He was President of the *Institute of Mathematical Statistics in 1946 and President of the *American Statistical Association in 1953. Cochran was the Association's *Wilks Medal winner in 1967 and Editor of the *Journal of the American Statistical Association* from 1946 to 1950. He was elected to the National Academy of Sciences in 1974.

Cochran's C test A test, introduced by *Cochran in a 1941 paper, for equality of *variance in m *normal populations having unknown *means and having sample variances denoted by $s_1^2, s_2^2, \ldots, s_m^2$. Let s_{max}^2 be the maximum of $s_1^2, s_2^2, \ldots, s_m^2$. Cochran's C is defined by

$$C = \frac{s_{max}^2}{\sum_{j=1}^{m} s_j^2}.$$

Unusually high values of C indicate unequal variances. The test is affected by departures from normality. *See also* BARTLETT'S TEST FOR HOMOGENEITY OF VARIANCES.

Cochran's Q test *See* MCNEMAR'S TEST.

Cochran's theorem A theorem concerning sums of *chi-squared variables. Let $\mathbf{y}$ represent an $n \times 1$ vector of *independent *standard normal random variables and let $\mathbf{A}_1, \mathbf{A}_2, \ldots, \mathbf{A}_k$ be symmetric *matrices such that

$$\sum_{j=1}^{k} \mathbf{y}'\mathbf{A}_j\mathbf{y} = \mathbf{y}'\mathbf{y},$$

where $\mathbf{y}'$ is the transpose of $\mathbf{y}$. Write $Q_j = \mathbf{y}'\mathbf{A}_j\mathbf{y}$. Cochran's theorem, published in 1934, states that, if one of the following three conditions is true, then so are the other two:

(i)   The ranks of $Q_1, Q_2, \ldots, Q_k$ sum to n.
(ii) Each of $Q_1, Q_2, \ldots, Q_k$ has a chi-squared distribution.
(iii) Each of $Q_1, Q_2, \ldots, Q_k$ is independent of all the others.

code; codebook; coding A number entered into a computer to signify each answer to a *questionnaire. The list of codes for each question is stored in the codebook. The process of translating the questionnaire responses into codes is called coding the data.

coefficient of alienation *See* CORRELATION COEFFICIENT, SAMPLE.

coefficient of concordance A *measure of agreement between m observers who *rank n items in order according to some characteristic. Let R_j be the total of the ranks assigned to the jth item and let S be given by

$$S = \sum_{j=1}^{n}\left\{R_j - \frac{1}{2}m(n+1)\right\}^2.$$

The coefficient of concordance, W, is given by

$$W = \frac{12S}{m^2(n^3 - n)},$$

which takes values from 0 (no general agreement) to 1 (complete agreement). The coefficient was proposed by Sir Maurice *Kendall and B. B. Smith in 1939.

coefficient of correlation *See* CORRELATION COEFFICIENT.

coefficient of determination *See* CORRELATION COEFFICIENT, SAMPLE.

coefficient of variation The ratio of the *standard deviation to the *mean. A term introduced by Karl *Pearson in 1896.

cofactor *See* MATRIX.

Cohen, Jacob (1923–98; b. New York City) An American psychologist and a pioneer of research methods. Cohen enrolled in the City College (now City University) of New York at the age of 15. After work in intelligence in World War II, he gained a PhD in clinical psychology at New York University, where he spent most of his teaching career. He was a co-author of a popular introductory statistics book for the behavioural sciences, first published in 1971.

Cohen's kappa (κ) A *measure of agreement between two observers, suggested by *Cohen in 1960. Suppose that the observers are required, independently, to assign items to one of m classes. Let f_{jk} be the number of individuals assigned to class j by the first observer and to class k by the second observer. Let $f_{j0} = \sum_{k=1}^{m} f_{jk}, f_{0k} = \sum_{j=1}^{m} f_{jk}$ and $f_{00} = \sum_{j=1}^{m}\sum_{k=1}^{m} f_{jk}$. Define the quantities O and E by

$$O = \sum_{j=1}^{m} f_{jj}, \quad E = \sum_{j=1}^{m} \frac{f_{j0}f_{0j}}{f_{00}},$$

so that O is the total number of individuals on which the observers are in complete agreement, and E is the expected total number of agreements that would have occurred if the observers had been statistically independent. The formula for Cohen's kappa is

$$\kappa = \frac{O - E}{f_{00} - E}.$$

A value of 0 indicates statistical independence, and a value of 1 indicates perfect agreement.

cohort study A *longitudinal study of the same group of people (the **cohort**) over time. By contrast with a *panel study, different members of the cohort may be studied at each time point. Usually the members of a cohort are of approximately the same age—for example, all those with 21st birthdays during the year 2000.

coiflet *See* WAVELET.

cokriging *See* KRIGING.

cold deck *See* IMPUTATION.

collapsing A term used in connection with the merging of neighbouring categories. For example, if the *random variable X takes values 0, 1, and 2 with high *probability and values 3, 4, and 5 with low probability, then it may be sensible to collapse the latter categories into a single '3 or more' category.

As a more extreme example, suppose that, in an $I \times J \times K \times L$ *contingency table, for variables A, B, C, D, it turns out that D is of no interest. All the categories of D may then be collapsed together to give an $I \times J \times K$ table—the original table has been collapsed over D.

collinearity *See* MULTIPLE REGRESSION MODEL.

column vector *See* MATRIX.

combination An unordered selection of r objects from a set of $n\,(\geq r)$ different objects. The number of different combinations is often denoted by ^nC_r. In fact

$$^nC_r = \binom{n}{r},$$

is the *binomial coefficient. Special values are $^nC_0 = 1, ^nC_n = 1, ^nC_1 = n$.

A frequently used relationship is

$$^{n+1}C_r = {}^nC_r + {}^nC_{r-1},$$

which is the defining relationship for *Pascal's triangle. For ordered selection, *see* PERMUTATION.

combinatorial optimization An optimization technique in which the values of the *variables are restricted to integers. Examples include the *knapsack and *travelling salesman problems.

combinatorics The study of the numbers of ways of selecting, or arranging, objects from a finite set. Its main applications in the theory of *probability are in calculating the probability of an event such as r of the objects chosen have a particular property, or the probability of a sequence of events with a particular property. *See* BINOMIAL DISTRIBUTION; HYPERGEOMETRIC DISTRIBUTION; NEGATIVE BINOMIAL DISTRIBUTION.

Combinatorial results often involve *factorials and *binomial coefficients. For example, the probability that, in a random deal of 13 cards from a normal pack, Player A receives 13 spades is

$$\frac{1}{\binom{52}{13}} \approx 1.57 \times 10^{-12},$$

and the probability that in a random deal to four players, each player receives 13 cards of the same suit is

$$\frac{4!}{\binom{52}{13}\binom{39}{13}\binom{26}{13}} \approx 4.47 \times 10^{-28}.$$

communicating states *See* MARKOV PROCESS.

complementary event The complementary event A' to an event A is the event 'A does not occur'. It satisfies $A \cup A' = S$, where S is the *sample space, and $A \cap A' = \phi$, where ϕ is the empty set. The complements of *intersections and *unions are given by de Morgan's laws:

$$(A \cap B)' = A' \cup B',$$
$$(A \cup B)' = A' \cap B'.$$

See BOOLEAN ALGEBRA; SAMPLE SPACE.

complementary log-log link *See* GENERALIZED LINEAR MODEL.

complete linkage clustering A method of collecting *multivariate data into *clusters. For a description, *see* AGGLOMERATIVE CLUSTERING METHODS.

completely randomized design An *experimental design in which *treatments are allocated to *experimental units at random.

complexity A measure of the computer time or space required to solve a problem by means of an *algorithm of interest, expressed as a function of the dimensions of the problem. If a problem with n dimensions can be solved in at most $P(n)$ time units, where P is a polynomial, then the algorithm is said to have **polynomial–time complexity**.

composite hypothesis *See* HYPOTHESIS TEST.

compositional variable A *variable describing the breakdown of some quantity into components that sum to the whole. An example is the outcome of the vote in a general election: in each constituency the separate votes for the competing parties sum to 100% of the votes cast. The composition varies from constituency to constituency. When the composition has just three parts a ternary diagram provides an appropriate display.

compound bar chart A bar chart used when a data set is cross-categorized according to two categorical variables. One variable is regarded as the main variable and the bars representing this variable are subdivided according to the other variable.

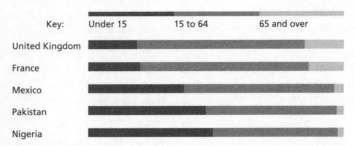

Compound bar chart. The diagram shows the similarity between the age distributions of two European countries and the difference between these and the age distributions of countries in other continents.

compound distribution A *distribution that results from allowing the *parameter of a distribution to vary. For example, if the success *probability of a *binomial distribution is not constant but has a *beta distribution, then the result is a *random variable having a *beta-binomial distribution.

compound Poisson process *See* POISSON PROCESS.

compound probability *See* MULTIPLICATION LAW FOR PROBABILITIES.

computer-intensive method A description of a method that relies on extensive computer calculations to obtain numerical results.

concentration *See* CYCLIC DATA.

conditional autoregressive model A model of a spatial process in which the *variance of the value at each point in space is equal and the expectation of the value at each point is a weighted sum of the observed values at neighbouring points.

conditional distribution; conditional expectation *See* BIVARIATE DISTRIBUTION.

conditional independence If, for each value of the *variable C, the variables A and B are *independent of one another, then they are said to exhibit conditional independence. If the variables are *categorical, with p_{jkl} denoting the *probability of an outcome in *cell (j, k, l), then A and B are conditionally independent, given the category of C, if and only if

$$p_{jkl} = \frac{p_{j0l} \, p_{0kl}}{p_{00l}},$$

for all j, k, and l, where $p_{j0l} = \sum_k p_{jkl}$, $p_{0kl} = \sum_j p_{jkl}$ and $p_{00l} = \sum_j \sum_k p_{jkl}$.

conditional probability If A and B are events (*see* SAMPLE SPACE), and $P(B) > 0$, the conditional probability of A given B is defined by

$$P(A|B) = \frac{P(A \cap B)}{P(B)},$$

or equivalently

$$P(A \cap B) = P(B) \times P(A|B).$$

If $P(A) > 0$ also, then $P(A \cap B) = P(A) \times P(B|A)$, so

$$P(A) \times P(B|A) = P(B) \times P(A|B).$$

The '$A|B$' notation was introduced by *Jeffreys and popularized by *Feller. *See also* INDEPENDENT EVENTS; BAYES'S THEOREM.

confidence ellipsoid The analogue of *confidence interval when a statement is to be made about the likely values of two or more unknown *parameters.

confidence interval An α% confidence interval for an unknown *population parameter θ, say, is an interval, calculated from *sample values by a procedure such that if a large number of *independent samples is taken, α% of the intervals obtained will contain θ. The term 'confidence interval' was introduced in 1934 by *Neyman.

A confidence interval can also be thought of as a single *observation of a random interval, calculated from a random sample by a given procedure, such that the *probability that the interval contains θ is α%. For example, if $X_1, X_2, \ldots, X_n$ is a random sample from a *normal distribution with unknown *mean μ and known *variance σ^2, and writing $\bar{X} = (X_1 + X_2 + \cdots + X_n)/n$,

$$P\left(-1.96\,\frac{\sigma}{\sqrt{n}} < \bar{X} - \mu < 1.96\,\frac{\sigma}{\sqrt{n}}\right) = 0.95,$$

so

$$P\left(\bar{X} - 1.96\,\frac{\sigma}{\sqrt{n}} < \mu < \bar{X} + 1.96\,\frac{\sigma}{\sqrt{n}}\right) = 95\%.$$

Hence, if $\bar{x}$ is the observed value of the sample mean, the end points of the corresponding 95% confidence interval for the mean, μ, are $\bar{x} \pm 1.96\sigma/\sqrt{n}$. This is a **symmetric confidence interval**. It is possible to have a **one-sided confidence interval**. For example $\mu > \bar{x} - 1.645\sigma/\sqrt{n}$ is a one-sided 95% confidence interval for the mean, μ.

If the population variance is not known then, to find a confidence interval for the mean, the *t-distribution can be used and the end points of the 95% confidence interval for the mean are

$$\bar{x} \pm t_{n-1}(0.025)\frac{s}{\sqrt{n}},$$

where $t_{n-1}(0.025)$ is the *critical value corresponding to an upper-tail probability of 2.5% for a t-distribution with $(n-1)$ *degrees of freedom, and s is the unbiased estimate of the population variance based on the sample values.

In the case when the population is not known to be normal the *central limit theorem may be used, provided n is reasonably large, to give the values $\bar{x} \pm 1.96\ s/\sqrt{n}$ as an approximate symmetric 95% confidence interval for μ.

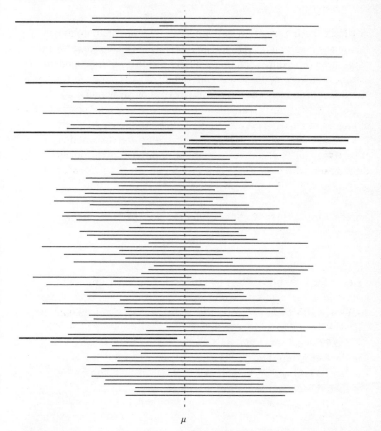

Confidence interval. The illustration shows one hundred 95% confidence intervals for the population mean. Each confidence interval is derived from a random sample from the same distribution. The intervals differ in width and location because of variations in the sample means and variances. On average 95% of such confidence intervals will include the true value of the population mean μ.

Finding a confidence interval for a population *proportion is difficult, owing to the discrete nature of the *binomial distribution, unless the *sample size n is large enough for the normal approximation to the binomial distribution to be valid. In this case the ends of the α%

symmetric confidence interval for the population proportion are the values p such that

$$p = \hat{p} \pm K\sqrt{\frac{p(1-p)}{n}},$$

where $\hat{p}$ is the sample proportion and K is the critical value corresponding to an upper-tail probability of $\frac{1}{2}(100 - \alpha)\%$ for a standard normal distribution.

Equivalently, the $\alpha\%$ confidence limits are the roots of the quadratic equation

$$p^2\left(1 + \frac{K^2}{n}\right) - 2p\left(\hat{p} + \frac{K^2}{n}\right) + \hat{p}^2 = 0.$$

An often used, but not recommended, approximate formula is

$$p = \hat{p} \pm K\sqrt{\frac{\hat{p}(1-\hat{p})}{n}}.$$

Since $\hat{p}(1 - \hat{p}) \approx \frac{1}{4}$ for values of $\hat{p}$ not too close to 0 or 1, an even simpler form is

$$p = \hat{p} \pm K\frac{1}{2\sqrt{n}}.$$

Hence for a sample of size 1 000 the 90% confidence limits are approximated by $p = \hat{p} \pm 0.03$, which is possibly the source of the oft-repeated statement that estimates of percentages from an opinion poll have a possible error of $\pm 3\%$.

A confidence interval for a *population variance σ^2 can be found, under the assumption that the population is normal. If s^2 is the unbiased estimate of the population variance, based on a sample of size n, then the $\alpha\%$ confidence for σ^2 is given by

$$\frac{(n-1)s^2}{U} < \sigma^2 < \frac{(n-1)s^2}{L},$$

where L is the critical value corresponding to a lower-tail probability of $\frac{1}{2}(100 - \alpha)\%$, for a *chi-squared distribution with $(n-1)$ degrees of freedom, and U is the critical value corresponding to an upper-tail probability of the same size (Appendix X).

confidence limit The end point of a *confidence interval.

confounded *See* FACTORIAL EXPERIMENT.

congruential generator method A method for generating a

sequence of *pseudo-random numbers. Let x_n, the nth number in the sequence, be an integer such that $0 \leq x_n \leq m - 1$. Then x_{n+1}, the next number in the sequence, is given by the relation

$$x_{n+1} = (a + bx_n) \qquad \mod (m),$$

where a and b are constants. The right-hand side of this equation should be interpreted as an instruction to subtract a suitable integer multiple of m from $(a + bx_n)$ so as to obtain a value x_{n+1} such that $0 \leq x_{n+1} \leq m - 1$. The values of a, b, and m have to be chosen carefully in order to get a useful sequence of x-values.

conjugate Latin squares Two *Latin squares are conjugate if the rows of one are the columns of the other in the same order.

conjugate prior *See* BAYESIAN INFERENCE.

connected graph *See* GRAPH.

consequent variable *See* ANTECEDENT VARIABLE.

consistency The property of a method that always produces a consistent estimator. The word 'consistency' was first used in this way by Sir Ronald *Fisher in 1922.

consistent estimator *See* ESTIMATOR.

consumer's risk *See* ACCEPTANCE SAMPLING.

contingency table (**cross-classification; cross-tabulation**) A table displaying the *frequencies for each combination of two or more *variables. The variables are either *categorical variables or *numerical variables for which the possible outcomes have been arranged in groups. The term was first used by Karl *Pearson in 1904. Each location in a table is called a **cell**, and the corresponding frequency is the **cell frequency**.

Suppose A and B are two categorical variables having J and K categories, respectively. There are therefore JK possible category combinations. The table described would be called a $J \times K$ table. One simple *model for such a table is the *independence model (*see also* CHI-SQUARED TEST). **Multidimensional contingency tables** summarize information from more than two categorical variables. A three-variable table might be called a

$J \times K \times L$ table. Models used include *logit models and, most commonly, *log-linear models.

continuity correction A correction term used when the distribution of a *discrete random variable is approximated by that of a *continuous random variable. For a discrete random variable, X, taking values $\ldots, x-1, x, x+1, \ldots$, the *probability of X taking the value x would be approximated by

$$P\left(x - \frac{1}{2} < Y < x + \frac{1}{2}\right),$$

where Y is the approximating continuous random variable. *See* BINOMIAL DISTRIBUTION; POISSON DISTRIBUTION; YATES-CORRECTED CHI-SQUARED TEST.

continuous distribution The *probability distribution of a *continuous random variable.

continuous random variable A *continuous variable subject to random variation. The *probability distribution is defined by the *probability density function. It is usual also to require that, for each real number x, $P(X = x) = 0$. As a result it is immaterial whether '<' or '≤' is used in probability statements, since e.g. $P(X \leq x) = P(X < x)$ and $P(x_1 \leq X \leq x_2) = P(x_1 < X < x_2)$.

continuous-time Markov chain *See* MARKOV PROCESS.

continuous variable A *variable whose set of possible values is a continuous interval of real numbers x, such that $a < x < b$, in which a can be $-\infty$ and b can be ∞.

contrasts *See* ANOVA.

control (control treatment) The standard treatment against which new treatments are compared. *See also* RETROSPECTIVE STUDIES.

control chart *See* QUALITY CONTROL.

controlled variable *See* REGRESSION.

control treatment *See* CONTROL.

control variable *See* COVARIATE.

convex hull The convex hull of a set of points in $\mathbb{R}^n$ is the smallest
*convex polyhedron (polygon when $n = 2$) that contains all the points.

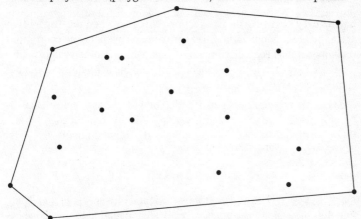

Convex hull. The hull is defined by the positions of the extreme points. In two
dimensions the hull is a convex polygon.

convex polyhedron A finite region bounded by a finite number of
*hyperplanes, in the sense that the interior of the region lies entirely on
one side of each hyperplane.

convolution Given two *independent *random variables X and Y, the
*probability distribution of their sum, Z, is called the convolution of the
distributions of X and Y. If $f_X(x)$, $f_Y(y)$, and $f_Z(z)$ denote the *probability
density functions, then

$$f_Z(z) = \int_{-\infty}^{\infty} f_X(z - y) f_Y(y) dy, \quad \text{or, equivalently,} \quad \int_{-\infty}^{\infty} f_Y(z - x) f_X(x) dx.$$

For *discrete random variables the equivalent result is

$$P(Z = z) = \sum_y P(X = z - y) P(Y = y) = \sum_x P(Y = z - x) P(X = x).$$

Cook, R. Dennis (1944–) An American Statistician. Cook gained his
PhD in 1971 from Kansas State University. His research interests include
*experimental design and *regression diagnostics. He is currently
Professor of Applied Statistics at the University of Minnesota.

Cook's statistic *See* REGRESSION DIAGNOSTICS.

copula A function that relates a joint *cumulative distribution function to the distribution functions of the individual variables. If the individual distribution functions are known, but the joint distribution is unknown, then a copula can be used to suggest a suitable form for the joint distribution.

Let F be the *multivariate distribution function for the *random variables $X_1, X_2, \ldots, X_n$ and let the cumulative distribution function of X_j be F_j (for all j). Define random variables $U_1, U_2, \ldots, U_n$ by $U_j = F_j(X_j)$ for each j, so that the marginal distribution of each U_j has a continuous *uniform distribution in the interval $(0, 1)$. Assume that for each value u_j there is a unique value $x_j = F^{-1}(u_j)$ and let the joint cumulative distribution function of $U_1, U_2, \ldots, U_n$ be C. Then

$$C(u_1, u_2, \ldots, u_n) = P(U_j < u_j \text{ for all } j) = F\{F_1^{-1}(u_1), F_2^{-1}(u_2), \ldots, F_n^{-1}(u_n)\},$$

for all $u_1, u_2, \ldots, u_n$ in $(0, 1)$, since $U_j < u_j$ if and only if $X_j < F_j^{-1}(u_j)$. The function C is called the copula. An equivalent equation to the above is

$$C\{F_1(x_1), F_2(x_2), \ldots, F_n(x_n)\} = F(x_1, x_2, \ldots, x_n),$$

for all $x_1, x_2, \ldots, x_n$, where $u_j = F_j(x_j)$ for each j. **Sklar's theorem**, formulated by Abe Sklar of the Illinois Institute of Technology and published in 1959, states that, for a given F, there is a unique C such that this equation holds.

Note that it may well be that it is not possible to express the inverse functions F_j^{-1} in a simple form (an example is the *multivariate normal distribution).

Assuming that the copula and the marginal distribution functions are differentiable, the corresponding result for *probability density functions is that

$$f(x_1, x_2, \ldots, x_n) = c\{F_1(x_1), F_2(x_2), \ldots, F_n(x_n)\} f_1(x_1)f_2(x_2) \ldots f_n(x_n).$$

The trivial case where $c\{F_1(x_1), F(x_2), \ldots, F(x_n)\} = 1$ corresponds to the case where the n X-variables are *independent. Thus the copula encapsulates the interdependencies between the X-variables and is therefore also known as the **dependence function**. If $c(u_1, u_2, \ldots, u_n)$ is the joint probability density function of $U_1, U_2, \ldots, U_n$, then

$$c(u_1, u_2, \ldots, u_n) = f(x_1, x_2, \ldots, x_n)/\{f_1(x_1)f_2(x_2) \ldots f_n(x_n)\},$$

where $x_j = F_j^{-1}(u_j)$, for each j.

Cornish, Edmund Alfred (1909–73; b. Perth, Australia) An Australian statistician and biometrician. Cornish was educated at the University of Melbourne, where he graduated in agricultural biochemistry. His first job was at an agricultural research institute where he was confronted by the need for statistics. His earliest work was concerned with the 23-year rainfall cycle in Adelaide (related to the well-known sunspot cycle). In 1937, at his own expense, he visited Sir Ronald *Fisher in England. This visit initiated fundamental work on approximations to *distributions. Subsequently Cornish headed the Mathematical Statistics division of the CSIRO (the Commonwealth Scientific and Industrial Research Organization), Australia's largest scientific and industrial research agency. He was President of the *International Biometric Society from 1970 to 1972.

Cornish–Fisher expansion A form of the *Edgeworth expansion, introduced by *Cornish and Sir Ronald *Fisher in 1937. In its most used inverse form, it relates the *cumulative distribution function of a *normal distribution to some distribution of interest.

Denote the $100p$ *percentiles of the normal distribution and of the distribution of interest by u_p and x_p, respectively, and let k_r be the rth *cumulant of the distribution of interest. Then, for all p,

$$
\begin{aligned}
x_p = u_p &+ \frac{1}{6}\left(u_p - 1\right)k_3 + \frac{1}{24}\left(u_p^3 - 3u_p\right)k_4 - \frac{1}{36}\left(2u_p^3 - 5u_p\right)k_3^2 \\
&+ \frac{1}{120}\left(u_p^4 - 6u_p^2 + 3\right)k_5 - \frac{1}{24}\left(u_p^4 - 5u_p^2 + 2\right)k_3k_4 \\
&+ \frac{1}{324}\left(12u_p^4 - 53u_p^2 + 17\right)k_3^3 + \frac{1}{720}\left(u_p^5 - 10u_p^3 + 15u_p\right)k_6 - \cdots.
\end{aligned}
$$

corrected moment *See* MOMENT.

correlated variables *Variables that display a non-zero *correlation. Correlated variables are not statistically independent.

correlation A general term used to describe the fact that two (or more) *variables are related. *Galton, in 1869, was probably the first to use the term in this way (as 'co-relation'). Usually the relation is not precise. For example, we would expect a tall person to weigh more than a short person of the same build, but there will be exceptions.

Although the word 'correlation' is used loosely to describe the existence of some general relationship, it has a more specific meaning in the context of linear relations between variables. *See* CORRELATION COEFFICIENT.

correlation coefficient, population A measure of the *linear dependence of one numerical *random variable on another. The phrase 'coefficient of correlation' was apparently originated by *Edgeworth in 1892. It is usually denoted by ρ (rho). The value of ρ, which lies between -1 and 1, inclusive, is defined as the ratio of the *covariance to the square root of the product of the *variances of the *marginal distributions of the individual variables:

$$\rho = \frac{\mathrm{Cov}(X, Y)}{\sqrt{\mathrm{Var}(X)\mathrm{Var}(Y)}}.$$

If the correlation coefficient between the random variables X and Y is equal to 1 or -1 then this implies that $Y = a + bX$, where a and b are constants. If b is positive then $\rho = 1$ and if b is negative then $\rho = -1$. The converse statements are also true.

If X and Y are completely unrelated (i.e. are *independent) then $\rho = 0$. If $\rho = 0$ then X and Y are said to be **uncorrelated**. However, ρ is concerned only with linear relationships, and the fact that $\rho = 0$ does not imply that X and Y are independent.

correlation coefficient, sample (product-moment correlation coefficient) If the n pairs of values of *random variables X and Y in a random sample are denoted by $(x_1, y_1), (x_2, y_2), \ldots, (x_n, y_n)$, the sample correlation coefficient r is given by

$$r = \frac{S_{xy}}{\sqrt{S_{xx}S_{yy}}},$$

where

$$S_{xy} = \sum_{j=1}^{n} x_j y_j - \frac{1}{n}\left(\sum_{j=1}^{n} x_j\right)\left(\sum_{j=1}^{n} y_j\right), \qquad S_{xx} = \sum_{j=1}^{n} x_j^2 - \frac{1}{n}\left(\sum_{j=1}^{n} x_j\right)^2,$$

and S_{yy} is defined analogously to S_{xx}. If the sample means are denoted by $\bar{x}$ and $\bar{y}$, alternative definitions are

$$S_{xy} = \sum_{j=1}^{n} x_j y_j - n\bar{x}\bar{y}, \qquad S_{xx} = \sum_{j=1}^{n} x_j^2 - n\bar{x}^2.$$

The coefficient r can take any value from -1 to 1, inclusive. When increasing values of one variable are accompanied by generally increasing values of the other variable then $r > 0$ and the variables are said to display **positive correlation**. If $r < 0$ then the variables display **negative correlation**.

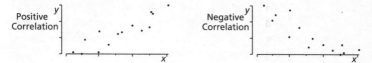

Sample correlation coefficient. When the correlation between two variables is positive, the values of one variable rise as the values of the other variable rise. The correlation is negative if the values of one variable rise as the values of the other fall.

The idea of correlation was put forward by *Galton in 1869, and it was Galton who was the first to denote it by the symbol r in 1888. The formulae given here were introduced by Karl *Pearson in 1896.

The sample correlation coefficient r is an estimate of the population *correlation coefficient ρ.

Correlation is closely linked to *linear regression. If the least squares regression lines of y on x and of x on y for the sample $(x_1, y_1), (x_2, y_2), \ldots, (x_n, y_n)$ are, respectively, $y = a + bx$ and $x = c + dy$ then $r^2 = bd$.

In a *hypothesis test, to test for significant evidence of a linear relationship between X and Y, we compare the *null hypothesis that $\rho = 0$ with the *alternative hypothesis that $\rho \neq 0$, rejecting the null hypothesis if $|r|$ is too large. Tables of critical values are given in Appendix XIII. If the assumption of normality cannot be made, then a *rank correlation coefficient may be preferred.

The square of r^2 is the **coefficient of determination**, and $1 - r^2$ is the **coefficient of alienation**.

correlation matrix A square symmetric *matrix in which the element in row j and column k is equal to the *correlation coefficient between *random variables X_j and X_k. The diagonal elements are each equal to 1.

correlogram *See* AUTOCORRELATION.

correspondence analysis A technique, originating in the 1930s, that results in the display of *data from a *contingency table in a *scatter diagram that includes points representing row categories and points representing column categories. If row points are positioned near each other in the diagram then this implies that the patterns of *counts along those rows are very similar. The same applies for groups of column points. If a row point and a column point are positioned close to one another then this implies a positive *association between the two. The calculations involved resemble those for *principal components analysis.

count A synonym for *frequency.

countable (denumerable) A set is countable if its members can be listed as a finite or infinite sequence, $x_1, x_2, \ldots$. The rational numbers are countable, but the irrational numbers, even in a finite interval, are not.

counting process A *stochastic process, $X_1, X_2, \ldots$, in which X_t is the number of *events (for some definition of an event) that have occurred by time t. One example is a *Poisson process.

coupon-collecting distribution *See* ARFWEDSON DISTRIBUTION.

covariance The covariance of two *random variables is the difference between the *expected value of their product and the product of their separate expected values. For random variables X and Y,

$$\mathrm{Cov}(X, Y) = \mathrm{E}(XY) - \mathrm{E}(X)\mathrm{E}(Y).$$

If X and Y are *independent then $\mathrm{Cov}(X, Y) = 0$. However, if $\mathrm{Cov}(X, Y) = 0$ then X and Y may not be independent. The term was used by Sir Ronald *Fisher in 1930. *See also* CORRELATION.

covariance models *See* ANOCOVA.

covariate (control variable) A *variable that has an effect that is of no direct interest. The analysis of the variable of interest is made more accurate by controlling for variation in the covariate.

covariogram *See* AUTOCORRELATION.

Cox, Sir David Roxbee (1924–; b. Birmingham, England) An English statistician knighted for his services to *Statistics. Cox was an undergraduate at Cambridge University and gained his doctorate at Leeds University. After employment in the Royal Aircraft Establishment and the Wool Industries Research establishment, he joined the statistics faculty at Cambridge University. In 1955 he moved to Birkbeck College,

London and in 1966 he was appointed Professor of Statistics at Imperial College, London. He was Warden of Nuffield College, Oxford from 1989 to 1994. He was Editor of *Biometrika* from 1965 to 1991. He has been President of the *Bernoulli Society (1979), the *Royal Statistical Society (1980), and the *International Statistical Institute (1995). He received the Royal Statistical Society's *Guy Medal in Silver in 1961 and in Gold in 1973. He was elected a Fellow of the Royal Society in 1973 and was knighted in 1985. He was elected to the National Academy of Sciences in 1988.

Cox, Gertrude Mary (1900–78; b. Dayton, Iowa; d. Durham, North Carolina) An American biometrician. Cox initially intended to be a deaconess in the Methodist Episcopal Church and did not start her studies at Iowa State College, until 1927. By 1933, however, she was a faculty member specializing in *experimental design. In 1940 she became head of the new Department of Experimental Statistics in the School of Agriculture at North Carolina State College. Cox was a pioneer in the use of computer programs, her staff developing many of the early *SAS algorithms. She was joint author, with William *Cochran, of the statistical classic *Experimental Designs*. She was President of the *American Statistical Association in 1956. She was the founding Editor of the journal *Biometrics* in 1945, remaining Editor until 1955. She was President of the *International Biometric Society in 1968 and was elected to the National Academy of Sciences in 1975.

Cox–Mantel test A *non-parametric test for comparing two *survival curves, which results from the work of Sir David *Cox and Nathan *Mantel. Denote the total number of deaths in the second group by D_2, and let the ordered survival times of the combined group be $t_{(1)} < t_{(2)} < \cdots < t_{(k)}$. The test statistic, C, is given by $C = U/\sqrt{I}$, where

$$U = D_2 - \sum_{j=1}^{k} m_{(j)} p_{(j)},$$

$$I = \sum_{j=1}^{k} \frac{m_{(j)}\big(d_{(j)} - m_{(j)}\big)}{d_{(j)} - 1} p_{(j)}\big(1 - p_{(j)}\big),$$

and $m_{(j)}$ is the number of survival times equal to $t_{(j)}$, $d_{(j)}$ is the total number of individuals who died (or were censored) at time $t_{(j)}$, and $p_{(j)}$ is the proportion of these who were in the second group. If the differences between the survival curves are attributable to random variation then C has a *standard normal distribution.

Cox process (doubly stochastic point process) A *Poisson process, introduced in 1955 by Sir David *Cox, in which the *mean is not constant but varies randomly in space or time.

Cox–Snell residuals Residuals introduced in 1968 by Sir David *Cox and E. Joyce Snell, for assessing the validity of a *survivor function that has been proposed for a set of survival *data. The value of the survivor function depends on the time, t, and on one or more *parameters estimated by the *vector $\hat{\boldsymbol{\theta}}$. The Cox–Snell residual r_j, corresponding to time t_j, is given by

$$r_j = -\ln\left\{ S\left(t_j;\, \hat{\boldsymbol{\theta}} \right) \right\},$$

where $S(t_j;\, \hat{\boldsymbol{\theta}})$ is the value of the estimated survivor function at time t_j. If the model is correct, then the residuals should have an *exponential distribution with mean 1.

Cox's regression model A *model, proposed by Sir David *Cox in 1972, for the lifetime of, for example, industrial components or medical patients. The model has the form

$$\ln\{h(t)\} = \alpha + \mathbf{x}'\boldsymbol{\beta},$$

where $h(t)$ is the *hazard rate at time t, α is a constant *parameter, $\boldsymbol{\beta}$ is a column *vector of slope parameters, and $\mathbf{x}'$ is a row vector of values of *background variables.

C_p See MALLOWS'S C_p.

Cramér, Carl Harald (1893–1985; b. Stockholm, Sweden; d. Stockholm, Sweden) A Swedish mathematical statistician who spent his entire career at Stockholm University. He entered as a student in 1912 and retired as its President in 1961. His research centred on probability, risk theory, and the mathematical underpinnings of Statistics. His best known work is *Mathematical Methods in Statistics*, published in 1945. He received the *Royal Statistical Society's *Guy Medal in Gold in 1972. He was elected to membership of the National Academy of Sciences in 1984.

Cramér–Rao inequality; Cramér–Rao lower bound See FISHER'S INFORMATION.

Cramér's V A *measure of association for a $J \times K$ *contingency table. Let f_{jk} denote a *cell frequency, with f_{j0}, f_{0k}, and f_{00} denoting the jth row total, the kth column total, and the grand total, respectively. For the *null hypothesis of *independence between the two *categorical variables, the *chi-squared test statistic, X^2, is given by

$$X^2 = \sum_{j=1}^{J} \sum_{k=1}^{K} \frac{(f_{jk} - e_{jk})^2}{e_{jk}},$$

where

$$e_{jk} = \frac{f_{j0}f_{0k}}{f_{00}}.$$

Cramér's V represents an attempt to create a function of X^2 that takes values in the interval between 0 and 1:

$$V = \sqrt{\frac{X^2}{Mf_{00}}},$$

where M is the smaller of $(I - 1)$ and $(J - 1)$.

Cramér–von Mises test An alternative to the *Kolmogorov–Smirnov test for testing the *hypothesis that a set of *data come from a specified *continuous distribution. The *test was suggested independently by *Cramér in 1928 and *von Mises in 1931. The test statistic W (sometimes written as W^2) is formally defined by

$$W = n \int_{-\infty}^{\infty} \{F_n(x) - F_0(x)\}^2 f_0(x) \mathrm{d}x,$$

where $F_0(x)$ is the *distribution function specified by the *null hypothesis, $F_n(x)$ is the *sample distribution function, and $f_0(x) = F'_0(x)$. In practice the statistic is calculated using

$$W = \frac{1}{12n} + \sum_{j=1}^{n} \left(t_j - \frac{2j-1}{2n} \right)^2,$$

where

$$t_j = F_0\left(x_{(j)}\right),$$

and $x_{(j)}$ is the jth ordered observation ($x_{(1)} \leq x_{(2)} \leq \cdots \leq x_{(n)}$).
    The test has been adapted for use with *discrete random variables, for cases where *parameters have to be estimated from the data, and for

comparing two samples. A modification leads to the *Anderson–Darling test.

critical path analysis A method of analysis aimed to schedule a set of tasks so that they are all completed in the shortest possible overall time. The difficulty is that the various tasks take different lengths of time to complete and that each task cannot be started until certain prerequisite other tasks have been completed. *See also* NETWORK FLOW PROBLEMS.

critical region The set of values of the *statistic, in a *hypothesis test, which lead to rejection of the *null hypothesis.

critical value An end point of a *critical region. In a *hypothesis test, comparison of the value of a test statistic with the appropriate critical value determines the result of the test. For example, 1.96 is the critical value for a two-tailed test in the case of a *normal distribution and a 5% significance level: thus if the test statistic z is such that $|z| > 1.96$ then the *alternative hypothesis is accepted in preference to the *null hypothesis.

Cronbach, Lee Joseph (1916–2001; b. Fresno, California; d. Palo Alto, California) An American psychologist. Cronbach graduated from Fresno State College in 1934, gaining his MS from the University of California at Berkeley in 1934 and his PhD in educational psychology from the University of Chicago in 1940. After posts at several universities, Cronbach joined the faculty at Stanford University in 1964, where he became Professor Emeritus of Education. His classic book *Essentials of Psychological Testing* was published in 1949; a fifth edition came out in 1990. He was elected to the National Academy of Sciences in 1974.

Cronbach's alpha *See* RELIABILITY.

cross-classification *See* CONTINGENCY TABLE.

cross-correlation The *correlation between a selected set of *data and the corresponding set displaced in time and/or space.

crossed design An *experimental design in which every *level of one *variable occurs in combination with every level of every other variable. An example is a *randomized block design, in which each *treatment occurs once within each block. *See also* NESTED DESIGN.

Cross–Fratar procedure *See* DEMING–STEPHAN ALGORITHM.

crossover trial An *experimental design in which each *experimental
unit is used with each *treatment being studied. The simplest crossover
trial uses two groups of experimental units (e.g. hospital patients), 1 and
2, and two treatments (e.g. medicines), A and B. The trial uses two equal-
length time periods. In the first period, group 1 is assigned treatment A
and group 2 is assigned treatment B. In the second period, the
assignments are reversed.

A complication in crossover trials, unless there is a protracted gap
between the two periods, is that the treatment allocated in the first period
may continue to have an effect (the **carry-over effect**) during the second
period. More complex allocations aim to estimate these effects. An
example—involving two treatments, four groups, and three time
periods—is designed to help with the estimation of the carry-over effects.

	GROUP 1	GROUP 2	GROUP 3	GROUP 4
time 1	A	A	B	B
time 2	B	B	A	A
time 3	A	B	A	B

The analysis of such a design is not simple; from the statistician's
viewpoint (though not the patient's) it would be preferable to minimize
the carry-over effects by allowing an interval (the **wash-out period**)
between successive treatments.

cross-sectional study A study of a human *population by means of a
*sample that includes representatives of all sections of society. An
alternative to a *longitudinal study.

cross-tabulation *See* CONTINGENCY TABLE.

cross-validation A method of assessing the accuracy and validity of a
statistical *model. The available *data set is divided into two parts.
Modelling of the data uses one part only. The model selected for this part
is then used to predict the values in the other part of the data. A valid
model should show good predictive accuracy.

cube law An empirical law relating to the outcome of two-party
multiple constituency elections. It states that if the votes gained by the
parties are in the ratio p to $(1 - p)$, then the numbers of constituencies

won by the parties will be in the ratio p^3 to $(1-p)^3$. A slight majority of votes (e.g. $p = 55\%$) leads to a much larger imbalance in constituencies won, since $0.55^3/(0.55^3 + 0.45^3) = 65\%$. In the United Kingdom the law worked well for many years, though more recently the imbalance has been less extreme.

cumulant-generating function (cgf) *See* CUMULANTS.

cumulants An alternative to *moments as a summary of the form of a *distribution. If the *moment-generating function of a distribution exists, then its natural logarithm exists and is called the **cumulant-generating function (cgf)**. The coefficient of $t^r/r!$ in the *Taylor expansion of the cgf is called the rth cumulant and is denoted by κ_r (where κ is kappa). The cumulants can be expressed in terms of the central moments, and vice versa. In particular, denoting the *mean by μ and the central moments by $\mu_2, \mu_3, \ldots$

$$\kappa_1 = \mu,$$
$$\kappa_2 = \mu_2,$$
$$\kappa_3 = \mu_3,$$
$$\kappa_4 = \mu_4 - 3\mu_2^2,$$
$$\kappa_5 = \mu_5 - 10\mu_3\mu_2.$$

For an example of the application of cumulants, *see* CORNISH–FISHER EXPANSION.

cumulative distribution function (distribution function)
The function F, for a *random variable X, defined for all real values of x by

$$F(x) = P(X \leq x).$$

Clearly, $F(-\infty) = 0$, and $F(\infty) = 1$, where $F(-\infty) = 0$ and $F(\infty)$ are the limits of $F(x)$ as x tends to $-\infty$ and ∞, respectively. This function is a non-decreasing function such that if $x_2 > x_1$ then $F(x_2) \geq F(x_1)$. If X is a *continuous random variable then F is a continuous function, and conversely. If X has *probability density function f then

$$F(x) = \int_{-\infty}^{x} f(t)dt$$

and $f(x) = F'(x)$, where $F'(x)$ denotes the derivative of $F(x)$.

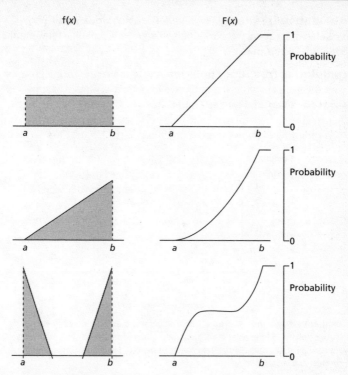

Cumulative distribution function. In each example the probability density function f defined in the interval (a, b) is shown on the left, with the corresponding cumulative distribution function F on the right. The scale from 0 to 1 refers to F.

A useful property of F is that, for any value of x, there is a corresponding value u, $0 \leq u \leq 1$, such that

$$u = F(x).$$

In the case where F is a continuous and increasing function for $a \leq x \leq b$, the random variable U defined by $U = F(X)$ has a continuous *uniform distribution on the interval $0 \leq u \leq 1$, and, for a given value of U, the corresponding value of X is given by $F^{-1}(U)$. *See* SIMULATION.

In the case of a *discrete random variable, the distribution function is a step function, in which the step at x_j is $P(X = x_j)$, and $F(x) \to F(x_j)$ as $x \to x_j$ from above, but $F(x) \to F(x_{j-1})$ as $x \to x_j$ from below.

cumulative frequency For a *sample of numerical data the cumulative frequency corresponding to a number x is the total number of *observations that are $\leq x$.

cumulative frequency polygon A diagram representing grouped numerical data in which *cumulative frequency is plotted against upper *class boundary, and the resulting points are joined by straight line segments to form a polygon. The polygon starts at the point on the x-axis corresponding to the lower class boundary of the lowest class. *See* OGIVE.

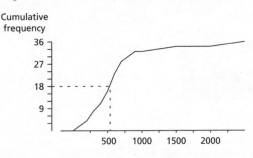

Cumulative frequency polygon. The diagram refers to the distances between the breeding colony and the point of recovery for a group of razorbills. The outline is typical, with slow increases at the left and right ends of the polygon indicating the scarcity of the corresponding values.

cumulative odds ratio A function used in models for *ordinal variables. Let j denote a possible value of the ordinal variable, Y, and let E_1 and E_2 be two possible events. The ratio R_j, given by

$$R_j = \frac{P(Y \leq j|E_1)}{P(Y > j|E_1)} \Big/ \frac{P(Y \leq j|E_2)}{P(Y > j|E_2)},$$

is called a cumulative *odds ratio. If E_1 and E_2 represent different values for a *vector of explanatory variables, the model $R_j = R$, where R is a constant, is a **proportional-odds model**.

cumulative probability For a *discrete random variable X that can take only the ordered values $x_{(1)} < x_{(2)} < \cdots$, the cumulative probability of a value less than or equal to $x_{(k)}$ is

$$\sum_{j=1}^{k} P\big(X = x_{(j)}\big).$$

cumulative probability function *See* CUMULATIVE DISTRIBUTION FUNCTION.

cumulative relative frequency *Cumulative frequency divided by total sample size.

cumulative relative frequency diagram (cumulative relative frequency graph) A *cumulative frequency polygon or a *step diagram in which the vertical axis has been scaled by dividing by the sample size so that the maximum *ordinate value is 1.

cumulative sum chart *See* QUALITY CONTROL.

curvilinear relation A relation between two *variables that is not linear but appears as a curve when the relation is graphed.

cusum chart *See* QUALITY CONTROL.

cut *See* NETWORK FLOW PROBLEMS.

cycle A repeating pattern in a *time series; examples are annual and daily patterns.

cyclic data *Data consisting of directions or times in which the measurement scale is cyclic (after 23.59 comes 00.00, after 359° comes 0°, after 31 December comes 1 January). Special techniques are required for summarizing and modelling all types of cyclic data. For example, the *histogram is replaced by the *circular histogram or the *rose diagram, and the principal distribution used to model the data is not the *normal distribution but the *von Mises distribution.

For *observations $\theta_1, \theta_2, \ldots, \theta_n$, the variability of the data is measured by the **concentration**, $\bar{R}$, defined by

$$\bar{R} = \frac{1}{n} \sqrt{\left(\sum_{j=1}^{n} \sin \theta_j \right)^2 + \left(\sum_{j=1}^{n} \cos \theta_j \right)^2}.$$

Thus $\bar{R} = R/n$, where R is the length of the **resultant vector**.

The **circular mean**, $\bar{\theta}$, is defined only when $\bar{R} \neq 0$ and is then the angle $(0° \leq \bar{\theta} < 360°)$ such that

$$n\bar{R}\sin(\bar{\theta}) = \sum_{j=1}^{n}\sin\theta_j, \qquad n\bar{R}\cos(\bar{\theta}) = \sum_{j=1}^{n}\cos\theta_j.$$

One way of representing cyclic data is to regard each observation as a move of length 1 unit in the stated direction. The complete sequence of such moves, taken in any order, will end at a finishing point that is a distance $n\bar{R}$ from the start. The direction of this finishing point from the start will be the angle $\bar{\theta}$.

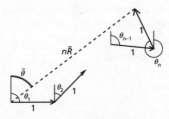

Cyclic data. Each data item is represented by a unit vector. The diagram shows the first two such vectors, and also the last two. The resultant vector, with length $n\bar{R}$, connects the start of the first unit vector to the end of the last unit vector. The direction of the resultant vector is $\bar{\theta}$.

D'Agostino, Ralph Benedict (1940–; b. Somerville, Massachusetts)
An American biostatistician. D'Agostino obtained his first degrees from
Boston University. After obtaining a PhD from Harvard University in
1968, he returned to Boston where *inter alia* he is now Executive Director
of the biostatistics programme.

D'Agostino's test A *test for *normality suggested by *D'Agostino in
1971. With ordered *observations $x_{(1)} \leq x_{(2)} \leq \cdots \leq x_{(n)}$, the test statistic
is D, given by

$$D = \frac{\sum_{j=1}^{n}\left\{j - \frac{1}{2}(n+1)\right\}x_{(j)}}{n\sqrt{n\sum_{j=1}^{n}\left(x_{(j)} - \bar{x}\right)^2}},$$

where $\bar{x}$ is the sample mean. Special tables are required.

damned lies In his autobiography Mark Twain attributed to Disraeli
the statement 'There are three sorts of lies: lies, damned lies and
statistics.' This might be interpreted as a comment on the difficulty of
interpreting data.

Daniell weights *See* MOVING AVERAGE.

Dantzig, George Bernard (1914–; b. Portland, Oregon) An American
*operational researcher. Dantzig is the originator of the *simplex
method and has been described as the father of *linear programming. He
studied mathematics at the Universities of Maryland and Michigan just
before the Second World War, during which he was head of the Combat
Analysis Branch of the United States Air Force's statistics section. His
work in the air force involved 'programming', which in those days
meant the production of efficient scheduling for the training and
deployment of men. He was awarded his PhD at the University of
California, Berkeley in 1946. He devised the simplex method in August
1947 while working at the Pentagon. Subsequent appointments
included chairs at the University of California at Berkeley and at Stanford
University. He was awarded the National Medal of Science in 1975.

data Information, usually *numerical or *categorical.

database A record, kept in a suitably accessible form on a computer, in which the values of several variables, which may be *categorical or *numerical, are separately recorded for each sampling unit.

data mining The exploration of a large set of *data with the aim of uncovering relationships between *variables.

data reduction The process of reducing an unmanageably large amount of data to produce a few *summary statistics (and graphical displays such as *histograms and *scatter diagrams) that encapsulate the useful content of the original data.

datum A single item of *data.

Daubechies, Ingrid (1954–; b. Houthalen, Belgium) A Belgian mathematician now working in the United States. Daubechies was educated at the Free University, Brussels, where she studied physics, gaining her doctorate in 1980 and joining the staff in 1984. She has developed an international reputation for her work on *wavelets. In 1987 she moved to AT&T Bell Laboratories in the United States. In 1991 she became Professor of Mathematics at Rutgers University, moving to her present post at Princeton University in 1993. She was elected to the National Academy of Sciences in 1998.

death rate *See* MORTALITY RATE.

decile An approximate value for the rth decile of a *data set can be read from a *cumulative frequency graph as the value of the *variable corresponding to a *cumulative relative frequency of $10r\%$. So the 5th decile is the *median and the 2nd decile is the 20th *percentile. The term decile was introduced by *Galton in 1882.

decision rule *See* GAME THEORY.

decision theory A method of deciding between various alternative experiments and actions. The decision process is informed by assigning *probabilities to the various alternatives and assigning costs (or benefits or *utilities) to the alternative outcomes. The decision process is governed by the wish to minimize the expected cost or to maximize the expected benefit or utility.

decision tree A graphical representation of the alternatives in a decision-making problem. As an example, suppose that we are choosing between two machines. One costs 100 units and has a 20% probability of breaking down within a year. The other costs 120 units and has a 5% breakdown probability. A breakdown costs 60 units. Ignoring the possibility of multiple breakdowns, which machine should we buy? Reading the decision tree from right to left, we see that it is cheaper to buy the 100-unit machine.

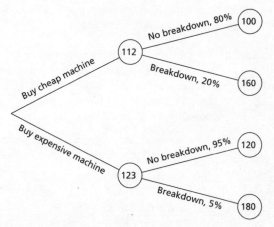

Decision tree. A circle at the right contains the total cost if this path occurs. An intermediate circle gives the expected costs at this stage. Thus, for example, $112 = 100 \times 80\% + 160 \times 20\%$. If cost is the only consideration, then in the case illustrated the cheaper machine is the better buy.

decomposable models *See* DIRECT MODELS.

defective An item that does not meet the required standard of a production process.

defender's fallacy A mis-statement of a *probability as a result of a misunderstanding of *conditional probability. *See also* PROSECUTOR'S FALLACY. As an example, suppose that a blood type possessed by only 1% of the population is found at a crime scene. The accused has blood of this type. The defender argues that, since the population size is one million, there are 10 000 with this blood type and therefore the probability that it is the accused who is guilty is 1 in 10 000.

For a randomly chosen person, let A be the *event 'person chosen has blood of this type' and let B be the event 'person chosen is guilty'. The figure quoted by the defender is $P(B|A)$, whereas the probability of interest is P(accused is guilty).

defining contrast *See* FACTORIAL EXPERIMENT.

degree *See* POLYNOMIAL.

degrees of freedom (df) A *parameter that appears in some *probability distributions used in *statistical inference, particularly the *t-distribution, the *chi-squared distribution, and the *F-distribution. The phrase 'degrees of freedom' was introduced by Sir Ronald *Fisher in 1922.

In the case of the t-distribution, the term usually reflects the fact that the population variance has been estimated. The number of degrees of freedom is equal to the number of independent pieces of information concerning the variance. In the most familiar case of n observations, $x_1, x_2, \ldots, x_n$, from a population with unknown mean and variance, there are $(n - 1)$ independent pieces of information, for example, $(x_1 - \bar{x})$, $(x_2 - \bar{x}), \ldots, (x_{n-1} - \bar{x})$, where $\bar{x} = (\sum_{j=1}^{n} x_j)/n$. In this case, therefore, there are $(n - 1)$ degrees of freedom.

In the case of a *random variable with a chi-squared distribution, if it can be expressed as the sum of squares of m *independent *standard normal variables, then the distribution has m degrees of freedom.

In the case where the chi-squared distribution is used as a *goodness-of-fit test, each independent parameter estimated from the data represents another constraint, so the number of degrees of freedom is $(c - 1 - p)$, where c is the number of cells, p is the number of parameters estimated from the sample data, and there is the constraint that the sum of the observed frequencies is the sum of the expected frequencies.

In the case where the chi-squared distribution is used to test the *null hypothesis of *independence in a $J \times K$ *contingency table, there are $(J - 1)(K - 1)$ degrees of freedom.

Delaunay, Charles Eugene (1816–72; b. Lusigny-sur-Barse, France; d. Cherbourg, France) A French mathematician. Delaunay studied at the Sorbonne and taught at the École Polytechnique. His theory of lunar motion advanced the development of theories of planetary motion. He drowned in a boating accident. A street in Paris and a lunar crater are named after him.

Delaunay triangles Triangles that provide a tessellation of the plane that is related to the *Dirichlet tessellation. If two points define the edge of a Dirichlet tile, then they form a side of a Delaunay triangle.

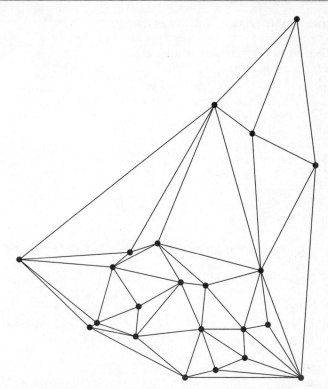

Delaunay triangles. The diagram shows the tessellating triangles corresponding to the locations of 22 rain-gauges in the Bolton area, England.

deletion residual A *statistic that provides an effective way of detecting an *outlier observation. The deletion residual corresponding to the jth of n *observations is calculated by comparing its value with the corresponding *fitted value based on the *model with the *parameters estimated from the remaining $(n - 1)$ observations. *See* REGRESSION DIAGNOSTICS.

Deming, Wiliiam Edwards (1900–93; b. Sioux City, Iowa; d. Washington, DC) An American engineer and statistician. Deming was a pioneer of *quality control. He was voted by business staff of the *Los Angeles Times* as being one of the 50 most influential business people of the century. He studied electrical engineering at the University of

Wyoming, graduating in 1921. As a summer job he worked for the Western Electric Company in Chicago where he encountered *Shewhart's work on quality control. He obtained his MS in mathematics and mathematical physics from the University of Colorado in 1925 and his PhD from Yale University in 1928. Deming began working first for the US Department of Agriculture and then for the US Bureau of the Census. In 1947 he spent three months in Japan helping with the Japanese census. On his return to Japan in 1950 he gave an extended course in quality control. The course was so successful and influential that he was invited back on many occasions, being received by Emperor Hirohito and awarded the Second Order of the Sacred Treasure. He was President of the *Institute of Mathematical Statistics in 1945. In 1983 he was awarded the *Wilks Medal of the *American Statistical Association.

Deming–Stephan algorithm (iterative proportional fitting) An algorithm, given by *Deming and *Stephan in 1940, that can be used to fit *log-linear models to *contingency tables. The algorithm matches marginal totals by using iterative scaling of a table whose entries are initially all equal to unity. Essentially the same algorithm allows one to revise sample margins to preserve known population values: depending on context this is known as the **Cross–Fratar procedure**, **raking**, or **structure-preserving estimation (SPREE)**.

demography The study of human populations, particularly with respect to births, marriages, deaths, employment, migration, health, etc.

de Moivre, Abraham (1667–1754; Vitry, France; d. London, England) A French mathematical prodigy. De Moivre studied at the University of Sedan when aged eleven, and later at the University of Saumur and the Sorbonne. In 1688 he emigrated to London to avoid persecution (he was a Protestant). In London he initially made a living by giving advice to gamblers and stock-exchange workers in the coffee-houses. In 1697 he was elected a Fellow of the Royal Society. In his 1711 paper *de Mensura Sortis* he derived both the distribution now called the *Poisson distribution and the *hypergeometric distribution. He is also famed for his work on *probability, encapsulated in his book *The Doctrine of Chances*, and he is credited with the first derivation of the *probability density function of the *normal distribution.

de Morgan, Augustus (1806–71; b. Madura, India; d. London, England) A British mathematician. De Morgan established the logical footing of mathematical induction, and his interest in algebra led to the formulation of *de Morgan's laws. De Morgan did not excel at school, having lost the sight of an eye shortly after birth, but nevertheless went to Cambridge University when aged sixteen. He refused to take the theological test required for an MA and, as a result, was not eligible for a Fellowship. After studying law at Lincoln's Inn for a year, he applied for the chair of mathematics at the newly founded University College, London. Although he had no mathematical publications he was appointed in 1828. De Morgan was a major contributor to the popular *Penny Cyclopedia*, contributing over 700 articles. He was described as a 'dry, dogmatic pedant' and during his career he twice resigned his chair on matters of principle—being re-appointed each time—and he refused to allow his name to be put forward for election to the Royal Society. He was the first President of the London Mathematical Society. A lunar crater is named after him.

de Morgan's laws *See* BOOLEAN ALGEBRA.

Dempster, Arthur Pentland (1929–) An American statistician. An inaugural member of the Statistics Department at Harvard University in 1957, Dempster was Department Chair from 1969 to 1982 and is currently Professor of Theoretical Statistics. In 1977 he was a co-author of the paper introducing the *EM algorithm that deals with the problem of the analysis of incomplete *multivariate *data.

dendrogram A diagram used in the context of *cluster analysis to trace the stages in the aggregation (or disaggregation) of clusters. Thus, in the following diagram, reading from left to right, items 1 and 2 are the first to join together. Then 3 and 4 join together and are subsequently joined by 5. Finally, all five items join in a single cluster. The horizontal axis is used to indicate the distances at which the joins occur; in the example items 1 and 2 are much closer together than items 3 and 4.

density estimation The process of estimating the *population *probability density function from a *sample of *observations. The question is how best to estimate the function. One standard method is to build up the estimate by assigning a rectangular unit area to each observation: this leads to the *histogram. A more sophisticated approach uses *kernel methods.

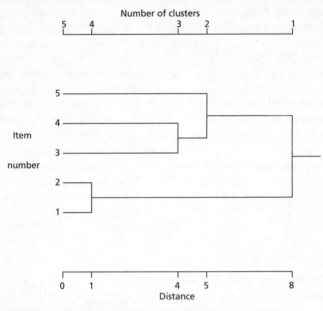

Dendrogram. The horizontal distances at which joins occur indicate the distances between the items.

density function *See* PROBABILITY DENSITY FUNCTION.

denumerable *See* COUNTABLE.

dependence function *See* COPULA.

dependent variable *See* REGRESSION.

Descartes, René (1596–1650; b. La Haye, France; d. Stockholm, Sweden) A French philosopher and mathematician. His birthplace is now named Descartes (a lunar crater and a street in Paris also bear his name). Descartes was initially educated at a Jesuit college in Anjou. At that time his health was poor and he was permitted to remain in bed until 11.00 a.m.—he continued this habit throughout his life. His treatise on universal science published at Leyden in 1637 included three appendices. One, '*La Géométrie*', introduced the ideas of coordinate geometry and in particular the use of *Cartesian coordinates.

deseasonalized data If a *time series exhibits regular seasonal fluctuations then for the purposes of analysis (for example, to estimate an underlying *trend) it is often necessary to remove the *seasonality to leave deseasonalized data.

design matrix *See* MULTIPLE REGRESSION MODEL.

determinant *See* MATRIX.

deterministic An outcome that can be predicted exactly (or a method with a perfectly predictable outcome).

detrending The removal of a *trend from a *time series.

Deutsche Statistiche Gesellschaft The German Statistical Society, founded in 1911.

deviance A measure for judging the extent to which a *model explains the variation in a set of *data when the *parameter estimation is carried out using the method of *maximum likelihood. The deviance, D, is given by

$$D = -2(\ln L_c - \ln L_s) \qquad \text{or, equivalently,} \, D = -2\ln\left(\frac{L_c}{L_s}\right),$$

where L_c is the likelihood of the current model, and L_s is the likelihood of the *saturated model. The deviance has an approximate *chi-squared distribution; the number of *degrees of freedom is given by the difference in the numbers of parameters in the two models.

deviance information criterion (DIC) A modification of the *AIC for use in cases where the *model parameters are not fixed but are subject to *random variation.

deviance residuals An alternative to *Anscombe residuals and Pearson residuals (*see* CHI-SQUARED TEST). In the case of a *generalized linear model with *random errors having a *Poisson distribution, the deviance residuals are given by

$$r = \frac{y - \hat{y}}{|y - \hat{y}|} \sqrt{2y\ln\left(\frac{y}{\hat{y}}\right) - y + \hat{y}},$$

where y and $\hat{y}$ are the corresponding *observed and *fitted values.

df The number of *degrees of freedom, often denoted by v (nu) or d.

DFBETA; DFFITS *See* REGRESSION DIAGNOSTICS.

diagonal matrix *See* MATRIX.

diagonals model *See* SQUARE TABLE.

DIC *See* DEVIANCE INFORMATION CRITERION.

dice problems *Probability problems concerning the outcomes of rolling dice. A problem considered at length by *Fermat and *Pascal concerned the number of times that one must throw a pair of dice before obtaining a double six. Their correspondence laid the foundation for the modern study of probability. *See* GEOMETRIC DISTRIBUTION.

dichotomous Adjective describing a *categorical variable with just two categories. The corresponding noun is **dichotomy**.

diffusion process *See* STOCHASTIC PROCESS.

directed divergence *See* KULLBACK–LEIBLER INFORMATION.

directed graphical model *See* GRAPHICAL MODELS.

direct models (decomposable models) *Log-linear models for which simple formulae exist for the *expected frequencies. The best known is the *independence model. A direct model that can be interpreted in terms of *conditional independence is a *graphical model.

Dirichlet, Johann Peter Gustav Lejeune (1805–59; b. Düren, Germany; d. Göttingen, Germany) A German mathematician. It is claimed that, as a 12-year-old, Dirichlet spent his pocket money on mathematics books. He graduated from the University of Cologne aged sixteen and went to Paris to act as a tutor. While there he published a paper on *Fermat's last theorem which brought him fame. Returning to Germany, he joined the staff at Cologne before moving to the University of Berlin in 1828. In 1855 he succeeded *Gauss as Professor at the University of Göttingen and, in the same year he was elected a Fellow of the Royal Society. He married a sister of the composer Mendelssohn and has a lunar crater named after him.

Dirichlet distribution The *multivariate distribution corresponding to the *beta distribution.

Dirichlet tessellation A tessellation that divides a region into n subregions, one for each measurement point, with the jth subregion consisting of all points in the region that are nearer to the *j*th measurement point than to any other.

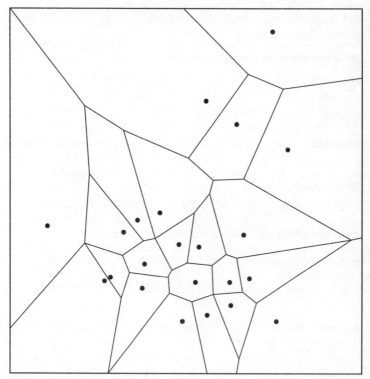

Dirichlet tessellation. The tessellation is defined by the locations of 22 rain-gauges in the Bolton area, England.

The Dirichlet tessellation has been rediscovered many times. Other names include **Voronoi polygons** and **Thiessen polygons**. Every edge of a Dirichlet subregion separates two of the original measurement points. Joining each such pair produces a new tessellation in which the subregions are *Delaunay triangles.

dirty data Data with unusual, missing, or incorrect values. Most real sets of data are dirty.

discrete distribution *See* DISCRETE RANDOM VARIABLE.

discrete random variable A *random variable whose set of possible values is a finite or infinite sequence of numbers $x_1, x_2, \ldots$. The set of values is often a subset of the non-negative integers $0, 1, \ldots$. The *probability distribution of a discrete random variable is referred to as a **discrete distribution**.

discrete-time Markov chain *See* MARKOV PROCESS.

discriminant analysis A procedure for the determination of the group to which an individual belongs, based on the characteristics of that individual. Suppose we have measurements on p characteristics for each of a *sample of individuals. We know that each individual belongs to one of g groups, but we do not know which. Discriminant analysis attempts to maximize the *probability of correct allocation. It differs from *cluster analysis in that we have an initial data set, the **training set**, whose group allocations are known.

For the case $g = 2$, suppose that the $p \times 1$ *vectors of sample means are $\bar{\mathbf{x}}_1$ and $\bar{\mathbf{x}}_2$. Let n_1 and n_2 be the numbers of members of the training set falling in the two groups and let $\mathbf{S}_1$ and $\mathbf{S}_2$ be the *variance–covariance matrices for the two parts of the training set. If we define the matrix $\mathbf{S}$ by

$$\mathbf{S} = \frac{(n_1\mathbf{S}_1 + n_2\mathbf{S}_2)}{(n_1 + n_2)},$$

a future individual with measurement vector $\mathbf{x}$ will be assigned to group 1 if and only if

$$\mathbf{a}'\mathbf{x} > \frac{1}{2}\mathbf{a}'(\bar{\mathbf{x}}_1 + \bar{\mathbf{x}}_2),$$

where $\mathbf{a}'$ is the *transpose of the vector $\mathbf{a}$ given by

$$\mathbf{a} = \mathbf{S}^{-1}(\bar{\mathbf{x}}_1 - \bar{\mathbf{x}}_2).$$

The function $\mathbf{a}'\mathbf{x}$ is called **Fisher's linear discriminant function**.

discrimination information *See* KULLBACK–LEIBLER INFORMATION.

disjunctive kriging *See* KRIGING.

dispersion *See* INDEX OF DISPERSION.

dispersion matrix *See* VARIANCE–COVARIANCE MATRIX.

dispersion test *See* INDEX OF DISPERSION.

distance measures Measures of distance between points in multidimensional space, used with techniques such as *cluster analysis and *multidimensional scaling. The two measures most commonly used are Euclidean distance and the city-block metric.

Euclidean distance

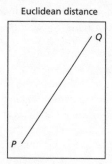

City-block distance

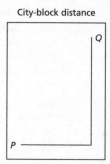

Distance measures. These are the two simplest of an infinite number of possible measures of distance.

Euclidean distance is the straight-line distance between two points, also called the l_2-**metric** (*see* l_p-**METRIC**). **In** *d* dimensions, if the positions of P and Q are given by the coordinates $(p_1, p_2, \ldots, p_d)$ and $(q_1, q_2, \ldots, q_d)$ then the Euclidean distance between P and Q is given by

$$\sqrt{\sum_{j=1}^{d} (p_j - q_j)^2}.$$

The **city-block metric** in two dimensions measures the distance between two points in a city if, for example, the only directions in which one could travel were north–south and east–west. It is also called the l_1-metric or **Manhattan distance**. In d dimensions the city-block distance between P and Q is given by

$$\sum_{j=1}^{d} |p_j - q_j|.$$

distributed lags models Models that relate y_t, the value of a dependent variable evaluated at time t, to the values taken by an explanatory variable x at times $t, t - 1, \ldots$. The general model is

$$y_t = \alpha + \sum_{s=0}^{\infty} \beta_s x_{t-s} + \varepsilon_t,$$

for suitable values of the constants $\alpha, \beta_0, \beta_1, \ldots$. Here, x_{t-s} is the value of x at lag s, and $\varepsilon_t, \varepsilon_{t-1}, \ldots$, are the (possibly *correlated) *random errors.

One simple model of this type is the three-parameter **Koyck model** for which $\beta_s = \gamma \beta^s$, where β and γ are positive constants. Another is the $(k+1)$-parameter **Almon model** in which

$$y_t = \sum_{s=0}^{n} \beta_s x_{t-s} + \varepsilon_t,$$

where n is known and

$$\beta_s = \sum_{j=0}^{k} \gamma_j s^j,$$

for constant $\gamma_0, \gamma_1, \ldots, \gamma_k$.

distribution The set of values of a set of *data, possibly grouped into classes, together with their *frequencies or relative frequencies. In the case of *random variables the distribution is the set of possible values together with their probabilities in the *discrete case and the *probability density function in the case of a *continuous variable.

distribution-free tests *See* NON-PARAMETRIC TESTS.

distribution function *See* CUMULATIVE DISTRIBUTION FUNCTION.

diversity index A measure of the extent to which the different types in a *population are unequally common. A value of 0 is attained if there is only one type in the population. One index is the **Shannon index**:

$$-\sum_{j=1}^{k} p_j \log(p_j),$$

where there are k types and p_j is the proportion of type j. The base of the logarithm is arbitrary, but the usual choice is 2 (so that $\log_2 (\frac{1}{2}) = -1$, $\log_2 (\frac{1}{4}) = -2$, etc.) in which case the index may alternatively be referred to as the **entropy** of the classifying *variable. An alternative is the **Gini index**, also called the **Simpson index**:

$$1 - \sum_{j=1}^{k} p_j^2.$$

DK Short for 'Don't know' as a *questionnaire answer.

D-optimal *See* OPTIMAL DESIGN.

dot plot An alternative to a *bar chart or line graph when there are very few *data values. Each value is recorded as a dot, so that the *frequencies for each value can easily be counted.

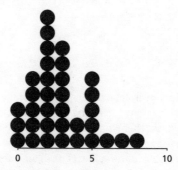

Dot plot. The diagram illustrates the numbers of goals scored in football games on one particular day. The modal number of goals was 2.

double-blind trial *See* BLINDING.

double bootstrap *See* BOOTSTRAP.

double exponential distribution An alternative name for the *Laplace distribution.

double sampling *Acceptance sampling in which the initial *sample leads either to a decision to immediately accept or reject the *batch being sampled, or to a decision to postpone the decision until a second sample has been taken.

doubly stochastic point process *See* COX PROCESS.

drunkard's walk *See* RANDOM WALK.

duality A property of an *optimization problem. Duality relates any linear maximization problem to an equivalent minimization problem. For example, with non-negative *x*-variables and *y*-variables, the maximum value for *A*, where

$$A = 3x_1 + 6x_2 + 2x_3,$$

subject to

$$3x_1 + 4x_2 + x_3 \leq 2,$$
$$x_1 + 3x_2 + 2x_3 \leq 1,$$

is equal to the minimum value of B, where

$$B = 2y_1 + y_2,$$

subject to

$$3y_1 + y_2 \geq 3,$$
$$4y_1 + 3y_2 \geq 6,$$
$$y_1 + 2y_2 \geq 2.$$

dummy variable A *variable, taking only the values 0 and 1, derived from a *polytomous *categorical variable. If the categorical variable has k categories then $(k - 1)$ dummy variables are required. For example, with four categories the three dummy variables (x_1, x_2, x_3) could be assigned the values (1, 0, 0) for category 1, (0, 1, 0) for category 2, (0, 0, 1) for category 3, and (0, 0, 0) for category 4. Dummy variables enable the inclusion of categorical information in *regression models.

Duncan's test A multiple comparison test proposed by D. B. Duncan in 1955. The test, based on the *Studentized range statistic, is a *multiple stage test for comparing sample means which aims to provide the specified significance level for each paired comparison. The **Newman–Keuls** test, which is also based on the Studentized range, aims to meet a specified significance level for the composite *null hypothesis that all the means are equal.

Dunnett, Charles William (1921–; b. Windsor, Ontario) A Canadian statistician. Dunnett obtained his BS in mathematics and physics from McMaster University in 1942, and an MA from the University of Toronto in 1946. From 1953 to 1974 he worked as a statistician for the American Cyanamid Co., obtaining his PhD from the University of Aberdeen in 1960. In 1974 he returned to McMaster University as Professor of Biostatistics, then as Professor of Applied Mathematics, and subsequently as Professor Emeritus.

Dunnett's test A *multiple comparison test for the special case where (typically) k new treatments are being compared with a *control.

Durbin, James (1923–) A statistician noted for his co-authorship of the 1950 paper introducing the *Durbin–Watson statistic. Durbin is a graduate of Cambridge University. He joined the London School of Economics in 1950; he was Professor of Statistics from 1961 to 1988 and then Professor Emeritus. He was awarded the *Royal Statistical Society's *Guy Medal in Silver in 1976 and was President of the Society in 1986.

Durbin–Watson statistic A statistic proposed by *Durbin and *Watson in 1950. Suppose the sequence of *observations $y_1, y_2, \ldots, y_t$ forms a *time series. Suppose also that there are k explanatory variables taking values $x_{1t}, x_{2t}, \ldots, x_{kt}$ at time point t. It is proposed that the variation over time in the value of y_t, at time t, may be explained by the *multiple regression model,

$$y_t = \beta_0 + \sum_{j=1}^{k} \beta_j x_{jt} + \varepsilon_t,$$

where ε_t is the *random error and $\beta_0, \beta_1, \ldots, \beta_k$ are unknown *parameters.

The Durbin-Watson statistic, d, given by

$$d = \frac{\sum_{m=2}^{t} \left(\hat{\varepsilon}_m - \hat{\varepsilon}_{m-1}\right)^2}{\sum_{m=1}^{t} \hat{\varepsilon}_m^2},$$

where $\hat{\varepsilon}_m$ is the *ordinary least squares estimate of ε_m, tests whether the sequence $\varepsilon_1, \varepsilon_2, \ldots, \varepsilon_t$ consists of *independent values (in which case d has expectation 2), or forms a *Markov process in which each random error is related to its predecessor.

dynamic programming The problem of optimizing a sequence of decisions in which each decision must be made after the outcome of the previous decision becomes known.

e *See* EXPONENTIAL.

ecological fallacy An error of reasoning brought about by arguing that the relationship between properties of groups of people must apply to the people themselves (or vice versa). The analogous situation with categorical data is described as *Simpson's paradox.

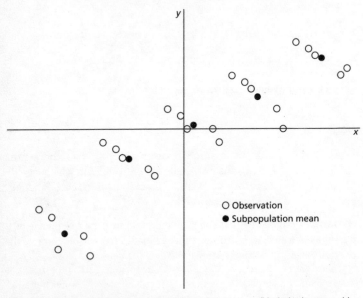

Ecological fallacy. Considering only the subpopulation means (black dots), we would conclude that the correlation between the random variables X and Y was positive. However, the data for each subpopulation (the groups of open circles) show that X and Y are in fact negatively correlated.

ecological inference A method for estimating probabilities in a *transition matrix. The method has mainly been used in the context of political data. It is assumed that for every pair of political parties the *probability of a supporter changing from one party to the other is the same in every constituency. If that is the case then, by studying the pairs of vote totals for two elections, it will be possible to identify the transition matrix.

Econometrica The journal of the *Econometric Society. The first issue appeared in 1933.

econometric models Models that describe the processes in an economic system. Typically an economic system is described by a set of simultaneous *multiple regression models that describe the links between a number of dependent (*endogenous) *variables and a number of predetermined (*exogenous) variables (which may include previous values of the endogenous variables).

Econometric Society The Society which publishes the journal *Econometrica*. It was founded by Irving *Fisher, *Hotelling, *Shewhart, *Wiener, and others in 1930.

EDA *See* EXPLORATORY DATA ANALYSIS.

edge effects Edge effects complicate the analysis of spatial processes. Many formulae relating to *spatial processes are affected by the finiteness of the region of interest. For example, suppose we look through a microscope at blood cells. The nearest neighbour to a cell of interest may be very close but out of the field of view. Statistical tests based on the distance to the apparent nearest neighbour must take account of this possibility.

Edgeworth, Francis Ysidro (1845–1926; b. Edgeworthstown, Ireland; d. Oxford, England) Irish econometrician, lawyer, and mathematician. His family had settled in Ireland in the sixteenth century, and had given their name to the town that subsequently developed. Edgeworth was educated at Trinity College, Dublin (studying modern languages) and Oxford University (studying classics). He then studied law, being called to the bar in 1877, and also mathematics, so that his first academic post, in 1880, was as Lecturer in Logic at King's College, London. In 1891 he became Professor of Political Economy at Oxford University, where he inaugurated the *Economic Journal*. Though much of his time was devoted to economics, he developed the early theory of *correlation, *estimation, and the application of the *normal distribution. Edgeworth lived extremely simply, with few personal possessions: an exaggerated claim was that these were limited to red tape and gum! He was President of the *Royal Statistical Society from 1912 to 1914 and was awarded its *Guy Medal in Gold in 1907.

Edgeworth expansion An expansion, derived by *Edgeworth in 1905, that relates the *probability density function, f, of a *random variable, X, having expectation 0 and *variance 1, to ϕ, the probability density function of a *standard normal distribution, using the *Chebyshev–Hermite polynomials. The first terms of the expansion are

$$f(x) = \phi(x)\left\{1 + \sum_{r=3}^{6} \frac{\kappa_r}{r!} H_r(x) + \frac{\kappa_3^2}{72} H_6(x)\right\},$$

where κ_r is the rth *cumulant of X, and $H_r(x)$ is the rth Chebyshev–Hermite polynomial. *See also* CORNISH-FISHER EXPANSION.

efficiency; efficient estimator *See* ESTIMATOR.

Efron, Bradley (1938–; b. St Paul, Minnesota) Professor of Statistics and Professor of Humanities and Sciences at Stanford University, where he obtained his PhD in 1964. His pioneering fundamental contributions, in particular to the theory of the *bootstrap, have resulted in several prestigious awards, including, in 1990, the *Wilks Medal of the *American Statistical Association, of which he was the 2003–5 President.

eigenvalue; eigenvector *See* MATRIX.

eighty/twenty rule A rule much quoted in business that states that eighty per cent of a firm's business is derived from twenty per cent of its customers; also that eighty per cent of its business is obtained by twenty per cent of its sales force. The rule derives from the work of *Pareto.

element *See* MATRIX.

EM algorithm An *algorithm for computing *maximum likelihood estimates of *parameters when some of the *data are missing. It is an *iterative algorithm that alternates two steps until convergence is attained to sufficient accuracy. Given some values assumed for the unknown parameters, the E step evaluates the joint *likelihood of the complete data set, suitably averaged over all values of the missing data. This is therefore an expectation of the likelihood that is *conditional on the observed data. The M step maximizes this expectation over the unknown parameter values. The values providing this maximization are used for the next E step. The algorithm was introduced by *Dempster, Laird, and Rubin in 1977.

empirical A statement based on a study of *data without the use of any mathematical *model.

empirical Bayes *See* BAYESIAN INFERENCE.

empty set *See* SAMPLE SPACE.

endogenous variable A dependent variable in the context of an *econometric model.

entropy *See* DIVERSITY INDEX.

entry *See* MATRIX.

Epanechikov kernel *See* KERNEL METHODS.

equilibrium A *stochastic process is said to be in equilibrium if, for every state of the process and for each time point, the *probability of being in that state at the next time point is equal to the probability for the current time point.

erf *See* NORMAL DISTRIBUTION.

ergodic state *See* MARKOV PROCESS.

Erlang, Agner Krarup (1878–1929; b. Lonborg, Denmark; d. Copenhagen, Denmark) A Danish engineer and probabilist. Erlang, the son of the local schoolmaster, learnt to read upside down, sitting opposite his elder brother. He graduated from the University of Copenhagen in 1901 and joined the Copenhagen Telephone Company in 1908, with whom he remained for the rest of his career. His first publication was entitled *The Theory of Probabilities and Telephone Conversations* and his subsequent research involved many trips down the manholes of Copenhagen.

Erlang distribution *See* GAMMA DISTRIBUTION.

error function *See* NORMAL DISTRIBUTION.

error of the first kind; error of the second kind *See* HYPOTHESIS TEST.

error sum of squares *See* ANOVA.

estimable Adjective used to describe a quantity that can be estimated.

estimate *See* ESTIMATOR.

estimation The process of using a mathematical formula, or formulae, to calculate from observed *data a value, or values, for an unknown *parameter, or parameters. Three common methods are the methods of *ordinary least squares and *maximum likelihood and the *method of moments. *See also* BAYESIAN INFERENCE; RESAMPLING.

estimator A *statistic used to estimate a *parameter. The realized value of an estimator for a particular *sample of *data is called the **estimate**.

If the expectation of the statistic is equal to the parameter then it is described as being an **unbiased estimator**. If T is an estimator of the parameter θ and the expectation of T is $\theta + b$, where $b \neq 0$, then b is the **bias** and the estimator is a **biased estimator**. If the bias tends to 0 as the *sample size increases, then the estimator is an **asymptotically unbiased estimator**.

The **efficiency** of an unbiased estimator is the ratio of its *variance to the *Cramér–Rao lower bound. For an **efficient estimator** the ratio is 1. The **relative efficiency** of two unbiased estimators is given by the inverse ratio of their variances.

Comparisons involving biased estimators are often based on the **mean squared error**, defined to be

$$E[(T - \theta)^2] = \text{Var}(T) + \{E(T) - \theta\}^2 = \text{Var}(T) + b^2,$$

where $E(T)$ and $\text{Var}(T)$ are, respectively, the expectation and variance of T. The **root mean squared error** is the square root of the mean squared error and has the same units as the original data. An estimator is said to be **consistent** if

$$\lim_{n \to \infty} P(|T - \theta| \geq c) = 0,$$

where n is the sample size, for all positive c.

A **sufficient estimator** or **sufficient statistic** is a statistic that encapsulates all the information about the unknown parameter.

The terms 'biased' and 'unbiased' appear in an 1897 text by *Bowley. The terms 'efficiency', 'estimate', 'estimation', and 'sufficiency' were introduced by Sir Ronald *Fisher in 1922. The term 'estimator' was introduced in a specialized sense by *Pitman in 1939.

Euclid (330?–275? BC; Athens?) Author of *Elements*, a rigorous mathematical treatise setting out, *inter alia*, the foundations of elementary geometry.

Euclidean distance The straight-line distance between two points. *See* DISTANCE MEASURES.

Euclidean space A term used to describe ordinary two- or three-dimensional space. Also used in connection with the n-dimensional space $\mathbb{R}^n$, in which a point is a real column vector $\mathbf{x} = (x_1 \ x_2 \ \ldots \ x_n)'$, the distance between two points $\mathbf{x}$ and $\mathbf{y}$ is

$$\sqrt{\sum_{k=1}^{n} (x_k - y_k)^2} = \sqrt{(\mathbf{x} - \mathbf{y})'(\mathbf{x} - \mathbf{y})},$$

and the angle θ between the vectors $\mathbf{x}$ and $\mathbf{y}$ is given by

$$\cos\theta = \frac{\mathbf{x}'\mathbf{y}}{\sqrt{\mathbf{x}'\mathbf{x}}\sqrt{\mathbf{y}'\mathbf{y}}}.$$

Euler, Leonhard (1707–83; b. Basel, Switzerland; d. St Petersburg, Russia) A Swiss mathematician. Euler attended Basel University when aged thirteen and gained a master's degree at sixteen. He was the author of more than 900 papers and introduced much of the notation in modern use (e.g. π, e, i, $\sum$). A street in Paris and a lunar crater are named after him.

Euler's constant The limit, as n approaches infinity, of

$$1 + \frac{1}{2} + \ldots + \frac{1}{n} - \ln n.$$

It is usually denoted by γ and its numerical value is $0.5772156649\ldots$.

event *See* SAMPLE SPACE.

event history analysis Analysis of the *recall data provided by individuals about their life. The life pattern is usually characterized as being in one of a number of states (e.g. education) and the aim is to study the pattern of change between these states using some form of *longitudinal analysis.

exact test Although there are many exact tests, this term is often intended to refer to *Fisher's exact test.

exclusive *See* MUTUALLY EXCLUSIVE EVENTS.

exclusive union *See* UNION.

exhaustive Events (*see* SAMPLE SPACE) $A_1, A_2, \ldots, A_n$, are exhaustive if $A_1 \cup A_2 \cup \ldots \cup A_n = S$, where S is the sample space. *See also* UNION.

exogenous variable An *independent variable in the context of an *econometric model.

exp *See* EXPONENTIAL.

expectation *See* EXPECTED VALUE.

expectation algebra The mathematical rules that govern expectations. For *random variables X, Y, X_j, with *expected values $E(X)$, $E(Y)$, and $E(X_j)$, and for any constants a, b, a_j, the following formulae hold, without any assumptions of *independence:

$$E(X + Y) = E(X) + E(Y),$$
$$E(X - Y) = E(X) - E(Y),$$
$$E(aX + b) = aE(X) + b,$$
$$E(aX + bY) = aE(X) + bE(Y),$$
$$E\left(\sum_{j=1}^{n} a_j X_j\right) = \sum_{j=1}^{n} a_j E(X_j).$$

The following law also holds, provided X and Y are independent:

$$E(XY) = E(X) \times E(Y).$$

expected frequency The *expected value for a *cell frequency for a *model of interest. Also called the *fitted frequency. In the context of a *chi-squared test the expected frequency for a particular outcome is the product of the *sample size and the *probability of the outcome under the *null hypothesis.

expected value The expected value of a *random variable X, which is denoted by $E(X)$, may be interpreted as the long-term average value of X. In the case of a *discrete random variable,

$$E(X) = \sum_j x_j P(X = x_j).$$

For a *continuous random variable with *probability density function f,

$$E(X) = \int_{-\infty}^{\infty} x f(x) \mathrm{d}x.$$

The expected value of X is often referred to as the **expectation** of X or as the *mean of X. The word 'expectation' has been used in this context since the use of 'expectatio' by *Huygens in his 1657 treatise on the results of playing games of chance: *De Ratiociniis in Ludo Alae.*

If g is any function, the expected value of $g(X)$, $E[g(X)]$ is defined by

$$E[g(X)] = \sum_j g(x_j) P(X = x_j)$$

in the case of a discrete random variable, and by

$$E[g(X)] = \int_{-\infty}^{\infty} g(x) f(x) \mathrm{d}x$$

in the case of a continuous random variable. Thus

$$E(X^2) = \sum_j x_j^2 P(X = x_j) \quad \text{or} \quad \int_{-\infty}^{\infty} x^2 f(x) \mathrm{d}x.$$

The expected value of a random variable may be thought of as the limiting value of the arithmetic mean of the observed values as the *sample size increases. It should be noted that in the case of a discrete random variable the expected value is generally not a possible value. For example, when an unbiased coin is tossed once, the expected value of the number of heads is $\frac{1}{2}$.

Some random variables (for example, those with a *Cauchy distribution) do not have an expected value. There are also discrete random variables without expected values, though this situation can only arise when the set of possible values is infinite and the sum involved is the sum of a series that does not converge. For example, if $P(X = r) = K/r^2$, for $r = 1, 2, \ldots$, where $1/K = \sum 1/r^2 (= \pi^2/6)$, then the expression for $E(X)$ is $\sum K/r$, and this series diverges to ∞. For the expected values of sums and products of random variables, *see* EXPECTATION ALGEBRA.

experimental design The branch of *Statistics concerned with the efficient *estimation of the unknown *parameters in a *linear model. Common experimental designs include *balanced incomplete blocks, *crossover trial, *factorial experiments, *Latin squares, *paired comparisons, *randomized blocks, *repeated measures, and *split plots. In each case a *linear model relates the *response variable to one or more explanatory variables. The results are usually summarized in an *ANOVA table.

The analysis of the ANOVA table is affected by the nature of the explanatory variables. For example, if a design compares treatment A with treatment B because these specific treatments are of interest, then they are said to have **fixed effects**. On the other hand, if treatments A and B have been chosen at random from a population of possible treatments with the intention of attempting to answer the general question 'Do treatments differ?', then they are said to have **random effects**. A design including both random and fixed effects is a **mixed effects design**.

experimental error The difference between an *observation and the value predicted by a mode.

experimental unit A person, a plot of land, a machine, etc. that forms the material on which an experiment is performed. In an agricultural context the term **plot** is usually preferred. *See* EXPERIMENTAL DESIGN.

expert system A computer application that can carry out the same problem-solving functions as a human expert in a particular area, for example, financial forecasts or medical diagnostics. Human expert knowledge in the area is collected and used as a base, with rules for solving relevant problems. The system is designed by studying how human experts make decisions and translating their methods into the software. The application may be interactive, such that the user is asked to clarify the problem.

explanatory variable *See* REGRESSION.

exploratory data analysis (EDA) The search for relationships between variables (as in *data mining) by the use of graphics such as *boxplots and techniques such as *classification trees.

exponent *See* INDEX.

exponential A function of a real variable defined for all real x by the sum to infinity of the (convergent) series

$$e^x = 1 + x + \frac{x^2}{2!} + \frac{x^3}{3!} + \ldots.$$

The notation reflects the **index property** that $e^x \times e^y = e^{x+y}$. The notation 'exp(x)' is sometimes used for e^x, particularly when x is a complicated algebraic expression. The number e is given by the above series with $x = 1$, so

$$e = 1 + 1 + \frac{1}{2!} + \frac{1}{3!} + \ldots \approx 2.718\,28.$$

The exponential and *natural logarithm functions are related by $\exp(\ln x) = x$ for $x > 0$ and $\ln(e^x) = x$ for all values of x.

exponential distribution A *random variable X with an exponential distribution has *probability density function f given by

$$f(x) = \lambda e^{-\lambda x}, \qquad x \geq 0,$$

where λ is a positive constant. The distribution function F is given by

$$F(x) = 1 - e^{-\lambda x}, \qquad x \geq 0.$$

The distribution has *mean $1/\lambda$ and *variance $1/\lambda^2$. The distribution has the *forgetfulness property. *See also* POISSON PROCESS.

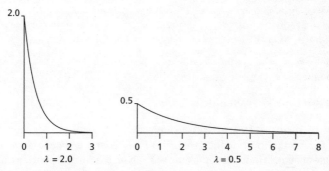

Exponential distribution. All exponential distributions have a range from 0 to ∞, a mode at 0, and a mean equal to the standard deviation.

exponential family If a *random variable X has a *probability distribution (*discrete case) or a *probability density function (*continuous case) that can be written in the form

$$\exp\{a(x)b(\theta) + c(\theta) + d(x)\},$$

where θ is a parameter and a, b, c, and d are known functions, then it is a member of the exponential family. The *Poisson distribution, the *binomial distribution, and the *normal distribution are members of this family. If $|a(x)| = x$, then the distribution is in **canonical form** and $|b(\theta)|$ is a **natural parameter** of the distribution. *See also* GENERALIZED LINEAR MODEL.

exponential smoothing A method of smoothing a *time series. Let $x_1, x_2, \ldots, x_n$ denote the observed values, and let $\hat{x}_1, \hat{x}_2, \ldots, \hat{x}_n$ be a corresponding set of smoothed values computed using the formulae

$$\hat{x}_1 = x_1,$$
$$\hat{x}_t = \alpha x_t + (1 - \alpha)\hat{x}_{t-1}, \qquad t = 2, 3, \ldots, n$$

where $\alpha\,(0 < \alpha < 1)$ is a constant. The formula can also be written as

$$\hat{x}_t = \alpha \sum_{k=1}^{t} (1 - \alpha)^{t-k} x_k + (1 - \alpha)^t x_1.$$

The value of α is usually chosen to minimize $\sum_{t=1}^{n} (x_t - \hat{x}_t)^2$.

With this value for α, the formula may be adapted for *forecasting a time series having neither *trend nor *seasonality by writing

$$\tilde{x}_{t+1} = \alpha x_t + (1 - \alpha)\hat{x}_t, \qquad t = n, n + 1, \ldots$$

where $\tilde{x}_{t+1}$ is the predicted value for time $(t + 1)$ and it is assumed that a good series of values exists.

extinction *See* BRANCHING PROCESS.

extra-binomial variation Greater variability in repeat estimates of a population proportion than would be expected if the population had a *binomial distribution. For example, suppose that n *observations are taken on *independent *Bernoulli variables that take the value 1 with *probability p, and the value 0 with probability $1 - p$. The *mean of the total of the observations will be np and the *variance will be $np(1 - p)$. However, if the probability varies from variable to variable, with overall mean p as before, then the variance of the total will now be $> np(1 - p)$.

extrapolation The use of a formula, deduced from an initial set of *data, for data values outside the range of the initial data set. *Compare* INTERPOLATION. Excessive extrapolation may give unreliable values and is not recommended.

For example, if we discover that 1 kg of fertilizer per ha results in a yield of 50 kg of produce per ha, that 1.5 kg gives 60 kg, and that 2 kg gives 70 kg, then we might reasonably extrapolate that 2.5 kg would give 80 kg. However, it would be excessive extrapolation to expect 10 kg to give 230 kg.

extreme-value distribution The *distribution of an extreme value. The best known distribution of this type is the *Gumbel distribution.

factor Depending on the context, *see* FACTOR ANALYSIS; FACTORIAL EXPERIMENT.

factor analysis A *multivariate method in which the basic *data consist of the observed values of the p characteristics of each of n individuals. The method, introduced in 1931 by *Thurstone, aims to explain the $p \times p$ *variance–covariance matrix for these data in terms of the relationships between a much smaller number (q) of unobserved variables (*latent variables) called **factors**. The underlying proposition is that the jth observed value, x_j, is the sum of an error term, ε_j, and a linear combination of the factors $f_1, f_2, \ldots, f_q$:

$$x_j = \lambda_{j1} f_1 + \lambda_{j2} f_2 + \cdots + \lambda_{jq} f_q + \varepsilon_j.$$

The multipliers $\lambda_{j1}, \lambda_{j2}, \ldots, \lambda_{jq}$ are called **factor loadings**.

Unlike *principal components analysis, which it resembles, the methodology makes many assumptions (for example, that factors exist and that there are exactly q of them). Its use has therefore attracted considerable criticism. Its popularity might be thought to be based on the ability of the method to produce the result desired by the researcher!

factorial If n is a positive integer, then factorial n, written as $n!$, is defined by

$$n! = 1 \times 2 \times \cdots \times n,$$

with 0! defined to be 1. The notation was introduced in about 1810 by the French mathematician Christian Kramp (1760–1826). Factorial n is the number of different orderings of n individuals. The numerical value of $n!$ increases rapidly with n:

$1! = 1, 2! = 2, 3! = 6, 4! = 24, 5! = 120, \ldots, 10! = 3\,628\,800, \ldots$.

For $n > 1$, successive values of $n!$ can be calculated using

$$n! = n \times (n-1)!$$

If n is not an integer, factorial n can be defined via the *gamma function Γ:

$$n! = \Gamma(n+1).$$

For large values of n an approximation is provided by *Stirling's formula.

factorial experiment An *experimental design to investigate the effects of several explanatory variables, in this context called **factors**, on a single response variable. Each factor takes only a small number of different values (typically two), which may not be quantitative and are called **levels**. For example, in an agricultural context the two levels might be two different varieties of a plant.

An experiment with k two-level factors (the upper and lower levels) is called a 2^k-factorial. The standard notation denotes factors by capital letters $A, B, \ldots$. Each **experimental unit** is subject to one or other of the levels of each factor. Lower-case letters are used to indicate factor combinations, so ac indicates that A and C occur at the higher level for that experimental unit, and other factors occur at the lower level. The combination with all factors at the lower level is indicated as (1), so the eight treatment combinations for a 2^3-factorial are

$$(1) \quad a \quad b \quad ab \quad c \quad ac \quad bc \quad abc$$

A 2^3-factorial enables one to estimate the three main effects (see below) as well as the three *interactions involving two variables and the three-variable interaction. The **main effect** of A compares the mean value of the observations at the higher level (a, ab, ac, abc), for example, with those at the lower level ($(1), b, c, bc$).

When 2^k is large it may not be feasible to maintain constant testing conditions. One solution is to test fractions at a time using *randomized blocks:

Block 1	Block 2
(1)　b　bc　c	a　ab　abc　ac

Suppose that the values in the second block of experiments are much larger than those in the first block. It is impossible to tell whether this is because of a difference between the two blocks or a difference between the two levels of *A*: the effects of *A* and of the blocks are said to be **confounded**. A better treatment allocation, in which it is the three-variable interaction that is confounded, is

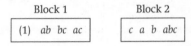

Block 1	Block 2
(1) *ab bc ac*	*c a b abc*

It will be seen that each of the treatment combinations in block 1 has an even number of letters in common with *abc* while each of those in block 2 has an odd number of letters in common with *abc*. In the previous example the reference combination was *a*. The reference combination, which determines the composition of the blocks, is called the **defining contrast**.

An alternative to performing a complete factorial using blocks is to use a **fractional replicate**, also called a **fractional factorial design**. As an example, a **half-replicate** with four factors would consist of eight rather than sixteen treatment combinations; these might be as follows:

(1) *ab ac ad bc bc cd abcd*

Since only a fraction of the full number of treatment combinations is tested, only a fraction of the full number of main effects and interactions can be estimated. In effect, what happens is that main effects and interactions are confounded with one another. Two such confounded effects are said to be **aliased** with each other. In the example no main effects are aliased with each other; the main effect of *A*, for example, is aliased with the *BCD* interaction. Fractional replicates are used in the exploratory stages of investigations with factors that are *continuous variables, where the aim is to find the global maximum (or minimum) of a *response surface.

If it can be assumed that there are no interactions then **Plackett and Burman designs** are appropriate. These designs, introduced in a 1946 paper by *Plackett and *Burman, are very economical in terms of the number of observations that need be performed. For example, with seven factors (*A–G*) each having two levels, the seven main effects can be estimated with the following set of treatment combinations:

dfg, aeg, abf, bcg, acd, bde, cef, abcdefg.

A design that has more factors than observations may be described as a **supersaturated design**.

factor loadings *See* FACTOR ANALYSIS.

fast Fourier transform (FFT) A numerical method for rapid computation of the coefficients in a finite *Fourier transform. The method was introduced in 1965 by *Tukey and Cooley. If the (possibly complex) values of a function f(x) are known at N equally spaced points, $x_0, x_1, \ldots, x_{N-1}$, where $x_k = 2\pi k / N$ for $k = 0, 1, \ldots, N-1$, the coefficients c_n are such that

$$f(x_k) = \sum_{n=0}^{N-1} c_n\, e^{ink}, \qquad k = 0, 1, \ldots, N-1,$$

where $i = \sqrt{-1}$.

father wavelet *See* WAVELET.

***F*-distribution** If Y_1 and Y_2 are *independent chi-squared random variables with v_1 and v_2 *degrees of freedom, respectively, then the ratio X, given by

$$X = \frac{Y_1}{v_1} \Big/ \frac{Y_2}{v_2},$$

is said to have an *F*-distribution with v_1 and v_2 degrees of freedom. This may be written as the F_{v_1, v_2}-distribution. Evidently $1/X$ will have a F_{v_2, v_1}-distribution.

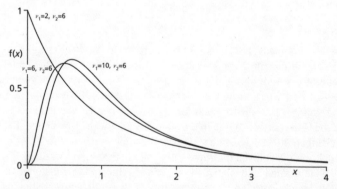

F-distribution. All *F*-distributions take values from 0 to ∞ and, for $v_1, v_2 > 4$, have mean and mode near 1.

The form of the distribution was first given in 1922 by Sir Ronald *Fisher, and it is sometimes still referred to as **Fisher's *F*-distribution**. In

1934 the distribution was tabulated (*see* APPENDIX IX) by *Snedecor, who used the letter F in Fisher's honour. The *probability density function f of the F_{v_1, v_2}-distribution is given by

$$f(x) = \frac{B\left(\frac{1}{2}v_1, \frac{1}{2}v_2\right) v_1^{\frac{1}{2}v_1} v_2^{\frac{1}{2}v_2} x^{\frac{1}{2}v_1 - 1}}{(v_2 + v_1 x)^{\frac{1}{2}(v_1 + v_2)}}, \qquad x > 0,$$

where B is the *beta function.

The distribution has *mean $v_2/(v_2 - 2)$ provided that $v_2 > 2$. The distribution has variance

$$\frac{2v_2^2(v_1 + v_2 - 2)}{v_1(v_2 - 2)^2(v_2 - 4)},$$

provided that $v_2 > 4$. If $v_1 \leq 2$ there is a mode at 0, otherwise the mode is at

$$\frac{v_2(v_1 - 2)}{v_1(v_2 + 2)}.$$

If X has an $F_{1, v}$-distribution, then $\sqrt{X}$ has a *t-distribution with v degrees of freedom.

feasible region In the context of *linear programming, the region in which all the constraints are satisfied. All feasible solutions must lie in this feasible region.

feasible solution *See* LINEAR PROGRAMMING.

Feller, William (1906–70; b. Zagreb, Croatia; d. New York City) A Croatian probabilist who spent most of his career in the United States. Feller was author of the influential 2-volume *An Introduction to Probability Theory and its Applications*. Volume 1 was published in 1950 and remained the standard introductory text for several decades. He obtained his MSc at the University of Zagreb in 1925 and his PhD at the University of Göttingen in 1926. Following posts at Kiel, Copenhagen, and Stockholm, he moved to Brown University in the United States in 1939, to Cornell University in 1945, and to Princeton University in 1950. He was President of the *Institute of Mathematical Statistics in 1947, and was elected to membership of the National Academy of Sciences in 1960.

Feller's coin-tossing constants Constants related to $w(n)$, the *probability that no *run of k consecutive heads is obtained when a fair coin is tossed n times. In a 1949 paper, *Feller showed that $w(n)$ obeyed the condition

$$\lim_{n \to \infty} w(n)\alpha^{n+1} = \beta,$$

where α is the smaller $(1 < \alpha < 2)$ positive root of the equation

$$1 - x + \left(\frac{1}{2}x\right)^{k+1} = 0,$$

and

$$\beta = \frac{2 - \alpha}{k + 1 - k\alpha}.$$

Fermat, Pierre de (1601–65; b. Beaumont-de-Lomagne, France; d. Castres, France) A French mathematician. He is particularly famous for his 'last theorem', which he discovered in about 1637, and of which he claimed he had a 'marvellous demonstration'. He died without revealing his proof and it was not until 1994 that the English mathematician Andrew Wiles gave a full proof. Fermat's father was a wealthy leather merchant. Fermat was educated at a Franciscan monastery and studied law at the University of Toulouse. He became a judge, but he had a passion for mathematics and obtained many mathematical theorems, which he communicated to fellow mathematicians, always remaining very secretive about his proofs. The correspondence between Fermat and *Pascal laid the foundations of the modern theory of *probability. A lunar crater and a street in Paris are named after him.

FFT *See* FAST FOURIER TRANSFORM.

fiducial inference A method of *statistical inference proposed by Sir Ronald *Fisher in 1930. Fisher's aim was to prescribe an entirely objective procedure that avoided prior assumptions or hypotheses. Fisher's approach is not easy to understand and continues to be the subject of discussion.

Fieller's theorem A theorem first stated by E. C. Fieller in 1940. Suppose that the *random variables X and Y have *means μ_x and μ_y, respectively, and a *bivariate normal distribution. The quantity of interest is γ, the ratio of the two means:

$$\gamma = \frac{\mu_x}{\mu_y}.$$

Let $\bar{x}$ and $\bar{y}$ be estimates of μ_x and μ_y, respectively, and let the respective estimates of the *variances and *covariance be denoted by s_x^2, s_y^2 and c_{xy}, with ν *degrees of freedom. Fieller's theorem states that

$$\frac{\bar{x} - \gamma\bar{y}}{s_x^2 - 2\gamma c_{xy} + \gamma^2 s_y^2}$$

has a *t-distribution with ν degrees of freedom.

filter A procedure that converts one *time series into another. A simple example is a *moving average, which, because it is a linear combination of the terms in the time series, is described as a **linear filter**. A filter that removes short-term random fluctuations is called a **low-pass filter**, and one that removes long-term fluctuations is called a **high-pass filter**.

filter coefficient *See* WAVELET.

finite mixture model In a finite mixture model the *distribution of the *random variable X has a known form (*e.g.* *normal), but the values of the *parameters of the distribution are not known. Instead, it is known that the *vector of the parameters takes one of the values $\boldsymbol{\theta}_1, \boldsymbol{\theta}_2, \ldots, \boldsymbol{\theta}_m$ with associated probabilities $w_1, w_2, \ldots, w_m$. If, for a *continuous random variable, the *probability density function of the kth of the possible distributions is denoted by $f(x, \boldsymbol{\theta}_k)$, then the finite mixture distribution of X is given by

$$\sum_{k=1}^{m} w_k f(x, \boldsymbol{\theta}_k).$$

An equivalent result holds for a *discrete random variable.

finite population correction *See* SAMPLING FRACTION.

Finnish Statistical Society This society was founded in 1920 and has around 500 members.

Fisher, Irving (1867–1947; b. Saugerties, New York; d. New York City) America's first mathematical economist. Co-founder and first President of the *Econometrics Society. Fisher's academic life was spent at Yale University, where he studied mathematics and then economics, gaining his PhD (the first in economics at Yale) in 1891. He then joined the mathematics faculty before moving into the social sciences, becoming

Professor of Political Economy from 1898 until his retirement as
Professor Emeritus in 1935.

Fisher, Sir Ronald Aylmer (1890–1962; b. East Finchley, England; d.
Adelaide, Australia) An English statistician; arguably the most influential
statistician of the twentieth century. Fisher was educated at Harrow and
at Cambridge University, where he studied mathematics. His initial
interest in statistics developed from his interest in genetics, and he
pursued both subjects for the rest of his life. Fisher's first paper
introduced the method of *maximum likelihood, his second the
mathematical derivation of the *t-distribution, and his third the
distribution of the *correlation coefficient. Following the First World
War (which saw Fisher employed as a teacher because his dreadful
eyesight prevented him from fighting), he joined the agricultural
research station at Rothamsted. During his time there he virtually
invented the subjects of *experimental design and *ANOVA, which
motivated his derivation of the *F-distribution. In 1925 the first edition of
his *Statistical Methods for Research Workers* appeared. The extent of the
influence of this work can be gauged from the fact that there was a new
edition about every three years until 1958, and a posthumous fourteenth
edition in 1970. At a conference in India he said, 'To call in the
statistician after the experiment is done may be no more than
asking him to perform a post-mortem examination: he may be able
to say what the experiment died of.' In 1933 Fisher became
Professor of Eugenics at University College, London and in 1943
Professor of Genetics at Cambridge University. He was elected a
Fellow of the Royal Society in 1929 and was awarded the Society's
Copley Medal in 1955. Fisher was elected to membership of the
National Academy of Sciences in 1948 and knighted in 1952. He was
President of the *Royal Statistical Society in 1952 and was awarded its
*Guy Medal in Gold in 1946.

Fisher–Behrens problem *See* BEHRENS–FISHER PROBLEM.

Fisher distribution *See* LANGEVIN DISTRIBUTION.

Fisher's exact test A test of *independence between two
*dichotomous *categorical variables that was introduced by Sir Ronald
*Fisher in 1935. Let a, b, c, and d denote the *frequencies of the four
(2×2) possible outcomes, with m, n, r, s, and N the totals of these
frequencies as indicated in the table

$$\begin{array}{cc|c} a & b & m \\ c & d & n \\ \hline r & s & N \end{array}$$

The test involves the use of the *hypergeometric distribution to calculate the *probability of the observed outcome, given the observed values of m, n, r, and s, under the *null hypothesis of independence. The probability of the observed outcome is given by

$$\frac{m!n!r!s!}{a!b!c!d!N!}.$$

As an example, suppose that five out of six patients treated with drug A recover, whereas only three out of five patients treated with drug B recover. The null hypothesis is that the outcome was independent of the drug used. The possible outcomes (and their probabilities) are as follows:

Actual Outcome

	R	R̄			R	R̄			R	R̄			R	R̄	
A	6	0	6		5	1	6		4	2	6		3	3	6
B	2	3	5		3	2	5		4	1	5		5	0	5
	8	3	11		8	3	11		8	3	11		8	3	11

$p = 2/33$	$p = 12/33$	$p = 15/33$	$p = 4/33$

where R denotes recovery and R̄ denotes non-recovery. Given the fixed marginal totals (6, 5, 8, 3), the probability of the observed outcome, or one in which drug A is more successful, is $\frac{12}{33} + \frac{2}{33} = \frac{14}{33}$. Similarly, the probability of the observed outcome, or one in which drug B is more successful, is $\frac{12}{33} + \frac{15}{33} + \frac{4}{33} = \frac{31}{33}$. Since neither $\frac{14}{33}$ nor $\frac{31}{33}$ is unusually small, the null hypothesis of independence is accepted.

Fisher's F-distribution *See* F-DISTRIBUTION.

Fisher's index (ideal index) An index, defined as the *geometric mean of *Laspeyres's price index and *Paasche's price index, that was suggested by Irving *Fisher in 1922. If p_{0j} and q_{0j} represent the price and quantity, respectively, of item j at time 0, with p_{nj} and q_{nj} representing the corresponding quantities at time n, then the index is given by

$$\sqrt{\frac{\sum_j p_{nj} q_{0j}}{\sum_j p_{0j} q_{0j}} \times \frac{\sum_j p_{nj} q_{nj}}{\sum_j p_{0j} q_{nj}}}.$$

Fisher's information; Fisher's information matrix The amount of information that a *sample provides about the value of an unknown *parameter. Writing L as the *likelihood for n *observations from a *distribution with parameter θ, Sir Ronald Fisher, in 1922, defined the information, I, as being given by

$$I = E\left\{ \left(\frac{\partial \ln L}{\partial \theta} \right)^2 \right\}.$$

If T is an unbiased estimator of θ then a lower bound to the variance of T is

$$\frac{1}{I}.$$

Since 1948 this result has generally been referred to as the **Cramér-Rao lower bound** or the **Cramér-Rao inequality** in recognition of the independent proofs by *Cramér and *Rao. If the distribution involves several parameters, $\theta_1, \theta_2, \ldots, \theta_p$, then Fisher's information matrix has elements given by

$$I_{jk} = E\left\{ \left(\frac{\partial \ln L}{\partial \theta_j} \right) \left(\frac{\partial \ln L}{\partial \theta_k} \right) \right\}.$$

Fisher's LSD *See* MULTIPLE COMPARISON TESTS.

Fisher's z-transformation A transformation of the sample *correlation coefficient, r, suggested by Sir Ronald *Fisher in 1915. The statistic z is given by

$$z = \frac{1}{2} \ln\left(\frac{1+r}{1-r} \right).$$

For samples from a *bivariate normal distribution with sample sizes of 10 or more, the *distribution of z is approximately a *normal distribution with *mean and *variance given by

$$\frac{1}{2} \ln\left(\frac{1+\rho}{1-\rho} \right) + \frac{\rho}{2(n-1)} \qquad \text{and} \qquad \frac{1}{n-3},$$

respectively, where n is the sample size and ρ is the population *correlation coefficient. The transformation is remarkably *robust.

fitted distribution *See* FITTED FREQUENCY.

fitted frequency The frequency of a value or category that results from some *model. When a distribution of a specified type is matched to a set of observed frequencies, the value corresponding to a particular observed frequency is the fitted frequency and the set of fitted frequencies is the **fitted distribution**. Usually the observed and fitted frequencies will be compared using a *goodness-of-fit test. The *parameters of the fitted distributions will often have been estimated using the *method of maximum likelihood. *See also* EXPECTED FREQUENCY.

fitted values The values predicted by a *model fitted to a set of *data.

five-barred gate *See* TALLY CHART.

five-number summary For a set of numerical *data, the least value, the lower *quartile, the *median, the upper quartile, and the greatest value, in that order. *See* BOXPLOT.

fixed effects *See* EXPERIMENTAL DESIGN.

flowchart A diagram showing the structure of a computer program, and the different pathways that can be taken.

forecasting The prediction of future values in a *time series. Methods include *exponential smoothing, the *Holt–Winters procedure, and the *Box–Jenkins procedure.

forgetfulness property (lack of memory) The property possessed by the *Poisson process, and by its discrete analogue (a sequence of *Bernoulli trials), that *probabilities of future occurrences are uninfluenced by past events. This is just a restatement of *independence. The corresponding *distributions of the time to the first observation (either *exponential or *geometric

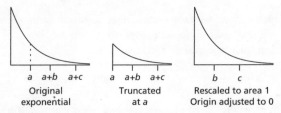

a *a+b* *a+c*	*a* *a+b* *a+c*	*b* *c*
Original exponential	Truncated at *a*	Rescaled to area 1 Origin adjusted to 0

Forgetfulness property. For the exponential distribution, if we know that the observed value is greater than *a*, then the resulting conditional distribution is a facsimile of the original distribution.

distributions) and their extensions to the time to the nth observation (either *gamma or *negative binomial distributions) all have the forgetfulness property.

FORTRAN Abbreviation for Formula Translator—one of the earliest computer programming languages.

forward selection *See* STEPWISE PROCEDURES.

Fourier series If f is an integrable function on $(-\pi, \pi)$ then the Fourier series for f is

$$\frac{1}{2}a_0 + \sum_{j=1}^{\infty}\left\{a_j\cos(jx) + b_j\sin(jx)\right\},$$

where

$$a_j = \frac{1}{\pi}\int_{-\pi}^{\pi} f(t)\cos(jt)\mathrm{d}t \quad \text{and} \quad b_j = \frac{1}{\pi}\int_{-\pi}^{\pi} f(t)\sin(jt)\mathrm{d}t, \qquad j = 0, 1, \ldots.$$

Fourier transform If f is a function of the real variable x, its Fourier transform F is defined by

$$F(t) = \frac{1}{\sqrt{2\pi}}\int_{-\infty}^{\infty} \mathrm{e}^{\mathrm{i}tx}f(x)\mathrm{d}x,$$

where $\mathrm{i} = \sqrt{-1}$ and t is a real variable. The **inverse Fourier transform** is given by

$$f(x) = \frac{1}{\sqrt{2\pi}}\int_{-\infty}^{\infty} \mathrm{e}^{-\mathrm{i}tx}\, F(t)\mathrm{d}t.$$

fractal An infinite set of points, usually in a plane, whose *fractal dimension is not an integer. Fractals are usually self-similar, *i.e.* any part of the fractal contains scaled-down versions of the whole fractal. The word 'fractal' was coined by *Mandelbrot in 1975. For a well-known example of a fractal, *see* SNOWFLAKE CURVE.

fractal dimension A generalization of the usual idea of dimension that permits non-integer values. For a plane set of points, the fractal dimension is estimated as follows. A lattice of square boxes of side s is superimposed over the set. The number, $N(s)$, of these boxes that contain a portion of the set is determined. This is repeated for a range of values of s. When $\ln\{N(s)\}$ is plotted against $-\ln(s)$ the result will be an approximate straight line. The fractal dimension is the slope of this line. An ordinary curve has a fractal dimension equal to 1. Any set of points in a plane has a fractal dimension of ≤ 2.

fractional factorial design *See* FACTORIAL EXPERIMENT.

fractional replicate *See* FACTORIAL EXPERIMENT.

Freeman–Tukey deviate *See* FREEMAN–TUKEY TRANSFORMATION.

Freeman–Tukey test A *goodness-of-fit test for models concerning *counts, introduced in 1950. If there are n counts, $f_1, f_2, \ldots, f_n$ with, according to a p-parameter model, corresponding *expected frequencies $e_1, e_2, \ldots, e_n$, then the Freeman–Tukey statistic, T^2, is given by

$$T^2 = \sum_{k=1}^{n} \left(\sqrt{f_k} + \sqrt{f_k + 1} - \sqrt{4e_k + 1} \right)^2.$$

If the model is correct then the distribution of T^2 is approximately *chi-squared with $(n - p)$ *degrees of freedom.

Freeman–Tukey transformation If X is a *binomial variable with *parameters n and p, then the transformed variable Y, given in radians by

$$Y = \frac{1}{2} \left\{ \sin^{-1}\left(\sqrt{\frac{X}{n+1}} \right) + \sin^{-1}\left(\sqrt{\frac{X+1}{n+1}} \right) \right\},$$

has an approximate *normal distribution with *mean $\sin^{-1}(\sqrt{p})$ and *variance $1/(4n + 2)$. The approximation improves as np increases and should not be used if $np < 1$.

If X is a *Poisson variable with expectation μ, then, for $\mu > 1$, the transformed variable Z, given by

$$Z = \sqrt{X} + \sqrt{X+1} - \sqrt{4\mu + 1}$$

has an approximate *standard normal distribution. Observed values of Z are referred to as **Freeman-Tukey deviates**. Both transformations were proposed in a 1950 paper by M. F. Freeman and *Tukey.

frequency 1. The number of times that a particular data value is obtained in a *sample. For example, the frequency of 5 in the sample 4, 6, 5, 7, 4, 5, 2, 5 is 3. The sum of the frequencies is the *sample size. The term is also used in connection with a set of values. For example, the number of people aged between 20 and 30, or the number of people with blue or green eyes. The term **relative frequency** is used for the ratio (frequency) ÷ (sample size). In the context of a *goodness-of-fit test, a frequency is referred to as an **observed frequency** and usually denoted by O, or f_O.

2. In the context of a *time series, the number of *cycles per unit time.

frequency polygon A graphical method for displaying *data that is an alternative to the *bar chart for displaying data on a *discrete random variable, and an alternative to the *histogram as a method of approximating the graph of a *probability density function.

Suppose a set of discrete numerical *data consists of the observed value x_1 with *frequency f_1, the observed value x_2 with frequency f_2, and so on. The frequency polygon consists of the straight line segments joining the points with coordinates (x_1, f_1), $(x_2, f_2), \ldots$. It is conventional to start and finish the polygon on the x-axis at appropriate points.

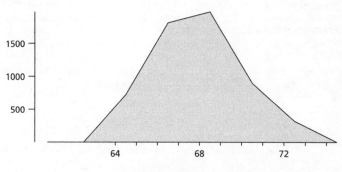

Height of militia man (inches)

Frequency polygon. This polygon illustrates the frequencies of the heights of 5732 Scottish militiamen in 1817.

In the case of grouped numerical data with classes of equal width d, say, the frequency polygon consists of the straight-line segments joining the points with coordinates $(x_1, f_1), (x_2, f_2), \ldots, (x_n, f_n)$, where $x_1, x_2, \ldots, x_n$ are now the midpoints of the class intervals and $f_1, f_2, \ldots, f_n$ are the corresponding class frequencies. The polygon starts at $(x_1 - d, 0)$ and finishes at $(x_n + d, 0)$. A suitable change of scale gives the **relative frequency polygon**. As the sample size increases, the relative frequency polygon approaches the graph of the corresponding probability density function.

frequentist One who views *probability as being equal to the limiting relative *frequency as the sample size increases.

Friedman, Milton (1912–; b. Brooklyn, New York) The American Nobel Prize winner for Economics in 1976. He was the son of an immigrant merchant who died when Milton was fifteen. Friedman obtained a BA in economics at Rutgers University in 1932, an MA from the University of Chicago in 1933, and a PhD from Columbia University in 1946, by which time he was Professor of Economics at the University of Chicago. From 1937 to 1981 Friedman was associated with the National Bureau of Economics Research. During this time he wrote many books on aspects of economics. In 1988 he was awarded the Presidential Medal of Freedom and the National Medal of Science. He is currently Professor Emeritus at the University of Chicago.

Friedman's two-way analysis of variance *Non-parametric *analysis of variance for a two-way *crossed design in which the *observations are replaced by their *ranks. Suggested by *Friedman in a paper published in 1937.

F-test A test of whether two *variances are equal. Denoting the estimates of the two variances by s_1^2 and s_2^2, with v_1 and v_2 *degrees of freedom, respectively, the test compares the ratio

$$\frac{s_1^2}{s_2^2}$$

with the *critical values of an *F-distribution with v_1 and v_2 degrees of freedom. The F-test is most frequently encountered in the context of an analysis of variance table (*See* ANOVA), where it is often referred to as a **variance-ratio test**.

F to enter; F to remove *See* STEPWISE PROCEDURES.

full rank *See* MATRIX.

G

Gabriel biplot *See* BIPLOT.

Galton, Sir Francis (1822–1911; b. Birmingham, England; d. Haslemere, England) An English doctor, explorer, meteorologist, biometrician, and statistician. Galton was also first cousin of Charles Darwin, the author of *The Origin of Species*. Galton studied medicine at Cambridge University. On coming into money, he abandoned this career and spent the period 1850–2 exploring Africa; he received the gold medal of the Royal Geographical Society in recognition of his achievements. In the 1860s he turned to meteorology and devised an early form of the weather maps used by modern meteorologists. He coined the term 'anticyclone'. Subsequently, perhaps inspired by Darwin's work, Galton turned to inheritance and the relationships between the characteristics of successive generations. In his 1869 book *Hereditary Genius* he used the term *correlation in its statistical sense. His best known work, published in 1889, was entitled *Natural Inheritance*. He made great use of the *normal distribution and illustrated it in a lecture to the Royal Institution in 1874 using a *quincunx. He is quoted as saying 'Whenever you can, count.' He was elected a Fellow of the Royal Society in 1860.

gambler's fallacy *See* LAW OF AVERAGES.

gambler's ruin A classic *random walk problem that can be solved using the theory of *Markov processes. At each play a gambler is supposed to have *probability p of winning one unit and probability q ($= 1 - p$) of losing one unit. The gambler starts with j units. Betting continues until either the gambler's fortune reaches N (in which case the gambler retires) or the gambler is ruined. If $p = \frac{1}{2}$ then the probability of ruin is $1 - (j/N)$, and otherwise it is

$$\left\{ \left(\frac{q}{p}\right)^{j} - \left(\frac{q}{p}\right)^{N} \right\} \Big/ \left\{ 1 - \left(\frac{q}{p}\right)^{N} \right\}.$$

game theory A theory that deals with the determination of the optimal strategies for each of the players in a game with well-defined rules. In a typical scenario, two players, A and B, are playing a game. Player A is required to make a decision in ignorance of a simultaneous decision made by player B. The outcome is a consequence of the two decisions. In a *zero-sum game A wins what B loses. Many real-life situations can be modelled in game-theoretic terms. The rules used by the players to determine their strategies are called **decision rules**.

gamma (γ) A *measure of association, of the type known as *proportional reduction in error, suitable for examining the association between two *ordinal variables.

gamma distribution The general form of the gamma distribution has *probability density function f given by

$$f(x) = \frac{(x-\gamma)^{\alpha-1} \exp\left[\frac{-(x-\gamma)}{\beta}\right]}{\beta^{\alpha}\Gamma(\alpha)}, \qquad x > \gamma,$$

where α and β are positive *parameters, and Γ is the *gamma function. The distribution has *mean $\alpha\beta + \gamma$ and *variance $\alpha\beta^2$. If $\alpha > 1$ then the distribution has *mode at $x = \gamma + \beta(\alpha - 1)$; otherwise the mode is at $x = \gamma$.

The case ($\alpha = \frac{1}{2}v$, $\beta = 2$, $\gamma = 0$) corresponds to the *chi-squared distribution with v *degrees of freedom. The case ($\beta = 1$, $\gamma = 0$) gives the standard form of the distribution:

$$f(x) = \frac{x^{\alpha-1}e^{-x}}{\Gamma(\alpha)}, \qquad x > 0.$$

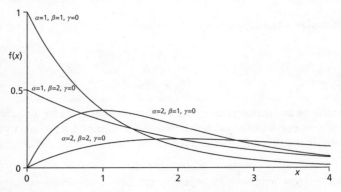

Gamma distribution. For simplicity $\gamma = 0$ for all the illustrated cases, which have mean $\alpha\beta$. When $\alpha \geq 1$ the mode is at $x = \beta(\alpha - 1)$, otherwise it is at $\alpha = 0$.

The case ($\alpha = 1, \beta = 1, \gamma = 0$) gives the *exponential distribution, and the case ($\alpha = k, \beta = 1, \gamma = 0$), where k is a positive integer, gives the **Erlang distribution**, which is the distribution of the time to the kth event in a *Poisson process. If Z is a *standard normal variable then $\frac{1}{2}Z^2$ has a gamma distribution with parameters $\alpha = \frac{1}{2}, \beta = 1, \gamma = 0$.

gamma function The gamma function, Γ, is defined by

$$\Gamma(x) = \int_0^\infty s^{x-1}e^{-s}ds.$$

In general $\Gamma(x + 1) = x\Gamma(x)$. If x is a positive integer then $\Gamma(x) = (x - 1)!$ (*see* FACTORIAL). If x is positive and an odd multiple of $\frac{1}{2}$ then the value of $\Gamma(x)$ is given by

$$\Gamma\left(\frac{1}{2}\right) = \sqrt{\pi},$$

$$\Gamma\left(\frac{2k+1}{2}\right) = \frac{2k-1}{2} \times \frac{2k-3}{2} \times \cdots \times \frac{1}{2}\sqrt{\pi}, \qquad k = 1, 2, \ldots.$$

Gantt chart A project-planning tool that can be used to represent the scheduling of tasks required to complete a project. Each task corresponds to a row. Dates run along the top of the chart in suitable increments of time. The expected time for each task is represented by a horizontal bar whose left end marks the expected beginning of the task and whose right end marks the expected completion date. As the project progresses, the chart is updated by filling in the bars to a length proportional to the fraction of work that has been accomplished on the task. Completed tasks lie to the left of the current date-line and are completely filled in. Current tasks cross the date-line and are behind schedule if their filled-in section is to the left of the line and ahead of schedule if the filled-in section stops to the right of the line. Future tasks lie completely to the right of the line. The chart was devised in 1917 by Henry L. Gantt, an American engineer and social scientist.

garbage in, garbage out A phrase that reminds us that if the basic *data are untrustworthy (for example, if most of the observed variation results from *experimental error) then the results of any analysis will be valueless.

GARCH (Generalized Autoregressive Conditional Heteroscedasticity) GARCH *models are used to predict the *volatility of financial *random variables.

Gauss, Carl Friedrich (1777–1855; b. Brunswick, Germany; d. Göttingen, Germany) A German mathematician and astronomer. Gauss was responsible, in a paper published in 1809, for developing the statistical theory underlying the method of least squares. He was educated at the Universities of Göttingen and Helmstedt, obtaining his doctorate from the latter in 1799. His work on least squares, which was a consequence of his appointment as director of the observatory at Göttingen in 1807, also entailed his deriving an appropriate error distribution—the distribution now called the *normal or Gaussian distribution. Gauss was elected a Fellow of the Royal Society in 1804 and awarded its Copley Medal in 1838. A lunar crater is named after him.

Gaussian distribution *See* NORMAL DISTRIBUTION.

Gaussian kernel *See* KERNEL METHODS.

Gauss–Markov theorem *See* MULTIPLE REGRESSION MODEL.

Gehan, Edmund Alpheus (1929–; b. Brooklyn, New York) An American biostatistician. Gehan obtained his BA in mathematics at Manhattan College, New York City, before moving to North Carolina State University where he obtained his MS in experimental statistics in 1953 and his PhD in experimental statistics and public health in 1957. He is currently Director of Biostatistics at the Georgetown University Medical Center in Washington.

Gehan test A test, introduced in 1965, as an alternative to the *Mantel–Haenszel test for comparing two *survivor functions using *data that include *censored observations (because, for example, some patients are still living). With two samples, of sizes m and n, the test statistic is W, given by

$$W = \frac{1}{mn}(N_{A>B} - N_{A<B}).$$

The calculation of $N_{A>B}$ and $N_{B>A}$ requires examination of the mn pairs of observations (with one member of each pair from each sample). The quantity $N_{A>B}$ is the number of pairs in which the member chosen from sample A is certain to have a longer lifetime than the member chosen from sample B.

generalized inverse *See* MATRIX.

generalized least squares (GLS) The extension of the *ordinary least squares procedure to the case where the *observations have been taken on *random variables that are not all *independent of one another. The GLS estimate of the $(p + 1) \times 1$ parameter vector $\boldsymbol{\beta}$ in the *multiple regression model

$$E(\mathbf{Y}) = \mathbf{X}\boldsymbol{\beta}$$

is given by

$$\hat{\boldsymbol{\beta}} = (\mathbf{X}'\Sigma^{-1}\mathbf{X})^{-1}\mathbf{X}'\Sigma^{-1}\mathbf{y},$$

where $\mathbf{y}$ is an $n \times 1$ column *vector of observations, Σ^{-1} is the inverse of the *variable-covariance matrix, and, with p explanatory variables and a constant term in the model, $\mathbf{X}$ is the $n \times (p + 1)$ *design matrix. *See also* WEIGHTED LEAST SQUARES.

generalized linear model (GLM) Model that relates μ, the expectation of the response variable Y, to a linear combination of the explanatory variables $x_1, x_2, \ldots, x_p$ using the model

$$g(\mu) = \beta_0 + \beta_1 x_1 + \cdots + \beta_p x_p,$$

where $\beta_1, \beta_2, \ldots, \beta_p$ are unknown *parameters and g is the **link function**. Examples of link functions are

$g(\mu) = \mu$ (the **identity link**),

$g(\mu) = \log(\mu)$ (the **logarithmic link**),

$g(p) = \log\left(\dfrac{p}{1 - p}\right)$ (the **logistic link**),

$g(p) = \log\{-\log(1 - p)\}$ (the **complementary log-log link**).

The first two correspond to random variables having, respectively, *normal distributions and *Poisson distributions. The last two provide different families of models for dealing with, for example, variation in the parameter of a binomial distribution.

generating function A function of an arbitrary variable (usually t) which, when expanded as a power series (in t), yields coefficients of interest to statisticians and others. Examples are the *moment-generating function and the *probability-generating function.

genetic algorithm An *algorithm used in particular for *combinatorial optimization. Later search locations are derived by **reproduction**, **mutation**, or **crossover**, which represent three different rules for adjusting or combining the coordinates of earlier search locations.

GENSTAT (Generalized Statistical Package) A statistical computer package particularly used in analysing natural science data.

geometric distribution A particular discrete distribution. When an experiment or trial with two possible results, usually classified as 'success' or 'failure', is repeated independently until the first success is obtained, then, if the *probability of a success, $p(\neq 0, 1)$, is the same for each trial, then the *distribution of the total number of trials up to and including the one in which the first success is obtained is a geometric distribution with *parameter p. The *probability function is given by

$$P(X = r) = p(1 - p)^{r-1}, \qquad r = 1, 2, \ldots.$$

The reason for the term 'geometric' is that the sequence of probabilities $P(X = r)$, for $r = 0, 1, 2, \ldots$, forms a geometric progression with first term p and common ratio $(1 - p)$. The *mean is $1/p$ and the *variance is $(1 - p)/p^2$. The *mode is 1 for all values of p.

The fact that the mode is always 1 means that, even for a very rare event (*e.g.* a lottery win), the most probable number of trials necessary to obtain the first success is 1. This is sometimes regarded as a paradox.

If we write $q = 1 - p$, cumulative probabilities are given by $P(X \geq r) = q^{r-1}$ and $P(X \leq r) = 1 - q^r$. The geometric distribution is a discrete analogue of the *exponential distribution and shares the 'non-ageing' or *forgetfulness property:

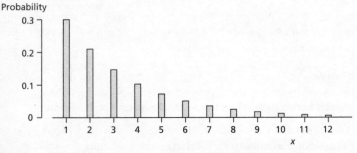

Geometric distribution. All geometric distributions have their mode at the lowest possible value, with successive probabilities having the common ratio p.

$$P(X = r + s | X \geq s) = P(X = r), \qquad s \geq 0.$$

If $X_1, X_2, \ldots, X_n$ are independent random variables each having a geometric distribution with parameter p, then $\sum_{k=1}^{n} X_k$, which is the number of trials up to and including the nth success, has a *negative binomial distribution with parameters n and p.

Some authors take the definition of a geometric distribution to be the number of trials before the first successful trial. This leads to $P(X = r) = p(1 - p)^r$, for $r = 0, 1, \ldots$, for which the mean is $(1 - p)/p$ and the mode is 0, though the variance is unchanged.

geometric mean For a set of positive observations $x_1, x_2, \ldots, x_n$, the geometric mean, g, is given by

$$g = (x_1 \times x_2 \times \cdots \times x_n)^{\frac{1}{n}}.$$

geometric variable A *random variable having a geometric distribution.

geostatistics A subject involving ideas from geology, mathematics, mining engineering, and statistics. It is principally concerned with methods of analysis for *spatial processes, the motivation being the accurate detection of mineral deposits. Its scope was well defined by the work of *Matheron.

Gibbs, Josiah Willard (1839–1903; b. New Haven, Connecticut; d. New Haven, Connecticut) An American pioneer of statistical mechanics. Gibbs entered Yale University in 1854 to study mathematics and Latin. He remained at Yale, becoming its first doctorate student in engineering (and the first in the United States) in 1863. During the 1870s Gibbs worked on thermodynamics and in the 1880s on vector analysis. He was elected to membership of the National Academy of Sciences in 1879. In 1897 he was elected a Fellow of the Royal Society and was awarded the Society's Copley Medal in 1901. His *Elementary Principles of Statistical Mechanics* was published in 1902. A lunar crater is named after him.

Gibbs sampler *See* MARKOV CHAIN MONTE CARLO METHODS.

Gini, Corrado (1884–1965; b. Motta di Livenza, Italy; d. Rome, Italy) An Italian statistician. Gini studied statistics as a compulsory part of his study of law at the University of Bologna. He realized that statistics provided a means of measuring social phenomena and this became his specialism. He was the founder of the first Department of Statistics in Italy (at Rome), in 1936.

Gini index *See* DIVERSITY INDEX.

Gini's statistic A measure of the spread or dispersion of a *probability distribution. Suppose that X_1 and X_2 are *independent observations from the distribution; then Gini's statistic is the expectation of $|X_1 - X_2|$.

Gittins, John Charles (1938–; b. Durham, England) An English statistician. Gittins was educated at St John's College, Cambridge and University College, Aberystwyth, where he gained his PhD on optimal resource allocation. He was appointed to the Mathematics Department at Oxford in 1974, where he is now Professor of Statistics. He was awarded the *Royal Statistical Society's *Guy Medal in Silver in 1984.

Gittins index *See* BANDIT PROBLEMS.

GLIM (Generalized Linear Interactive Modelling) A statistical computer package especially suitable for fitting *generalized linear models. The package was commissioned by the *Royal Statistical Society in 1972 with the guidance of *Nelder.

GLM *See* GENERALIZED LINEAR MODEL.

global maximum *See* LOCAL MAXIMUM.

GLS *See* GENERALIZED LEAST SQUARES.

Gompertz, Benjamin (1779–1865; b. London, England; d. London, England) An English actuary. Gompertz's parents were Jewish Dutch merchants. Because of his faith he was denied entry to university, but he nevertheless became a member of the London Mathematical Society when aged eighteen. He joined the Stock Exchange in 1810 and was elected a Fellow of the Royal Society in 1819. As an actuary he was concerned with the expectation of life. Statisticians remember Gompertz for his 'Law of Mortality' (*see* GOMPERTZ DISTRIBUTION), while animal rights campaigners recall that Gompertz, struck by 'the similitude between man and other animals' was one of the founders, in 1824, of the Society for Prevention of Cruelty to Animals (now the RSPCA).

Gompertz curve *See* GROWTH CURVE.

Gompertz distribution A probability distribution now better known as the *Gumbel distribution. In 1825 Gompertz proposed a probability model for a 'Law of Mortality', assuming the 'average exhaustion of a man's power to avoid death to be such that at the end of equal infinitely small intervals of time he lost equal portions of his remaining power to oppose destruction which he had at the commencement of these intervals'.

Goodman, Leo A. (1928–; b. New York City) An American statistician and sociologist. Goodman was a graduate of mathematics and sociology at Syracuse University in 1948. After obtaining his MA and PhD degrees from Princeton University in 1950, he joined the faculty at the University of Chicago, where he was appointed full professor in 1955. In 1986 he moved to the University of California, Berkeley, where he is now Professor Emeritus. He was awarded the *Wilks Medal of the *American Statistical Association in 1985.

Goodman and Kruskal's gamma (γ) *See* ORDINAL VARIABLE.

Goodman and Kruskal's lambda (λ) A *measure of association between two *categorical variables A and B, suggested by *Goodman and *Kruskal in 1954. Let A have J categories and B have K categories. A *sample contains f_{jk} occurrences of the combination of category j for A and k for B. Define $f_{j0}, f_{0k},$ and f_{00} by

$$f_{j0} = \sum_{k=1}^{K} f_{jk}, \qquad f_{0k} = \sum_{j=1}^{J} f_{jk}, \qquad f_{00} = \sum_{j=1}^{J} \sum_{k=1}^{K} f_{jk}.$$

Suppose that we are asked to guess the category of B for the next *observation. An intelligent guess would be the category that was the commonest so far (with *marginal total equal to f_{0m}, say). Judging by the past data, the *probability that our guess will be correct will be P, given by $P = f_{0m}/f_{00}$.

Suppose now that the next observation belongs to category j of variable A. Our best guess for the category of B now corresponds to the maximum of $f_{j1}, f_{j2}, \ldots, f_{jK}$. The revised probability of our being correct will be estimated by

$$\frac{\max_k (f_{jk})}{f_{j0}}.$$

Since the estimated probability of the next observation belonging to category j of A is estimated as f_{j0}/f_{00}, the estimated probability of being correct taking into account the category of A for the next observation is P_A, given by

$$P_A = \frac{\sum_{j=1}^{J} \max_k (f_{jk})}{f_{00}}.$$

The statistic λ, given by

$$\lambda = \frac{P_A - P}{1 - P},$$

is a measure of the *proportional reduction in error resulting from knowledge about A.

As an illustration consider the following 3×4 table in which the relevant cell and marginal frequencies are shown in bold.

	B_1	B_2	B_3	B_4	TOTAL
A_1	10	5	18	**20**	53
A_2	8	13	5	**16**	42
A_3	**11**	10	3	1	25
Total	29	28	26	**37**	**120**

Here $\lambda = \{(20 + 16 + 11) - 37\}/(120 - 37) \approx 12\%$, so the benefit of knowing the category of A for the next observation is slight. This suggests that the association between A and B is not great.

goodness-of-fit test A test of the fit of some *model to a set of *data. The most commonly used tests are the *chi-squared test and the *likelihood-ratio test. *See also* INDEX OF DISPERSION; KOLMOGOROV–SMIRNOV TEST.

gooseberry bushes An example of *nonsense correlation. If the number of gooseberry bushes in London gardens is plotted against the number of births in London in successive decades of the twentieth century, then the resulting graph is likely to show a strong positive nonsense correlation, since both are a reflection of increasing population size. In the same way, increases in births can be linked to the numbers of storks nesting on the rooftops of European cities.

Gosset, William Sealy (1876–1937; b. Canterbury, England; d. Beaconsfield, England) An English statistician. Gosset was a pioneer of the statistical theory associated with small *sample sizes. He was educated at Winchester College and Oxford University, where he studied chemistry. In 1899 he joined the staff of Arthur Guinness Son & Co. Ltd as a 'brewer'. An early task was to investigate the relationship between the quality of the final product and that of the raw materials (such as barley and hops). The difficulty with this task was the expense and time involved in obtaining an observation, so large samples were not available. Gosset correctly mistrusted the existing theory and, in a paper published in 1908, entitled *The Probable Error of a Mean*, he conjectured the form of the *t-distribution relevant for small samples. Guinness company policy at the time meant that Gosset was obliged to publish under a pseudonym: being naturally modest he chose the pen-name 'Student', and the t-distribution is still sometimes referred to as 'Student's t-distribution'.

Graeco-Latin square *See* LATIN SQUARE.

Gram–Charlier expansion An expansion relating a *probability density function, f, to ϕ, the probability density function of a *standard normal distribution, using *Chebyshev–Hermite polynomials. The alternative *Edgeworth expansion is usually preferred.

graph A graph consists of a set of *n* **nodes** represented by small circles, and a set of **arcs** represented by lines. If a path can be found that connects all the nodes, then a graph is said to be a **connected graph**.

graphical models *Graphs in which nodes represent *random variables. Pairs of nodes connected by arcs correspond to variables that are not *independent of one another. If two sets of nodes A and B are not connected, except via paths that pass through a node in a third set, C, then the variables represented by the nodes in the set A are *conditionally independent of the variables represented by the nodes in the set B, given the values of the variables represented by the nodes in the set C.

If an arc has no direction attached, then this implies that there is *association but not causation. An **undirected graphical model** (also called a *Markov random field) is a model in which no arcs have directions attached. A model in which all arcs have directions attached is called a **directed graphical model** (or **Bayes network** or **Bayes net**). If we consider a directed path as referring to 'generations' of variables, then a simple example of conditional independence (not the only possibility) occurs when a 'child variable' is conditionally independent of earlier 'ancestor variables', given the values of the 'parent variables'. A special case is the *hidden Markov model. A graph containing a mixture of directed and undirected arcs is a **chain graph**.

Graunt, John (1620–74; b. London, England; d. London, England) A prosperous London haberdasher and a Freeman of the Draper's Company. Graunt became interested in the information implicit in the weekly 'Bills of Mortality' for London and, in 1662, he published *Natural and Political Observations Mentioned in a following Index and Made Upon the Bills of Mortality*. This was well received by the Royal Society, which duly elected him to be a Fellow. At one time he had a fine collection of paintings, described by the diarist Samuel Pepys as 'the best collection of anything almost I saw'. However in 1666 he lost much property in the Great Fire of London and he died in poverty. His book contains the foundations of *demography, showing understanding of the concepts of *sample and *population and introducing the *life table.

Graybill, Franklin Arno (1927–) An American statistician. Graybill obtained his PhD from Iowa State University in 1952. In 1960 he moved to Colorado State University, founding, in 1961, the statistical laboratory that bears his name and where he is now Professor Emeritus. He was co-author with *Mood of the 1963 textbook *Introduction to the Theory of Statistics* which for many years was the recommended introductory textbook. In 1976 Graybill became President of the *American Statistical Association after a contested election (an unusual occurrence—the loser was *Kruskal).

Greenhouse, Samuel W. (1918–2000; b. New York City; d. Rockville, Maryland) An American biometrician. Greenhouse was educated at City College (now City University), New York, gaining his BS in mathematics in 1938. On graduating he joined the Bureau of Census, working with *Deming. After serving in the army during the war, Greenhouse was recruited, in 1948, to be an inaugural member (with *Mantel) of the first *biometry group in the National Cancer Institute. His subsequent career took him through the leadership of several sections of the National Institute of Health (NIH). At the same time, from 1946 he had been a part-time staff member at George Washington University (GWU), where he also studied under *Kullback for his PhD. On his retirement from NIH in 1974 (as an Associate Director), he took a full-time post at GWU, retiring as Professor Emeritus in 1988.

Greenhouse–Geisser correction An adjustment made to the

numbers of *degrees of freedom, in *analysis of variance, when it is known that the *observations do not obey the usual assumptions of being *uncorrelated with constant *variance.

group-average clustering A method for collecting *multivariate data into *clusters. *See* AGGLOMERATIVE CLUSTERING METHODS.

grouped data The *frequencies of occurrence of values in specified intervals. *Histograms and *frequency polygons are used to illustrate grouped data.

grouped mean *See* MEAN (DATA).

growth curve A curve describing the growth of a population. If the population size at time t is y then possibilities include the **Gompertz curve** given by

$$y = ae^{-bt},$$

and the **logistic curve** given by

$$y = \frac{a}{1 + e^{-bt}},$$

where, in each case, a and b are constants.

G^2 *See* LIKELIHOOD-RATIO GOODNESS-OF-FIT STATISTIC.

Gumbel, Emil Julius (1891–1966; b. Munich, Germany; d. New York City) A German Jewish statistician who spent most of his career in exile. He was educated at Munich University, obtaining a PhD in population statistics in 1914. In 1923 he joined the faculty of the University of Heidelberg but he was an outspoken pacifist and his political publications led to exile, first in France in 1932 where he worked at the University of Lyon, and then in the United States in 1940. It was while in France that he published the definitive study of the *extreme-value distribution that bears his name.

Gumbel distribution An *extreme-value distribution, introduced by *Gumbel in 1935, that has *probability density function f given by

$$f(x) = \frac{1}{\beta}\exp\left\{-\frac{(x-\alpha)}{\beta} - \exp\left(-\frac{x-\alpha}{\beta}\right)\right\}, \qquad -\infty < x < \infty.$$

The distribution has *mode α, *mean $\alpha + \gamma\beta$ (where $\gamma = 0.5772156649\ldots$ is *Euler's constant), and *variance $\frac{1}{6}\beta^2\pi^2$. *See following diagram.*

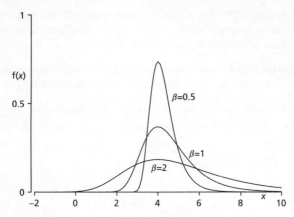

Gumbel distribution. All the distributions illustrated have $\alpha = 4$.

Guttman, Louis Eliahu (1916–87; b. Brooklyn, NY; d. Jerusalem, Israel) An American-born psychometrician who spent most of his working life in Israel. His PhD, at the University of Minnesota, was in *factor analysis and led to the development of the *Guttman scale. He was the founder of the Israel Institute of Applied Social Research.

Guttman scale A method of quantifying the strength of a person's opinion on a topic of interest. The idea is that a series of k related yes/no questions in a *questionnaire may be ordered in the sense that when the questions have been arranged in the appropriate order, an individual who answers 'Yes' to question j ($< k$) will probably answer 'Yes' to questions $j + 1, j + 2, \ldots, k$. For example, a Guttman scale is formed by the series of questions asking whether an individual deserves the death penalty for killing another when the killing was

1. by accident;
2. in self-defence;
3. in the heat of the moment;
4. premeditated.

Usually the ordering of the questions will be less obvious.

Guy, William Augustus (1810–85; b. Chichester, England; d. London, England) An English physician and statistician. Guy was appointed Professor of Medical Jurisprudence at King's College London in 1838. In 1844 he published a paper entitled 'On the Value of Numerical Methods as applied to Science, but especially to Physiology and Medicine'. He was a Fellow of the Royal Society and President of the *Royal Statistical Society from 1873 to 1875.

Guy Medal Medals founded by the *Royal Statistical Society in honour of their past President, William *Guy. Currently, the Guy Medal in Gold is awarded once every three years to individuals judged to have made innovative contributions to the theory or application of *Statistics. Guy Medals in Silver and Bronze are awarded annually to Fellows of the Society for papers published in the Society's journals or given at a meeting organized by the Society.

H

H₀;H₁ *See* HYPOTHESIS TEST.

Haar, Alfréd (1885–1933; b. Budapest, Hungary; d. Szeged, Hungary) A Hungarian mathematician. Haar obtained his PhD in mathematics from the University of Göttingen in 1909. In 1912 he returned to Hungary and after the First World War he founded the Mathematical Centre at Szeged University.

Haar wavelet *See* WAVELET.

Hadamard, Jacques Salomon (1865–1963; b. Versailles, France; d. Paris, France) A French mathematician. Hadamard was educated in Paris where, after graduating from the École Normale Supérieure in 1888, he obtained his doctorate in 1892. His wife was a relative of Alfred Dreyfus, who was convicted of selling military secrets to the Germans. It was subsequently discovered that the evidence against Dreyfus had been manufactured and Hadamard was instrumental in helping to clear his name. Hadamard's research spanned many areas of mathematics. He was elected a member of the National Academy of Sciences in 1926 and a Fellow of the Royal Society in 1932.

Hadamard matrix A square *matrix H_n with n rows and columns, in which every entry is either 1 or -1, arranged so that $H'_n H_n = nI$, where I is an $n \times n$ identity matrix. These matrices are useful in the construction of *experimental designs. The possible values of n are 1, 2, and multiples of 4.

Haenszel, William Manning (1910–98; b. New York City; d. Wheaton, Illinois) An American epidemiologist. Haenszel was a graduate of the University of Buffalo. He joined the New York State Department of Health in 1934. Much of his time there was spent in the National Cancer Institute, where he worked with *Mantel. He was co-author (with Mantel) of the 1959 paper which introduced the eponymous *Mantel–Haenszel test of *independence. From 1976 to 1995 he was Professor of Epidemiology at the University of Illinois.

half-normal plot *See* NORMAL PROBABILITY PAPER.

half-replicate *See* FACTORIAL EXPERIMENT.

halo effect A bias affecting judgements of performance. If an individual does well in one aspect then this is likely to trigger favourable reports on other aspects.

Hammersley, John Michael (1920–) An English Statistician. Educated at Cambridge and Oxford Universities, he was on the faculty of both. A Fellow of the Royal Society, he was co-author of the seminal 1964 book, *Monte Carlo Methods*.

Hammersley–Clifford theorem A theorem, formulated by *Hammersley and Peter Clifford, in 1971, that is concerned with the values at a set of interconnected locations. The theorem defines the class of *probability distributions that are consistent with the Markov property that the value at a location is dependent only on the values of its immediate neighbours. *See* MARKOV RANDOM FIELD.

Hamming, Richard Wesley (1915–98; b. Chicago, Illinois; d. Monterey, California) An American computing expert. Hamming studied at the Universities of Chicago and Nebraska before obtaining his PhD from the University of Illinois at Urbana-Champaign. From 1946 to 1976 he worked at Bell Telephones. He ended his career as Professor of Computer Science at the Naval Postgraduate School at Monterey, California.

Hamming window *See* PERIODOGRAM.

Hardy, Godfrey Harold (1877–1947; b. Cranleigh, England; d. Cambridge, England) An English mathematician. Hardy was the author of the inspirational *A Mathematician's Apology* and he is also remembered as the host of the Indian prodigy, Ramanujan. He was educated at Winchester and Cambridge University, graduating in 1898. Elected a Fellow of Trinity college in 1900, he remained at Cambridge until 1919, when he became Professor of Geometry at Oxford, where he wrote what was for many years the standard textbook on pure mathematics. Hardy made significant advances in the theory of numbers, particularly the prime number theorem.

Hardy was an eccentric—on entering hotel rooms, he covered all mirrors with a towel. He was a keen fan of cricket, which he watched on most days in the season; in order to ensure that the weather stayed fine he would arrive at the cricket ground with thick sweaters and an umbrella, which he referred to as his 'anti-God battery'. Among his sayings is the statistical observation that 'It is not worth an intelligent man's time to be in the majority. By definition, there are already enough people to do that.'

Hardy was elected a Fellow of the Royal Society in 1910 and was awarded its Copley Medal in 1947. He was elected to membership of the National Academy of Sciences in 1927. He was President of the London Mathematical Society in 1926 and again in 1939 and was awarded its De Morgan Medal in 1929.

Hardy–Weinberg law A law that states that under random mating in a *population the *proportions with particular genetic characteristics remain constant over the generations. Suppose that a large population contains individuals with gene pairs *AA*, *Aa*, and *aa*, and that individuals mate with randomly chosen partners. Their offspring receive one gene from each parent. If, from each gene pair, the gene is chosen at random, then the Hardy–Weinberg law, first published in 1908, states that the proportions of the three types of gene pair will remain constant. The population is in *equilibrium.

harmonic mean The *reciprocal of the *mean of the reciprocals of a set of non-zero *observations. For observations $x_1, x_2, \ldots, x_n$ the harmonic mean, h, is given by

$$h = n \left/ \sum_{j=1}^{n} \frac{1}{x_j} \right.$$

Hartley, Herman Otto (1912–80; b. Berlin, Germany; d. Durham, NC) A German statistician who spent the majority of his career in the United States. Hartley gained his PhD in mathematics at the University of Berlin in 1933. In the following year, he emigrated to England to work with Egon *Pearson, gaining a PhD in Statistics from Cambridge University in 1940. In 1946 he joined Pearson on the staff at University College, London, and he was co-author with Pearson of the widely used *Biometrika Tables for Statisticians*. In 1953, Hartley moved to the United States, with posts successively at Iowa State College, Texas A & M University (1963), and Duke University (1979). He was President of the *American Statistical Association in 1979.

Hartley's test A simple test of homogeneity of *variance, suggested by

*Hartley in 1950. The test is suitable when random *samples of equal size have been taken from k different *populations. The *null hypothesis is that all the populations have the same variance. If the unbiased sample estimates are denoted by $s_1^2, s_2^2, \ldots, s_k^2$, the test statistic is

$$\frac{\max(s_1^2, s_2^2, \ldots, s_k^2)}{\min(s_1^2, s_2^2, \ldots, s_k^2)}.$$

hat matching An intriguing *probability problem. At a birthday party n children are given hats. The hats are later shuffled and randomly reassigned. The probability that exactly k get the same hat twice is

$$\frac{1}{k!}\left\{\frac{1}{2!} - \frac{1}{3!} + \cdots + \frac{(-1)^{n-k}}{(n-k)!}\right\}.$$

As n increases, this probability approaches $(1/k!)e^{-1}$. The probability that none gets the same hat twice approaches $e^{-1} \approx 0.368$. Note that the result for $n = 5$ is 0.367, so that the probability of no matches is essentially the same whether there are 5 or 500 children.

hat matrix *See* REGRESSION DIAGNOSTICS.

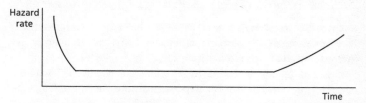

Hazard rate. The typical bathtub curve that results from a plot of the hazard rate against time. A high initial rate falls to a 'normal level' and then slowly increases as ageing sets in.

hazard rate (age-specific failure rate) A measure, $h(t)$, of the chance that a component, still working at age t, is about to fail. Formally, it is defined by

$$h(t) = \lim_{\delta t \to 0}\left\{\frac{P[\text{a component, working at } t, \text{fails in the interval}(t, t + \delta t)]}{\delta t}\right\}.$$

The hazard rate is related to the *probability density function, f, and the *survivor function, S, by the equation

$$f(t) = h(t)\,S(t).$$

For many situations the graph of a hazard rate is a **bathtub curve**: initially the rate is high as the component beds in, there is then a constant hazard rate, and finally the component starts to wear out.

Hellinger, Ernst David (1883–1950; b. Strzegom, Poland; d. Chicago, Illinois) A Polish mathematician. Hellinger was a Polish Jew, educated at several German universities, who obtained his doctorate from the University of Göttingen in 1907. In 1914 he was appointed Professor of Mathematics at the University of Frankfurt. In 1938 he was arrested (because he was a Jew) and taken to Dachau concentration camp. The following year he was allowed to emigrate to the United States, where he joined the staff at Northwestern University.

Hellinger distance A measure of the distance between populations with *multivariate distributions having *probability density functions f and g. The distance is given by $\sqrt{2(1 - \rho)}$, where

$$\rho = \int_{-\infty}^{\infty} \cdots \int_{-\infty}^{\infty} \sqrt{f(x_1, x_2, \ldots, x_n)g(x_1, x_2, \ldots, x_n)} \, dx_1 \, dx_2 \ldots dx_n.$$

See also BHATTACHARYA'S DISTANCE; KULLBACK–LEIBLER INFORMATION.

Helmert, Friedrich Robert (1843–1917; b. Freiburg, Germany; d. Potsdam, Germany) A German mathematical physicist. Helmert's interest in statistics resulted from his research in geodesy—he was Professor of Geodesy in Aachen in 1872, and in Berlin in 1887, where he became director of the Prussian Geodetic Institute. His research included fundamental work on the *chi-squared distribution.

Helmert contrast *See* ANOVA.

hessian matrix The *matrix of second-order (partial) derivatives of a multivariate function. If $y = f(x_1, x_2, \ldots, x_n)$ the element in row j and column k is

$$\frac{\partial^2 y}{\partial x_j \partial x_k}.$$

The matrix is $n \times n$ and symmetric (provided that its elements are continuous).

heteroskedastic A term used of a set of *random variables that have different *variances. By contrast, a set of variables all having the same variance is described as **homoskedastic**.

heuristic A general recommendation based on statistical evidence. For example, 'Smoking may severely damage your health.'

hidden Markov model A special type of *graphical model. The observed *variable, Y, has a *distribution that depends on the value of the corresponding X-variable. The idea is that the successive values of X obey the Markov property, with the value of each depending only on the value of its predecessor.

hierarchical models Two or more models that are nested in the sense that the more complex model includes all the *parameters of the simpler model together with at least one other parameter.

high-order interaction An *interaction involving more than two *variables.

high-pass filter *See* FILTER.

highspread The difference between the greatest value in a *sample and the sample *median.

hinge An alternative word for *quartile.

histogram A diagram that uses rectangles to represent *frequency. It differs from the *bar chart in that the rectangles may have differing

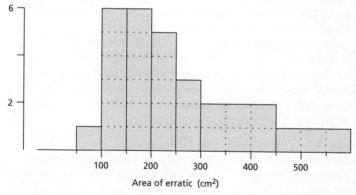

Histogram. This histogram represents data on the cross-sectional area of 30 erratics (boulders left behind by retreating glaciers). Note the use of wider intervals for the classes corresponding to the scarcer larger boulders. In a histogram, area is proportional to frequency.

widths, but the key feature is that, for each rectangle, the area is proportional to the frequency represented. The term 'histogram' was introduced by Karl *Pearson in his lectures prior to 1895.

Hodges, Joseph L., jun. (1922–2000; b. Shreveport, Louisiana; d. Berkeley, California) An American mathematical statistician. Hodges entered the University of California at Berkeley, aged sixteen, and graduated in 1942. After the Second World War, during which he served with *Lehmann in the operations analysis group in Guam, he obtained his doctorate at Berkeley (under the supervision of *Neyman) and promptly joined the faculty, retiring in 1991. He was Editor of the *Annals of Mathematical Statistics* during 1961–4.

Hodges–Lehmann estimators *Robust estimators suggested by *Hodges and *Lehmann in 1963. The estimate of location for a *sample of n observations is the *median of the averages of the $\frac{1}{2}n(n-1)$ possible pairs of observations. For example, with the four observations 22, 24, 26, and 60, the six possible averages are 23, 24, 41, 25, 42 and 43, so that the Hodges–Lehmann location estimate is 33 (equal to the *mean in this case).

In the case of samples (of sizes m and n) from two populations, where an estimate is required of the difference in their locations, the Hodges–Lehmann estimate is the median of the mn possible differences resulting from taking one observation from each sample.

Hoeffding, Wassily (1914–91; b. Mustamäki, Finland; d. Chapel Hill, North Carolina) A Finnish mathematical statistician who spent most of his career in the United states. After schooling in Denmark and Germany, Hoeffding gained his PhD from the University of Berlin in 1940. In 1946 he moved to the United States and joined the staff at the University of North Carolina, Chapel Hill, retiring in 1979. He was President of the *Institute of Mathematical Statistics in 1969 and was elected to membership of the National Academy of Sciences in 1976.

Hoeffding's D A rather complicated function of the *ranks of the *observations. It was suggested by *Hoeffding in 1948 as the basis of a *non-parametric test of the *independence of two *random variables.

Hölder's inequality For *random variables X and Y, and constants p and q both > 1 and satisfying $p^{-1} + q^{-1} = 1$,

$$E(XY) \leq \{E(|X^p|)\}^{\frac{1}{p}}\{E(|Y^q|)\}^{\frac{1}{q}}.$$

See also CHEBYSHEV'S INEQUALITY; MARKOV'S INEQUALITY; MINKOWSKI'S INEQUALITY.

Holt–Winters forecasting An application of *exponential smoothing to a *time series that displays a *trend and *seasonality.

homoskedastic *See* HETEROSKEDASTIC.

honestly significant difference test *See* TUKEY'S TEST.

Hooke–Jeeves pattern search A search procedure for finding the minimum (or maximum) of a multidimensional surface. An original base point is chosen. This is followed by exploratory moves, changing the values of one variable at a time, and resulting, in general, in a new base point. A pattern move is now made, the new point being dictated by the values of the preceding base points. The alternating sequence of exploratory and pattern moves continues until convergence.

hot deck *See* IMPUTATION.

Hotelling, Harold (1895–1973; b. Fulda, Minnesota; d. Chapel Hill, North Carolina) An American statistician and economist. Hotelling's first degree, a BA in Journalism at the University of Washington in 1919, was followed by an MA in mathematics at Washington in 1921, and a PhD in topology at Princeton University in 1924. His first job was in the Food Research Institute at Stanford University, where he became interested in *Statistics. In 1931 he was appointed Professor of Economics at Columbia University. In 1946 he was invited to found the Department of Statistics at the University of North Carolina, Chapel Hill. He was President of the *Econometrics Society in 1936 and President of the *Institute of Mathematical Statistics in 1941. He was elected to membership of the National Academy of Sciences in 1970.

Hotelling–Lawley trace *See* MULTIVARIATE ANALYSIS OF VARIANCE.

Hotelling's T^2 A *statistic having a multivariate distribution which is the analogue of the univariate *t-distribution. The statistic was introduced in 1931 by *Hotelling. A sample of size n is drawn from a *multivariate normal distribution with mean vector $\boldsymbol{\mu}$ and *variance–covariance matrix estimated as **S**. If the column vector of means is denoted by $\bar{\mathbf{x}}$ the statistic T^2 is given by

$$T^2 = n(\bar{\mathbf{x}} - \boldsymbol{\mu})'\mathbf{S}^{-1}(\bar{\mathbf{x}} - \boldsymbol{\mu}).$$

For $p \times 1$ vectors, $\{(n - p)T^2\}/\{(n - 1)p\}$ has an *F-distribution with p and $(n - p)$ *degrees of freedom. *See also* MAHALANOBIS'S D^2.

HSD (honestly significant difference) *See* TUKEY'S TEST.

Hsu's MCB test A *multiple comparison test. If t treatments are being compared, then the comparisons made are those between the best treatment and each of the remaining $(t - 1)$ treatments.

Huber function *See* M-ESTIMATES.

Huygens, Christiaan (1629–95; b. The Hague, Netherlands; d. The Hague, Netherlands) A Dutch mathematician, astronomer, and horologist. The son of a diplomat, Huygens studied law and mathematics, first at the University of Leiden and then at the College of Orange at Breda. His principal interest was astronomy, but this required precision instrumentation and led to his devising new methods for polishing lenses and keeping accurate time—in 1656 he patented the first pendulum clock. The previous year he had discovered the first of Saturn's moons. In 1657 he wrote the first printed work on probability, *Van Rekeningh in Spelen van Geluck* (*Calculation in Games of Chance*).

Huynh–Feldt correction A correction applied in *ANOVA calculations for *longitudinal data when the usual assumptions of constant *variance independent experimental errors is untrue.

hypergeometric distribution The *distribution of the number of 'successes' when *sampling without replacement from a finite *population each of whose members is classified as either a 'success' or a 'failure'. As an example, suppose that an urn contains N balls of which w are white. Suppose n balls are taken from the urn at random and without replacement, and let the *random variable X be the number of white balls obtained. Then X has the hypergeometric distribution given by

$$P(X = r) = \frac{\binom{w}{r}\binom{N-w}{n-r}}{\binom{N}{n}}, \quad 0 \le r \le w.$$

The initial derivation of the distribution was published by *de Moivre in 1711. The *mean of the distribution is nw/N and the *variance is

$$\frac{nw(N-n)(N-w)}{N^2(N-1)}.$$

For large values of N the hypergeometric distribution may be approximated by the *binomial distribution $B(n,p)$, where $p = w/N$, since in this case sampling without replacement is approximately equivalent to sampling with replacement.

hyperplane A hyperplane in $\mathbb{R}^n$ is the set of points $(x_1, x_2, \ldots, x_n)$ such that $a_1x_1 + a_2x_2 + \cdots + a_nx_n = c$, where $a_1, a_2, \ldots, a_n$ are constants, not all zero, and c is a constant. In the case $n = 2$ a hyperplane is a line, and in the case $n = 3$ it is a plane.

hypothesis test A procedure (based on the *Neyman–Pearson lemma) for deciding between two hypotheses on the basis of the value of a *statistic called the **test statistic**, which is a function of the *observations in a random *sample. In a test concerning the value of an unknown *parameter, the **null hypothesis** specifies a particular value for the parameter, whereas the **alternative hypothesis** specifies either an alternative value or, more usually, a range of alternative values. The null hypothesis is often denoted by H_0 or NH, and the alternative hypothesis by H_1 or AH.

A typical null hypothesis might state that the *population mean $\mu = 20$. The alternative hypothesis $\mu < 20$ is described as **one sided** and the test procedure is described as **one tailed**. By contrast, the alternative hypothesis $\mu \neq 20$ is **two sided** and the test is **two tailed**.

If necessary, assumptions are made about the *distribution type of the population, so that the *probability distribution of the statistic can be determined assuming H_0. The probability of obtaining a value for the test statistic that is as extreme, or more extreme (taking account of the alternative hypothesis), is called the **p-value**. If the actual value of the statistic is too far from its *expected value the test is deemed to be **significant** and the decision is to reject H_0 in favour of H_1. If the actual value of the statistic is close to its expected value the test is deemed to be **not significant** and the decision is not to reject H_0. The set of values of the statistic that lead to rejection of H_0 is called the **critical region** or **rejection region**, and the set of values that do not lead to rejection of H_0 is called the **acceptance region**.

There are two cases when the test leads to a correct result. These occur when H_0 is true and the test leads to its acceptance and when H_1 is true and the test leads to rejection of H_0. On the other hand there are two cases when the test leads to an incorrect result. These occur when H_0 is true but the test leads to its rejection (a **Type I error**, or **error of the first kind**) and when H_1 is true but the test leads to acceptance of H_0 (a **Type II error**, or **error of the second kind**).

The size of the critical region is determined by the desired **significance level** of the test, often denoted by α (alpha), which is the probability of making a Type I error. The smaller the significance level, the smaller the critical region. The significance level is usually expressed as a percentage. The word 'significance' seems to have been introduced into Statistics by *Edgeworth in 1885. The ideas concerning the two types of error were introduced by *Neyman and Egon *Pearson in 1928.

As an example, suppose X has a *normal distribution $N(\mu, 9)$ and it is desired to test $H_0: \mu = 20$ against $H_1: \mu > 20$, using a sample of size 25. An appropriate statistic is the sample mean $\bar{X}$, which has distribution $N(20, 9/25)$ under H_0, or its standardized value

$$Z = \frac{\bar{X} - 20}{\sqrt{9/25}},$$

which has distribution $N(0, 1)$. If the desired significance level is 5%, the critical region, from tables of **critical values** for the standard normal distribution (*see* APPENDIX VI) is $Z > 1.645$. In terms of $\bar{X}$, the critical region is $\bar{X} > 20 + 1.645\sqrt{9/25}$ or, equivalently, $\bar{X} > 20.99$.

The probability of making a Type II error is often denoted by β (beta). The **power** of the test, which is the probability of accepting the alternative hypothesis when it is in fact true, is $1 - \beta$ and its value depends on the value of the parameter under test. A plot of the relation between power and the parameter value is called the **power curve**. If there is a choice of test, with a predetermined significance level, it is usual to choose the test (if one exists) that maximizes the power.

In the previous example the variance of the normal distribution was known to be 9. Usually, though the distribution may be assumed to be normal, the variance will be unknown. Suppose (with the same hypotheses as previously) that the sample of 25 observations has sample mean and variance (using the $(n - 1)$ divisor) given, respectively, by $\bar{x} = 21.4$ and $s^2 = 12.25$; then a **t-test** is appropriate. The test statistic t is given by

$$t = \frac{21.4 - 20}{\sqrt{12.25/25}} = 2.0$$

The upper 5% point (*see* PERCENTAGE POINT) of a *t*-distribution with $24(= 25 - 1)$ *degrees of freedom is 1.711 (*see* APPENDIX VIII). Since $2.0 > 1.711$, the null hypothesis would be rejected in favour of the alternative hypothesis.

Frequently, hypothesis tests involve the comparison of two populations. The case where unrelated random samples are taken from the populations gives rise to **two-sample tests**. A two-sample test of the equality of two population means in which the populations have the same (unknown) variance requires the use of a *pooled estimate of variance. The case where the population variances are unknown and cannot be assumed to be equal is the *Behrens–Fisher problem.

If the two random samples have the same size and matched pairs of individuals are obtained, one from each population, then, for a test of equality of population means, a **paired-sample test** is appropriate. Here the differences within each pair of values constitute the observations. A *t*-test is performed, the null hypothesis being that the mean difference is 0, and the sample variance being the variance of the differences.

A hypothesis that specifies several simultaneous conditions is described as a **composite hypothesis**. For example, with samples from k populations, the null hypothesis might specify that all k population means are equal.

IBS *See* INTERNATIONAL BIOMETRIC SOCIETY.

icon plot A diagram for displaying multidimensional observations. The diagram consists of an array of individual diagrams (**icons**), one for each data item. In the individual diagrams a separate feature corresponds to each of the variables measured. One example of an icon plot is the use of *Chernoff faces.

ICSA *See* INTERNATIONAL CHINESE STATISTICAL ASSOCIATION.

ideal index *See* FISHER'S INDEX.

idempotent matrix *See* MATRIX.

identity link *See* GENERALIZED LINEAR MODEL.

identity matrix *See* MATRIX.

i.i.d. Abbreviation for 'independent and identically distributed'. Thus i.i.d. *random variables are *independent variables all having the same *distribution. The most common situation involving i.i.d. random variables arises when a random *sample of *observations is taken from a single *population.

ill-conditioned *See* MULTIPLE REGRESSION MODEL.

importance sampling A method of reducing the *variance of an estimate obtained via *simulation. Suppose, for example, that we wish to estimate the quantity I, given by

$$I = \int_0^1 g(x)dx,$$

where g is a given function. An obvious method is to generate uniform *pseudo-random numbers $u_1, u_2, \ldots, u_n$, in the interval (0, 1), and to calculate the estimate, I_1, given by

$$I_1 = \frac{1}{n} \sum_{j=1}^{n} g(u_j).$$

Importance sampling makes a more representative choice of values: suppose that f is a *probability density function that resembles g in its general shape, and let F be the corresponding *distribution function. Instead of working with $u_1, u_2, \ldots, u_n$, we work with $v_1, v_2, \ldots, v_n$, where $F(v_j) = u_j$. The resulting estimate, I_2, given by

$$I_2 = \frac{1}{n} \sum_{j=1}^{n} \frac{g(v_j)}{f(v_j)},$$

will have a smaller variance than I_1.

improper prior *See* BAYESIAN INFERENCE.

imputation The process of replacing missing values in (usually) a large-scale social survey. Suppose, for example, that the salary information is missing for an individual who is known to be a doctor aged 55. One approach would be to determine the *average salary of all 55-year-old doctors and to replace the missing value with this average value (or some estimate obtained, for example, by *multiple regression of salary on other *variables). There are two possible objections to this approach: (i) the imputed value might not be a salary actually obtained by any 55-year-old doctor, and (ii), if there are many doctors of this age with missing salary information and if each were given the same imputed salary, this would give a very misleading idea of salary variability.

The first objection can be met by insisting that the value imputed must be a real salary. Both objections can be met if the imputed value is taken to be the most recently encountered actual salary of a doctor of that age. This is called **hot deck** imputation, which refers to a time when the records on each individual were on a separate card. Imagine examining the deck of cards a card at a time, imputing missing values. The salary chosen for the doctor is that most recently encountered in the cards containing information on 55-year-old doctors. The alternative, in which all imputation takes place after all the cards have been examined, is called **cold deck** imputation—the most recently encountered value is used to replace all those missing (so that only the first objection is met).

IMS *See* INSTITUTE OF MATHEMATICAL STATISTICS.

inadmissible decision rule A *decision rule that is never superior to a given alternative rule. Suppose that δ_1 and δ_2 are two decision rules and that the outcome of each rule is dependent on the values of the *parameter vector $\boldsymbol{\theta}$. If, for all values of $\boldsymbol{\theta}$, the expected loss from using rule δ_2 is at least as large as the loss from using rule δ_1 and if it is greater for at least one value of $\boldsymbol{\theta}$, then the rule δ_2 is described as being inadmissible.

incidence rate The number of new cases of a disease occurring during a given period as a proportion of the number of people in the *population. It is usually expressed as cases per 1000 or 100 000 per annum. *See* MORBIDITY RATE; MORTALITY RATE.

independence For independence of events, *see* INDEPENDENT EVENTS; for independence of random variables, *see* INDEPENDENT RANDOM VARIABLES.

independence model A *model in which two (or more) *random variables are *independent of one another: the value taken by one variable is completely unaffected by the value taken by the other variable. Suppose, for example, that A, B, and C are three *categorical random variables, with p_{jkl} denoting the *probability of an outcome in cell (j, k, l); then the independence model states that

$$p_{jkl} = p_{j00}\, p_{0k0}\, p_{00l},$$

where

$$p_{j00} = \sum_k \sum_l p_{jkl}, \qquad p_{0k0} = \sum_j \sum_l p_{jkl}, \qquad p_{00l} = \sum_j \sum_k p_{jkl}.$$

With just two variables, A and B, the corresponding result would be

$$p_{jk} = p_{j0}\, p_{0k},$$

where

$$p_{j0} = \sum_k p_{jk}, \qquad p_{0k} = \sum_j p_{jk}.$$

independent events Two events (*see* SAMPLE SPACE) A and B are independent if

$$P(A \cap B) = P(A) \times P(B),$$

or, equivalently, if $P(A|B) = P(A)$, or if $P(B|A) = P(B)$. Three events A, B, and C are said to be independent (or **mutually independent**) if each pair is independent and if, in addition,

$$P(A \cap B \cap C) = P(A) \times P(B) \times P(C).$$

For a set of more than three events to be independent the *multiplication rule must hold for all possible subsets. *See also* INTERSECTION; CONDITIONAL PROBABILITY; PAIRWISE INDEPENDENT.

independent random variables Two *random variables are said to be independent if the value taken by one has no effect on the value taken by the other.

independent variable *See* REGRESSION.

index (exponent) In the term a^x, the index is x. The plural is **indices**.

index number A measure of the value of a variable relative to its value at some base date or state. The index is often scaled so that its base value is 100. *See* FISHER'S INDEX; LASPEYRES'S PRICE INDEX; PAASCHE'S PRICE INDEX; RETAIL PRICE INDEX.

index of dispersion A measure of the extent to which a set of observed *frequencies (for example the numbers of plants in randomly distributed *quadrats) follow a *Poisson distribution. For a sample of n observations, let $\bar{x}$ and s^2 denote, respectively, the sample mean and the sample variance (using the $(n - 1)$ divisor). Under the *null hypothesis of a Poisson distribution the quantity I (the index of dispersion), given by

$$I = \frac{(n - 1)s^2}{\bar{x}},$$

has an approximate *chi-squared distribution with $(n - 1)$ *degrees of freedom. According to the null hypothesis the value of I should be near $(n - 1)$. This test is called the **dispersion test**.

If I is unusually large then the data are described as being **overdispersed**, and, in the case of plants (or stars, or other point objects), the data are described as *clustered. If I is unusually small then the data are described as **regular**.

index of diversity *See* LOGARITHMIC DISTRIBUTION.

index property *See* EXPONENTIAL.

Indian Statistical Institute (ISI) The Institute, which is located at Kolkata, was founded by *Mahalanobis in 1931. It publishes the journal *Sankhya*.

indicator kriging *See* KRIGING.

indicator variable An alternative name for a *dummy variable.

indices *See* INDEX.

inference The process of deducing properties of the underlying *distribution by analysis of *data.

inflation The percentage increase in the *retail price index relative to its value one year earlier.

influence *See* REGRESSION DIAGNOSTICS.

information *See* FISHER INFORMATION.

inspection by attributes; inspection by variables *See* ACCEPTANCE SAMPLING.

Institute of Mathematical Statistics (IMS) The Institute is one of the major American societies for statisticians. Founded in 1935 at Ann Arbor in Michigan, it currently has about 3000 members worldwide. The IMS publishes the *Annals of Applied Probability*, *Annals of Probability* and *Annals of Statistics*. These three titles, which commenced in 1973, replaced a single predecessor, the *Annals of Mathematical Statistics*.

Institute of Statistical Mathematics Japanese research institute based in Tokyo. Its publications include the *Annals of the Institute of Statistical Mathematics*.

Institute of Statisticians A British society founded in 1948 for non-academic statisticians. It published a journal originally called *The Incorporated Statistician*, though the title was soon changed to *The Statistician*. The Institute merged with the *Royal Statistical Society in 1993, with *The Statistician* becoming the fourth of the Society's journals.

instrumental variable *See* MULTIPLE REGRESSION MODEL.

integer programming *See* LINEAR PROGRAMMING.

intelligence quotient (IQ) In 1905, the French psychologists Alfred Binet and Théodore Simon proposed the concept of an intelligence test that measured mental age. In 1912, the German psychologist William Stern proposed the division of mental age by chronological age, to give the intelligence quotient. In 1916 the test was introduced into the USA by Lewis Terman of Stanford University. Terman proposed scaling the quotient by 100, to remove awkward decimal places, and as a result the test became known as the Stanford–Binet test. The idea of a mental age to chronological age ratio only works well for the young, however. As a consequence the original ratio concept has been abandoned. In 1939 the Romanian-born US psychologist David Wechsler introduced a statistical definition of IQ as a *random variable having a *normal distribution with *mean 100 and *standard deviation 15, so that most people then have IQs in the range 70 to 130, with IQs over 150 being extremely rare.

interaction A term particularly used in the contexts of *factorial experiments and *log-linear models to describe cases where the combined effects of two *variables is not a simple sum of their separate effects. For example, we might write

$$E(Y) = \beta_0 + \beta_1 x_1 + \beta_2 x_2 + \beta_{1,2} x_1 x_2,$$

the final term quantifying the interaction between the two x-variables.

International Biometric Society (IBS) Society, founded in 1947, devoted to the mathematical and statistical aspects of biology. Through its regional organizations the Society sponsors regional and local meetings. Its principal journals are *Biometrics* and the *Journal of Agricultural, Biological and Environmental Statistics*.

International Chinese Statistical Association (ICSA) Association, founded in 1987, with a membership primarily consisting of Chinese in the United States, Hong Kong, and Taiwan. The ICSA has published *Statistica Sinica* since 1991.

International Statistical Institute (ISI) An autonomous society, founded in 1885, that seeks to develop and improve statistical methods and their application, through the promotion of international activity and co-operation. It has about 2000 members worldwide. Its principal publications are the *Bulletin of the International Statistical Institute, *International Statistical Review*, and *Statistical Theory and Methods Abstracts*.

International Statistical Review A journal of the *International Statistical Institute. Founded in 1933, the journal contains a high proportion of historical and review papers.

interpolation The use of a formula to estimate an intermediate data value. *Compare* EXTRAPOLATION. For example, if we discover that 1 kg of fertilizer per hectare results in a yield of 50 kg of produce per hectare, and that 2 kg gives 70 kg, then we might reasonably interpolate between these values to deduce that 1.5 kg would give a yield of about 60 kg ha^{-1}.

In general, if $x_1 < x < x_2$ and the values of y corresponding to x_1 and x_2 are y_1 and y_2, respectively, then the estimate of y_0, the value corresponding to x_0, is given by **linear interpolation** as

$$y_0 = \frac{y_2(x_0 - x_1) - y_1(x_0 - x_2)}{x_2 - x_1} = y_1 + \frac{x_0 - x_1}{x_2 - x_1}(y_2 - y_1).$$

interquartile range (midspread) If the lower and upper *quartiles are denoted by Q_1 and Q_3, respectively, the interquartile range is $Q_3 - Q_1$. This term can be used either for a set of *data or for a *probability distribution. The phrase 'interquartile range' was used by *Galton in 1882.

inter-rater reliability A measure of the extent to which raters (for example, judges of an ice-skating contest) agree.

intersection The intersection of two events (*see* SAMPLE SPACE) A and B is the event 'both A and B occur'. It is denoted by $A \cap B$. Clearly, for any events A, B, C,

$$A \cap A = A, \qquad A \cap B = B \cap A,$$
$$A \cap S = A, \qquad A \cap \phi = \phi,$$
$$A \cap (B \cap C) = (A \cap B) \cap C,$$

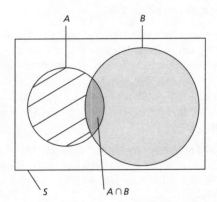

Intersection. This Venn diagram shows the intersection of two events as the overlapping part of the circles that represent the individual events.

where S is the sample space and ϕ is the empty set. *See also* BOOLEAN ALGEBRA; VENN DIAGRAM.

The intersection of the n events $A_1, A_2, \ldots, A_n$ is the event 'all of $A_1, A_2, \ldots, A_n$ occur'. It is denoted by $A_1 \cap A_2 \cap \cdots \cap A_n$.

interval estimate *See* CONFIDENCE INTERVAL.

interval scale A scale of measurement which can be used to measure the difference, or distance, between two general states or points. For example, use of a ruler to measure length, use of a stop-watch to measure a time interval, or measurement of a musical interval (octave, fifth, etc.).

In a **ratio scale** the difference, or distance, is measured between a state or point of interest and a standard state or point. For example, height above sea-level, distance from London, frequency of a musical note (in cycles per second), temperature in °Kelvin (above absolute zero), clock-time (13:05 on 5 November 2001 AD). Confusingly, comparison of ratio scale measurements gives an interval scale, and in music an interval is measured by the ratio of the frequencies.

inverse Fourier transform *See* FOURIER TRANSFORM.

inverse Gaussian distribution *See* INVERSE NORMAL DISTRIBUTION.

inverse matrix *See* MATRIX.

inverse normal distribution (inverse Gaussian distribution; Wald distribution) A continuous distribution. The *probability density function f (*see following diagram*) is given by

$$f(x) = \left(\frac{\lambda}{2\pi x^3} \right)^{\frac{1}{2}} \exp\left\{ -\frac{\lambda}{2\mu^2 x}(x - \mu)^2 \right\}, \qquad x > 0,$$

where $\mu > 0, \lambda > 0$. The distribution has *mean μ, *variance μ^3/λ, and *mode

$$\frac{\mu}{2\lambda} \left(\sqrt{4\lambda^2 + 9\mu^2} - 3\mu \right).$$

inverse transformation method A general method for the *simulation of *observations of a *continuous random variable X. Let u be a *pseudo-random number and let F be the *distribution function of X. Then x, given by

$$x = F^{-1}(u),$$

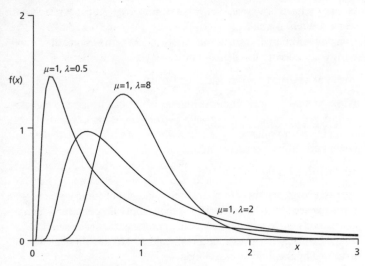

Inverse normal distribution. The distribution has two parameters: the mean μ and a parameter λ related to the variance.

is a random observation of X. For example, an *exponential random variable X, with mean μ, has distribution function $(1 - e^{-x/\mu})$; hence $-\mu \ln(1 - u)$ is an observation of X.

invertible *See* MATRIX.

IQ *See* INTELLIGENCE QUOTIENT.

IRLS *See* ITERATIVELY REWEIGHTED LEAST SQUARES.

irreducible *See* MARKOV PROCESS.

ISI An abbreviation used for both the *Indian Statistical Institute and the *International Statistical Institute.

Ising model A model originating in the context of statistical mechanics to predict the behaviour of a simple magnet. The model regards the magnet as consisting of a lattice of locations, each of which is either in the + state or the − state. The state of a location is affected by random fluctuations in the energy levels of the location itself and, crucially, its neighbours. Similar ideas underlie *Markov chain Monte Carlo processes.

isotropy If a *time series would have the same basic properties if time were to run backwards then it displays **isotropy** and is said to be **isotropic**. Similarly, if the properties of a *spatial process are the same in all directions then it too is isotropic. However, if, for example, the spatial *autocorrelation is different in the north–south direction and the east–west direction, then it displays **anisotropy** and is said to be **anisotropic**.

item non-response See NON-RESPONSE.

item reliability See RELIABILITY.

iteration See ITERATIVE ALGORITHM.

iterative algorithm A numerical method usually used when explicit formulae are unavailable. The idea is that a repetition of (usually simple) calculations will result in a sequence of approximate values for the quantity of interest. The differences between successive values will usually diminish rapidly and a usual termination rule is based on the difference having reached an acceptably small value. Each repetition is called an **iteration**. Examples include the *Deming–Stephan algorithm, the *EM algorithm, and *iteratively reweighted least squares.

iterative proportional fitting See DEMING–STEPHAN ALGORITHM.

iteratively reweighted least squares (IRLS) An *iterative algorithm for fitting a *linear model in the case where the *data may contain *outliers that would distort the *parameter estimates if other *estimation procedures were used. The procedure uses *weighted least squares, the influence of an outlier being reduced by giving that observation a small weight. The weights chosen in one iteration are related to the magnitudes of the residuals in the previous iteration—with a large residual earning a small weight. This is one of a number of methods for *robust regression, the weights being related to *M-estimates.

J

JABES *See* JOURNAL OF AGRICULTURAL, BIOLOGICAL AND ENVIRONMENTAL STATISTICS.

Jaccard's coefficient A measure of the similarity between two individuals in a population. Each individual is assessed with respect to q characteristics that are either present or absent. For each characteristic there are three possible outcomes: 'Absent for both individuals', 'Present for just one of the individuals', 'Present for both individuals'. Denoting the corresponding *counts by n_0, n_1, and n_2, respectively, with $n_0 + n_1 + n_2 = q$, Jaccard's coefficient is the proportion of characteristics present in at least one individual that are present in both individuals:

$$\frac{n_2}{n_1 + n_2}.$$

See also MATCHING COEFFICIENT.

jackknife A *computer-intensive *resampling method for *estimating some unknown *parameter of a *distribution while making minimal assumptions. In this respect it resembles the *bootstrap. Denote the parameter by θ and its usual estimate, based on a *sample of n *observations, by $\hat{\theta}$. For example, if the parameter were the *mean of a distribution then the usual estimate would be the sample mean.

The jackknife procedure produces an alternative estimate $\tilde{\theta}$, together with an estimate of the *bias (if any) of the usual estimate. Let $\hat{\theta}_{-j}$ be the usual estimate of θ calculated from the same sample but with observation j omitted. Now define the **pseudovalue** $\tilde{\theta}_j$ by

$$\tilde{\theta}_j = n\hat{\theta} - (n-1)\hat{\theta}_{-j}.$$

The jackknife mean and variance are given by

$$\tilde{\theta} = \frac{1}{n}\sum_j \tilde{\theta}_j, \qquad s_j^2 = \frac{1}{n-1}\sum_j \left(\tilde{\theta}_j - \tilde{\theta}\right)^2.$$

The estimated bias is $\hat{\theta} - \tilde{\theta}$. The ratio

$$\frac{\hat{\theta} - \theta}{s_j/\sqrt{n}}$$

has an approximate *standard normal distribution. The method also applies to the estimation of more complex characteristics, such as the *correlation in a set of *bivariate observations.

Japan Statistical Society The Society, which publishes a biennial English version of its journal, is based in Tokyo.

JASA See JOURNAL OF THE AMERICAN STATISTICAL ASSOCIATION.

Jeffreys, Sir Harold (1891–1989; b. Durham, England; d. Cambridge, England) An English astronomer, geophysicist, and mathematician. Jeffreys studied at Cambridge University, being made a Fellow of St John's College in 1914 and holding this title for a record 75 years. From 1946 to 1958 he was Plumian Professor of Astronomy and Experimental Philosophy. He was interested in scientific method and developed *Bayesian inference. He was elected a Fellow of the Royal Society in 1925 and was awarded its Copley Medal in 1960. He was elected to membership of the National Academy of Sciences in 1945 and was knighted in 1953. He was awarded the *Royal Statistical Society's *Guy Medal in Gold in 1962.

Jenkins, Gwilym Meirion (1933–82) A Welsh statistician. Jenkins was a graduate of University College, London, gaining his BSc in 1953 and his PhD in 1956. His knowledge of the theoretical side of the analysis of *time series was augmented by practical insights gained from two years at the Royal Aircraft Establishment. In 1957 he joined the Statistics faculty at Imperial College, London. From 1965 to 1974 he was Professor of Systems Engineering at Lancaster University. Jenkins was co-author with *Box of the influential 1970 book *Time Series Analysis: Forecasting and Control*, which set out what is now referred to as the *Box–Jenkins procedure.

jitter A procedure for improving the display of *bivariate data. If the variables are *discrete, or if they are *continuous, but the values have been rounded, then some value pairs may occur very often, but would be represented by a single point on a normal *scatter diagram. A better impression of the data can be obtained by adding small random quantities to all values before plotting. *See following diagram.*

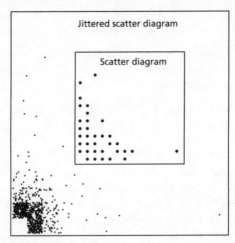

Jitter. The axes in the diagrams represent the numbers of tips reported by two tipping-bucket rain-gauges in September 2000. The main diagram illustrates the amounts of rain for the 813 2-minute periods for which at least one gauge reported a tip. The inner diagram shows the corresponding 34-point scatter diagram.

Johnson, Norman Lloyd (1917–; b. Ilford, England) An English statistician who spent most of his career in the United States. Johnson is a graduate of University College, London, gaining his MSc just before the Second World War and his PhD just afterwards. He joined the staff of University College in 1938 and remained there till 1962, except during the war, when he was seconded to the London Ordnance Board. In 1962 he moved to the University of North Carolina, Chapel Hill, where he is now Alumni Distinguished Professor Emeritus. He was awarded the *Wilks Medal of the *American Statistical Association in 1993.

Johnson distributions *Probability distributions, of widely differing shapes, that can be obtained by simple transformations of the *standard normal distribution. The relationship to the normal distribution makes it easy to calculate, for example, a *tail probability. The entire family is defined by

$$Y = g\left(\frac{Z - \alpha}{\beta}\right),$$

where α and β are constants, g is a function, Z is a standard normal *random variable and Y is a Johnson random variable.

There are three classes of Johnson distributions, referred to as the S_U, S_B, and S_L classes, which result from different choices for the function $g(x)$. In the order given, for a variable x, the functions are

$$\sinh x, \qquad \frac{1}{2}(1 + \tanh x), \qquad \text{and} \qquad e^x.$$

The last case gives a type of *lognormal distribution.

joint distribution *See* BIVARIATE DISTRIBUTION.

joint independence The *random variables A and B are jointly independent of C if their *joint probability distribution is unaffected by the value of C. For *categorical variables, with p_{jkl} denoting the probability of an outcome in *cell (j, k, l), the condition for A and B to be jointly independent of the category of C, is that

$$p_{jkl} = p_{jk0}p_{00l},$$

for all j, k, and l, where

$$p_{jk0} = \sum_l p_{jkl} \qquad \text{and} \qquad p_{00l} = \sum_j \sum_k p_{jkl}.$$

joint probability The joint probability of a set of *events is the *probability that all occur simultaneously.

joint probability density function *See* BIVARIATE DISTRIBUTION.

Journal de la SFdS The journal of the *Société Francaise de Statistique; first published (under a different title) in 1859.

Journal of Agricultural, Biological and Environmental Statistics (JABES) A publication of the *International Biometric Society.

Journal of the American Statistical Association (JASA) The principal journal published by the *American Statistical Association. It was first published under a slightly different name in 1888 and under its current name since 1922. Its emphasis is on developments in statistical theory.

Journal of the Italian Statistical Society An English language journal of the *Società Italiana di Statistica which was first published in 1994.

Journal of the Royal Statistical Society (*JRSS*) The first regularly published *Statistics journal in the world. Currently the Journal is published as four Series. Series A, now entitled **Statistics in Society**, was first published in 1839. Series B, **Statistical Methodology**, was first published in 1934. Series C, **Applied Statistics**, was first published in 1954. Series D, **The Statistician**, first appeared in 1950 as a publication of the Association of Incorporated Statisticians (later the *Institute of Statisticians).

JRSS *See* JOURNAL OF THE ROYAL STATISTICAL SOCIETY.

Kalman, Rudolf Emil (1930–; b. Budapest) A Hungarian electrical engineer. Kalman was a graduate of the Massachusetts Institute of Technology, gaining his DSci from Columbia University in 1957. Kalman introduced the widely used *Kalman filter in papers published in 1960 and 1961. At that time he was at the Research Institute for Advanced Study in Baltimore. He was appointed Professor of Engineering Mathematics at Stanford University in 1964, moving to the University of Florida in 1971 as Professor of System Theory. He is a member of the National Academy of Sciences, the National Academy of Engineering and the American Academy of Arts and Science.

Kalman filter A computationally efficient method of updating the estimates of the time-dependent *parameters of a *multiple regression model as successive values in the time series of values of the *dependent variable become available. *Exponential smoothing provides an extremely simple example of the recursive calculations involved. The procedure was introduced by *Kalman in 1960.

Kaplan, Edward Lynn (1920–) An American mathematician. Kaplan is the joint author, with *Meier, of the influential 1958 paper on the estimation of the *survivor function. He was elected Emeritus Professor of Mathematics at Oregon State University in 1981.

Kaplan–Meier estimate (product-limit estimate) A procedure for estimating the *survivor function from *observations of lifetimes (of people, machine components, etc.) when some observations are *censored (e.g. people move away from the observation site, or functioning components are removed before the end of the experimental period). The estimate is calculated as follows. First, order the *data by length of lifetime. Suppose there are m distinct lifetimes (where, for example, people alive at the time of calculation are given their current lifetime). Let $t_{(j)}$ be the jth of these ordered lifetimes, let n_j individuals have lifetimes of $t_{(j)}$ or more, and let d_j individuals have lifetimes of exactly $t_{(j)}$. The Kaplan–Meier estimate of the survivor function is given by

$$\hat{S}\big(t_{(j)}\big) = \prod_{k=1}^{j}\bigg(1 - \frac{d_k}{n_k}\bigg).$$

Closely related to the Kaplan–Meier estimate is the **Nelson–Aalen** estimate of the cumulative hazard rate, given by

$$\hat{\Lambda}\big(t_{(j)}\big) = \sum_{k=1}^{j}\frac{d_k}{n_k}.$$

Kendall, David George (1918–; b. Ripon, England) An English mathematical statistician and probabilist. A graduate of Oxford University, Kendall worked during the Second World War for the Ministry of Supply, receiving his statistics training in evening discussions with *Anscombe. In 1946 he was elected a Fellow of Magdalen College, Oxford. In 1962 he was appointed Professor of Mathematical Statistics at Cambridge University, where he is now Professor Emeritus. He was elected a Fellow of the Royal Society in 1964. He was the President of the London Mathematical Society in 1972 and President of the *Bernoulli Society in 1973. He was awarded the *Royal Statistical Society's *Guy Medal in Silver in 1955 and in Gold in 1981.

Kendall, Sir Maurice George (1907–83; b. Kettering, England; d. Redhill, England) An English statistician. The son of a publican, Kendall failed to get a place in his local grammar school, but nevertheless won a scholarship to study mathematics at Cambridge University. On graduating he initially worked at the Ministry of Agriculture, where his work involved the study of *time series. Kendall's work from that time underlies the modern approach to the analysis of time-series *data. He also became interested in methods for measuring *correlation; his initial paper on *rank correlation was published in 1938, though *Kendall's tau became widely used only after the publication in 1948 of his book *Rank Correlation Methods*. Kendall may be best known as the author of the two-volume *The Advanced Theory of Statistics* (the volumes appearing in 1943 and 1946). For many years these volumes were the first reference used whenever there was a statistical inquiry. Further editions, with other authors, have appeared regularly ever since (currently with Kendall's name transferred to the title of the book). Kendall became Professor of Statistics at the London School of Economics in 1949, where he remained until 1961. From 1960 to 1962 he was President of the *Royal Statistical Society and was awarded its *Guy Medal in Silver in 1945 and in Gold in 1968. From 1972 to 1980 he was Director of the World Fertility Study. He was knighted in 1974.

Kendall's tau (τ) A *correlation coefficient that can be used as an alternative to *Spearman's rho for *data in the form of *ranks. It is a simple function of the minimum number of neighbour swaps needed to produce one ordering from another. Its properties were analysed by Sir Maurice *Kendall in a paper published in 1938.

As an example, suppose that we have four objects (A–D) and the two orderings (D,B,A,C) and (A,B,D,C). To convert the first ordering into the second using neighbour swaps we could begin by swapping A and B to get (D,A,B,C). Now the swap of A and D gives (A,D,B,C) and then the swap of B and D gives the desired (A,B,D,C). The reordering cannot take fewer than the three swaps used. In a general case with n items to order and a minimum of Q swaps required, τ (which takes values in the interval -1 to 1, inclusive) is given by

$$\tau = 1 - \frac{4Q}{n(n-1)}.$$

An easy way of counting the minimum number of neighbour swaps is as follows. First, write down the items in the order specified by the first ordering. Next, write down the list of ranks assigned to these items by the second ordering. For each number in this list in turn, count how many subsequent numbers are smaller than it. The sum of these *counts is Q. Tables of critical values of Kendall's τ are given in Appendix XII.

kernel density estimation; kernel function *See* KERNEL METHODS.

kernel methods (kernel density estimation) Methods for the *estimation of *probability density functions; the process is also called **kernel density estimation**. Suppose X is a *continuous random variable with unknown probability density function f. A random *sample of *observations of X is taken. If the sample values are denoted by $x_1, x_2, \ldots, x_n$, the estimate of f is given by

$$f(x) = A \sum_{j=1}^{n} K(x, x_j),$$

where K is a **kernel function** and the constant A is chosen so that $\int_{-\infty}^{\infty} f(x)dx = 1$. The observation x_j may be regarded as being spread out between $x_j - a$ and $x_j + b$ (usually with $a = b$). The result is that the naive estimate of f being a function capable of taking values only at $x_1, x_2, \ldots, x_n$, is replaced by a continuous function having peaks where the data are densest. Examples of kernel functions are the **Gaussian kernel**,

$$K(x, t) = \exp\left\{ -\frac{(x - t)^2}{2h^2} \right\} \qquad -\infty < x < \infty,$$

and the **Epanechikov kernel**,

$$K(x, t) = \begin{cases} 1 - \dfrac{(x - t)^2}{5h} & -\sqrt{5}h < x - t < \sqrt{5}h, \\ 0 & \text{otherwise.} \end{cases}$$

The constant h is the **window width** or **bandwidth**. The choice of h is critical: small values may retain the spikes of the naive estimate, and large values may oversmooth so that important aspects of f are lost.

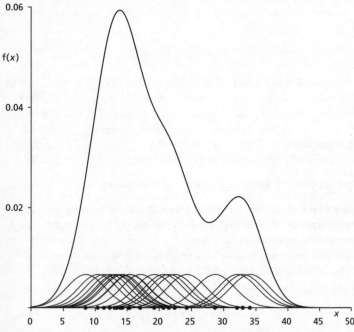

Kernel method. In this case a sample of twenty observations have been generated randomly from a chi-squared distribution with twenty degrees of freedom and a Gaussian kernel with $h = 3$ has been used to generate the kernel density estimate.

knapsack problem A *combinatorial optimization problem. Given a knapsack of limited capacity and various items that vary in value and size, the problem is to determine which items should be placed in the knapsack so as to maximize the value of its contents.

Kohonen networks (self-organizing feature maps) *Neural nets based on the topological properties of the human brain.

Kolmogorov, Andrei Nikolaevich (1903–87; b. Tambov, Russia; d. Moscow, Russia) An outstanding Russian mathematician and probabilist. Kolmogorov's mother died in childbirth and he was brought up by his aunt. Before enrolling at Moscow University in 1920, he spent some time as a railway conductor. A brilliant student, he obtained fundamental results in Fourier series (1922), logic (1925), and probability (1929). In 1929 he joined the faculty of Moscow University, becoming a professor in 1931. He was successively appointed Professor of Probability in 1938, of Statistical Methods in 1966, of Mathematical Statistics in 1976, and of Logic in 1980. Kolmogorov also worked on such practical projects as turbulence, the motion of the planets, the theory of fire, telecommunications scheduling, and the landing of planes on aircraft carriers. His work is now especially remembered by statisticians in the context of the *Chapman–Kolmogorov equations for *Markov processes and the *non-parametric *Kolmogorov–Smirnov test for a specified distribution. He was selflessly devoted to the development of talent in others, particularly the young, and worked tirelessly to improve the teaching of mathematics in secondary schools. He was elected a Fellow of the Royal Society in 1964 and a member of the National Academy of Sciences in 1967.

Kolmogorov's inequality A stronger form of *Chebyshev's inequality. Let $X_1, X_2, \ldots, X_n$ be *independent random variables with finite expectations and *variances. Define the partial sum S_k by

$$S_k = \sum_{j=1}^{k} X_j, \qquad k = 1, 2, \ldots, n.$$

Then Kolmogorov's inequality states that, for all positive values of ε,

$$P\left(\max_k \left\{ \frac{|S_k - \mu_k|}{\sigma_k} < \varepsilon \right\} \right) \geq 1 - \frac{1}{\varepsilon^2},$$

where, for each k, μ_k and σ_k^2 are, respectively, the expectation and variance of S_k.

Kolmogorov–Smirnov test A *non-parametric test for the *null hypothesis that a random *sample has been drawn from a specified *distribution (either *discrete or *continuous). There are several similar tests, each involving a comparison of the sample *distribution function with that hypothesized. For example, let the sample values, in increasing order, be $x_{(1)}, x_{(2)}, \ldots, x_{(n)}$. Let the hypothesized *probability of a value less than or equal to $x_{(j)}$ be p_j. Let u_j and v_j be defined by

$$u_j = \frac{j}{n} - p_j \qquad v_j = p_j - \frac{j-1}{n} = \frac{1}{n} - u_j.$$

The test statistic is the largest of the absolute magnitudes of these $2n$ differences. A two-sample version of the test compares the two sample distribution functions. In the single-sample case, approximate *critical values are $1.36/\sqrt{n+1}$ at the 5% level and $1.63/\sqrt{n+1}$ at the 1% level. The test was introduced by *Kolmogorov in 1933, and further developed by *Smirnov in 1939.

As described, the test refers to a fully prescribed distribution. However, by using special tables of critical values and estimating unknown parameters from the sample data, its use has been extended to testing for *exponential, *extreme-value, *logistic, *normal (the *Lilliefors test), or *Weibull distributions with unspecified parameters.

As an example, to test the hypothesis that the values 0.273, −1.184, 1.456, −0.655, −0.323, −0.733, −1.600, 0.819, 0.081, 0.971 have been drawn from a standard normal distribution the following results are required:

value	j/n	p_j	$(j-1)/n$	u_j	v_j
−1.600	0.1	0.055	0.0	0.045	0.055
−1.184	0.2	0.118	0.1	0.082	0.018
−0.733	0.3	0.232	0.2	0.068	0.032
−0.655	0.4	0.256	0.3	0.144	−0.044
−0.323	0.5	0.373	0.4	0.127	−0.027
0.081	0.6	0.532	0.5	0.068	0.032
0.273	0.7	0.608	0.6	0.092	0.008
0.819	0.8	0.793	0.7	0.007	0.093
0.971	0.9	0.834	0.8	0.066	0.034
1.456	1.0	0.927	0.9	0.073	0.027

The value of the test statistic is 0.144, which is much less than $1.36/\sqrt{11} = 0.41$: so the null hypothesis is acceptable.

Koyck model *See* DISTRIBUTED LAGS MODELS.

kriging A method of spatial prediction, named by *Matheron after the South African mining engineer Danie G. Krige. The values of some quantity, s, are known for n points in space. Estimates for intervening points are required—for example, we might wish to derive contour lines from a series of spot heights. The predictions are to be calculated as the weighted sum of the observed values. The question is what *weights to use so as to minimize the average squared difference between the predicted value and the true value.

If, at each spatial location, *data are collected on several *variables, then the *interpolation of values for the variable of prime interest may be improved by noting the variation in the other variables—this is called **cokriging**.

Other variants include **disjunctive kriging**, which uses linear combinations of functions of the data, **indicator kriging**, which uses *binary indicators (*e.g.* presence/absence) in place of the observed data, and **universal kriging**, in which the surface being estimated is an unknown linear combination of known spatial functions.

Kruskal, William Henry (1919–; b. New York City) An American statistician. Kruskal was a co-author (with *Wallis) of the 1952 paper that introduced one of the best known *non-parametric tests. Kruskal was educated at Harvard and Columbia Universities. He joined the Statistics Department at the University of Chicago in 1950, where he worked with *Goodman and Wallis. He retired as Professor Emeritus in 1990. He was Editor of the *Annals of Mathematical Statistics* during 1958–61. He was President of the *Institute of Mathematical Statistics in 1971 and President of the *American Statistical Association in 1982. He was awarded the Association's *Wilks Medal in 1978. He was made an Honorary Fellow of the *Royal Statistical Society in 1989.

Kruskal stress *See* MULTIDIMENSIONAL SCALING.

Kruskal–Wallis test A *non-parametric test, introduced in 1952 by *Kruskal and *Wallis; k-sample extension of the two-sample *Mann–Whitney test. It tests the *null hypothesis that the k sampled *populations have the same *distribution function.

With n_j *observations in the jth *sample, denote the total number of observations, $\sum_{j=1}^{k} n_j$, by N. Arrange these N observations in order of size and replace their values with the corresponding *ranks. Let the total of the ranks in sample j be R_j. The test statistic is H, given by

$$H = \frac{12}{N(N+1)} \sum_{j=1}^{k} \frac{R_j^2}{n_j} - 3(N+1).$$

Under the null hypothesis, for large N, with no n_j small, H has an approximate *chi-squared distribution with $(k-1)$ *degrees of freedom.

Kuder–Richardson formulae *See* RELIABILITY.

Kullback, Solomon (1907–94, b. Brooklyn, New York) A statistician and cryptographer. Kullback attended City College (now City University), New York (BA 1927) and Columbia University (MA 1929). He obtained his PhD in mathematical statistics in 1934 from George Washington University (GWU). By this time he had been a member of the United States Army's Signals Intelligence Service for four years. The war years were busy ones for code-breakers and Kullback is credited with being part of the three-man team that broke the Japanese codes. Kullback did not leave the National Security Agency until 1962, when he joined the faculty at GWU. Kullback was awarded the *Wilks Medal of the *American Statistical Association in 1976. Among his other honours was the Legion of Merit from the United States Army.

Kullback–Leibler information (directed divergence; discrimination information) A measure of the difference between two *probability density functions (f and g, say) taking non-zero values on the same interval (a, b). Usually denoted by I, it is related to *entropy and is defined as the expectation of the *logarithm of the ratio of *likelihoods. The information in favour of f is

$$I = \int_a^b \ln\left\{\frac{f(x)}{g(x)}\right\} f(x) dx.$$

An equivalent expression holds for the comparison of two *discrete distributions. The measure was introduced in 1951 by *Kullback and *Leibler. *See also* BHATTACHARYA'S DISTANCE; HELLINGER DISTANCE.

Kulldorff scan statistic A *statistic that evaluates reported spatial or space-time disease clusters to determine their *significance. Conversely, it can be used to search for clusters of events in space, time, or space-time.

kurtosis A measure of the peakedness of a *distribution popularized by Karl *Pearson before 1905. The usual measure is β_2 (also denoted by κ), given by

$$\beta_2 = \frac{\mu_4}{\mu_2^2},$$

where μ_4 is the fourth central *moment of the distribution and μ_2 is the *variance. It is invariant under a change of scale or origin.

For a *normal distribution $\beta_2 = 3$; an alternative definition of kurtosis reduces the ratio by 3 to give a value of 0 for the normal. With this adjustment, a distribution having negative kurtosis is described as being **platykurtic** (flatter), and a distribution having positive kurtosis is described as being **leptokurtic** (more peaked). A distribution having kurtosis zero is described as **mesokurtic**.

l_1-metric; l_2-metric *See* l_p-METRIC.

lack of memory *See* FORGETFULNESS PROPERTY.

lag *See* AUTOCORRELATION.

lag window *See* PERIODOGRAM.

lambda *See* GOODMAN AND KRUSKAL'S LAMBDA.

Langevin, Paul (1872–1946; b. Paris, France; d. Paris, France) A French physicist. Langevin derived the *Langevin distribution during a study in 1905 of the magnetic properties of molecules.

Langevin distribution (Fisher distribution) A *distribution used to model spherical data. The *probability density function f in the direction (θ, ϕ) (*see* SPHERICAL DATA) is given by

$$f(\theta, \phi) = c \exp[\kappa\{\cos\theta_0 \cos\theta + \sin\theta_0 \sin\theta \cos(\phi - \phi_0)\}],$$
$$0 \leq \theta \leq \pi, \; -\pi \leq \phi < \pi,$$

where c is a positive constant and the non-negative *parameter κ is a measure of the concentration of the distribution, which is symmetrical about the mode. The case $\kappa = 0$ corresponds to a uniform distribution over the entire sphere. If $\kappa \neq 0$ the distribution has *mode at (θ_0, ϕ_0).

Laplace, Marquis Pierre-Simon (1749–1827; b. Beaumont-en-Auge, France; d. Paris, France) A French mathematician. Laplace was born into the French middle class but died a marquis having prospered as well during the French Revolution as during the monarchy. The name of Laplace is well known to mathematicians both for his work with Laplace transformations and for his work on planetary motion. He had a deterministic view of the universe. Reputedly, his reply, when asked by Napoleon where God fitted in, was 'I have no need of that hypothesis.' Napoleon appointed him as Minister of the Interior—but removed him six weeks later for trying 'to carry the spirit of the infinitesimal into administration'. In statistics, besides the *Laplace distribution he worked on many probability problems: he is credited with independently discovering *Bayes's theorem. He was elected a Fellow of the Royal Society in 1789. A street in Paris is named after him, as is the Promontorium Laplace on the Moon.

Laplace distribution A distribution, first given by *Laplace in 1774, that has *probability density function f given by

$$f(x) = \frac{1}{2\phi} \exp\left(-\frac{|x|}{\phi}\right), \quad -\infty < x < \infty,$$

where ϕ is a positive *parameter. The distribution is also called the **double exponential distribution** and is the distribution of the difference of two independent *exponential random variables with the same mean. The distribution is symmetrical about 0, which is therefore both its *mean and its *mode. The distribution has *variance $2\phi^2$.

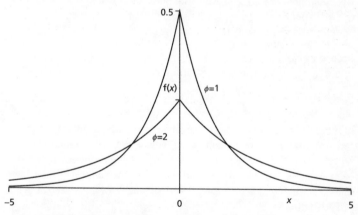

Laplace distribution. The distributions have variances equal to $2\phi^2$.

Laspeyres, Ernst Louis Étienne (1834–1913; b. Halle/Saale, Germany; d. Giessen, Germany) A German economist. Laspeyres was the son of a law professor. He studied law and public finance at a succession of universities, gaining his PhD in political science and public finance from the University of Heidelberg. From 1874 to 1900 he was Professor of Political Science at Giessen. He was an active member of the International Statistical Institute. He is remembered for *Laspeyres's price index, though he was not the first to suggest its use and apparently did not use it himself.

Laspeyres's price index A measure of the value of money in which the cost of a standard collection, or basket, of goods and services at some particular date is compared with the cost of the same basket at a base date. The basket is chosen to represent the expenditure of a typical household at the base date. The index, suggested by Laspeyres in 1871, is calculated as the ratio

$$\sum_j p_{nj}q_{0j} \Big/ \sum_j p_{0j}q_{0j},$$

where p_{0j} and p_{nj} are the prices, at times 0 and n respectively, of the jth item in the basket, and q_{0j} is the quantity of that item at time 0. *See also* FISHER'S INDEX; PAASCHE'S PRICE INDEX; RETAIL PRICE INDEX.

latent class model A model for cross-tabulated *data. The idea is that the apparently complex relationship between the observed *variables can be explained by a simple relationship that holds for unobserved *latent variables. *See* CONTINGENCY TABLE.

latent variable An unobserved *variable that may account for variation in the data and/or for apparent relations between observed variables.

Latin square An *experimental design that extends the *randomized blocks design. In a Latin square two types of *blocks are used. The numbers of both type of block must be equal to the numbers of *treatments (m, say) being compared. The design is conveniently represented as an $m \times m$ square in which each experimental unit has been assigned to one treatment, one row, and one column in such a way that each treatment occurs exactly once in each row and exactly once in each column.

3 × 3 Latin square

A	C	B
B	A	C
C	B	A

4 × 4 Latin square

A	D	C	B
B	C	A	D
D	A	B	C
C	B	D	A

An extension of the Latin square introduces a fourth effect, represented by a Greek letter. The resulting design, in which each pair of letters occurs just once, with both the Greek letters and the Roman letters forming Latin squares, is called a **Graeco-Latin square**. Removal of a row or column from a Latin square results in a so-called **Youden square** (actually a rectangle). Youden squares can be used to generate *balanced incomplete block designs.

Graeco-Latin square

Cδ	Bα	Dγ	Aβ
Aα	Dδ	Bβ	Cγ
Dβ	Aγ	Cα	Bδ
Bγ	Cβ	Aδ	Dα

Youden square

A	D	C	B
B	C	A	D
C	B	D	A

Latin squares were introduced into *Statistics by Sir Ronald *Fisher in 1925. Graeco-Latin squares were introduced by Fisher and *Yates in 1934. In 1938 Fisher and Yates introduced Youden squares which they named after *Youden.

lattice models Models describing the variation in the values at the rc nodes of an $r \times c$ lattice.

lattice square An *experimental design related to *Latin square and *balanced incomplete block designs. As an example, suppose there are nine treatments A–I. The design consists of two squares:

A	B	C
D	E	F
G	H	I

and

A	F	H
I	B	D
E	G	C

Considering the squares separately, we find that each pair of treatments occurs exactly once—either in a row or in a column. In general, with n^2 treatments, the design requires $(n-1)$ separate $n \times n$ squares.

law of averages As a *sample size increases so the *data will increasingly resemble the *population being sampled and will take on its characteristics (*see* LAWS OF LARGE NUMBERS). However, there is nothing *deterministic about this process. It is easy to be deluded by the **gambler's fallacy**, which advocates betting on numbers that 'should have come up'. As an example, suppose that a fair die is rolled 24 times and no six is obtained. According to the gambler's fallacy a six is long overdue and must be a good bet for the 25th roll—the truth is that its *probability will be the same as for every other roll, namely, 1/6.

law of compound probability *See* MULTIPLICATION LAW FOR PROBABILITIES.

law of total probability *See* ADDITION LAW FOR PROBABILITIES.

laws of large numbers Laws that describe the way in which the sample mean approaches the *population mean as the sample size increases. The phrase is due to Poisson, who, in 1835, referred to 'La loi des grands nombres'. The **weak law of large numbers** states that if $X_1, X_2, \ldots, X_n$ are a set of *independent identically distributed random variables, each with expectation μ, and if $\bar{X} = \frac{1}{n} \sum_{j=1}^{n} X_j$ then, for every positive ε,

$$\lim_{n \to \infty} P(|\bar{X} - \mu| \geq \varepsilon) = 0.$$

An equivalent statement is that

$$\bar{X} \to \mu \qquad \text{'almost surely', as } n \to \infty.$$

The **strong law of large numbers** states that, for every positive δ, it is always possible to find positive values ε and N such that

$$P(|\bar{X} - \mu| \geq \varepsilon) \leq \delta, \qquad n = N, N+1, \ldots.$$

An equivalent statement is that

$$\bar{X} \to \mu \qquad \text{'in probability', as } n \to \infty.$$

leading diagonal *See* MATRIX.

least significant difference *See* MULTIPLE COMPARISON TESTS.

least squares *See* ORDINARY LEAST SQUARES.

Legendre, Adrien-Marie (1752–1833; b. Paris, France; d. Paris, France) A French mathematician. Legendre was educated at the Collège Mazarin in Paris. In 1775 he taught with *Laplace at the École Militaire. Legendre's career covered many branches of mathematics and, in 1805, he published the first account of the method of least squares.

Lehmann, Erich Leo (1917–) A German mathematical statistician who has spent most of his career in the United States. Lehmann's family left Germany for Zurich in 1933, when Hitler came to power. After study at Cambridge University, Lehmann moved to the University of California, Berkeley, where he obtained his PhD in 1946. He promptly joined the staff at Berkeley, remaining there for the rest of his career—currently as Professor Emeritus. He was Editor of the *Annals of Mathematical Statistics* during 1953–5. He was created an Honorary Fellow of the Royal Statistical Society in 1986.

Lehmann–Scheffé theorem If T is a sufficient statistic for the *parameter θ, then the *minimum variance *unbiased *estimator of θ is given by $E(\hat{\theta}|T)$, where $\hat{\theta}$ is any unbiased estimator of θ. The theorem, published in 1950, is an extension of the *Rao–Blackwell theorem.

Leibler, Richard Arthur (1914–2003; b. Chicago, Illinois; d. Reston, Virginia) An American mathematician and cryptoanalyst. Leibler worked in the National Security Agency with *Kullback. In 1958 he was appointed to the Institute for Defence Analysis, becoming its Director in 1963. In 1977 he was appointed Chief of the Office of Research at the National Security Agency.

Leibniz, Gottfried Wilhelm (1646–1716; b. Leipzig, Germany; d. Hannover, Germany) A German philosopher and mathematician. Leibniz graduated from the University of Leipzig at the age of seventeen. In 1667 he was awarded a doctorate in law by the University of Altdorf. In 1673 he was elected a Fellow of the Royal Society. In 1675, in Paris, he developed the dx notation for differentials. In 1679 he developed the binary system. Leibniz was an assiduous letter writer whose more than 600 correspondents included all the leading mathematicians of the day. He studied all areas of mathematics and is credited with devising the diagrams popularized by *Venn. His later years were much taken up with disputes on priority concerning the work on differentiation and integration: he was accused of plagiarism by a supporter of Sir Isaac Newton (who had derived equivalent results independently of Leibniz).

leptokurtic *See* KURTOSIS.

Leslie matrix; Leslie model A *matrix that captures the evolution of the age distribution of the females in a *population. The female population is divided into g age-groups, each of width k years. The *probability that a female, in group j at time t, will be alive in group $(j + 1)$ at time $(t + k)$ is denoted by S_j. The mean number of female offspring born to a mother in group j between times t and $(t + k)$, and alive at time $(t + k)$, is denoted by M_j. The Leslie matrix, **L** is given by

$$\mathbf{L} = \begin{pmatrix} M_0 & M_1 & M_2 & M_3 & \cdots \\ S_0 & 0 & 0 & 0 & \cdots \\ 0 & S_1 & 0 & 0 & \cdots \\ 0 & 0 & S_2 & 0 & \cdots \\ \vdots & \vdots & \vdots & \vdots & \ddots \end{pmatrix}.$$

The entries in this matrix are the mean numbers at time $(t + k)$ resulting from single individuals in each group at time t. If $\mathbf{N}(t)$ is the $g \times 1$ column vector with entries being the mean numbers of females in the groups at time t then the Leslie model states that

$$\mathbf{N}(t + k) = \mathbf{L}\mathbf{N}(t).$$

level 1. In *significance testing, the designated probability for rejection of the *null hypothesis—the significance level (e.g. 5%).
2. In *experimental design, the category of a *treatment (typically, in a 2^n *factorial experiment, reference is made to upper and lower levels of a treatment).

Levene, Howard (1914–2003; d. New York City) An American biostatistician. After undergraduate study at New York University, Levene moved to Columbia University, gaining his PhD in 1947 and joining the Biostatistics Department. He is now Professor Emeritus in the Department of Statistics at Columbia University.

Levene's test A test for homogeneity of *variance that is insensitive to departures from *normality. Let the kth *observation in the jth of m *samples be denoted by y_{jk} and let the *mean of the n_j observations in this sample be denoted by $\bar{y}_j$. Define z_{jk} by

$$z_{jk} = |y_{jk} - \bar{y}_j|,$$

with the mean of the z-values in the jth sample being denoted by $\bar{z}_j$ and the overall mean of the $n(= \sum_j n_j)$ z-values being $\bar{z}$. Levene's test statistic is W given by

$$W = \frac{n - m}{m - 1} \frac{\sum_{j=1}^{m} n_j(\bar{z}_j - \bar{z})^2}{\sum_{j=1}^{m} \sum_{k=1}^{n_j} (z_{jk} - \bar{z}_j)^2}.$$

If the hypothesis of equality of variance is correct, then the distribution of W is approximately an *F-distribution with $(m - 1)$ and $(n - m)$ *degrees of freedom.

A drawback of Levene's test is that the sample means may be affected by *outliers. The **Brown–Forsythe test** avoids this problem by working with z'_{jk} instead of z_{jk}, where

$$z'_{jk} = |y_{jk} - M_j|,$$

and M_j is the *median of the jth sample.

leverage *See* REGRESSION DIAGNOSTICS.

Lexis, Wilhelm (1837–1914; b. Eschweiler, Germany; d. Göttingen, Germany) A German social scientist. Lexis graduated in 1859 from the University of Bonn, where he had studied mathematics and physics. His subsequent career focused on the social sciences: at various times he held chairs in economics, geography, and political science. In 1879 he published *On the Theory of the Stability of Statistical Series* which began the study of time series. Lexis was a pioneer of the analysis of demographic *time series.

Lexis diagram A plot of age against time, consisting of lines inclined at 45° representing the lifetimes of individuals. Each line starts with birth (or immigration into the population) and ends with death (or emigration from the population).

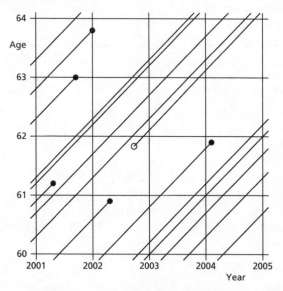

● Death or emigration   ○ Birth or immigration

Lexis diagram. Diagonal lines show lifetimes of individuals; vertical lines show points in time. There were three people aged 60 at the beginning of 2001. Of these, one died later that year aged 61. In 2002 a 61-year-old immigrant entered the population.

life table A table, first developed by *Graunt, that presents the *probability of dying in the next time interval, as a function of current age. The table provides a means of calculating life expectancy, conditional on current age.

AGE AT START OF DECADE	MALES		FEMALES	
	PROBABILITY OF DEATH IN NEXT DECADE	NUMBER ALIVE AT START OF DECADE	PROBABILITY OF DEATH IN NEXT DECADE	NUMBER ALIVE AT START OF DECADE
0	0.01050	100 000	0.00852	100 000
10	0.00660	98 950	0.00307	99 148
20	0.01458	98 297	0.00537	98 844
30	0.01964	96 864	0.00967	98 313
40	0.03791	94 962	0.02083	97 362
50	0.08286	91 362	0.05060	95 334
60	0.19825	83 792	0.12347	90 510
70	0.40089	67 180	0.26973	79 335
80	0.73074	40 248	0.58404	57 936
90	0.95709	10 837	0.91394	24 099
100	0.99963	465	0.99517	2074
110	1.00000	1	1.00000	10

Life table. The table records the probabilities for males and females in the United States, using 1997 information. As an example, the probability that, in the United States, a 50-year-old male lives to be at least 70 is $(1 - 0.08286)(1 - 0.19825) = 0.73532$.

likelihood For a random sample $x_1, x_2, \ldots$ of *observations on a *discrete random variable X, the likelihood (a term coined by Sir Ronald *Fisher) is proportional to the product of the *probabilities of the individual values:

$$\prod_j P(X = x_j).$$

When X is *continuous, a reported value x_j has to be regarded as meaning that the observed value is (in general) in the interval $(x_j - \delta, x_j + \delta)$, where δ represents the accuracy of measurement. The likelihood is then proportional to

$$\prod_j P(x_j - \delta < X < x_j + \delta).$$

If δ is very small then this is approximately proportional to

$$\prod_j f(x_j),$$

where f is the *probability density function of X.

likelihood-ratio goodness-of-fit statistic (G^2) A *statistic used in an alternative to the *chi-squared goodness-of-fit test. The two test statistics usually have similar values and both have approximate *chi-squared distributions when the *model under test is a correct description of the *data. The G^2 test is usually preferred when comparing *hierarchical models. If the *observed cell frequencies are denoted by $O_1, O_2, \ldots, O_k$ and the corresponding *expected cell frequencies by $E_1, E_2, \ldots, E_k$, respectively, G^2 is given by

$$G^2 = 2 \sum_{j=1}^{k} O_j \ln\left(\frac{O_j}{E_j}\right),$$

where ln is the *natural logarithm.

likelihood residuals An alternative name for *deletion residuals.

Likert, Rensis (1903–81; b. Cheyenne, Wyoming; d. Ann Arbor, Michigan) An American psychologist. Likert graduated from the University of Michigan with an AB in Economics and Sociology in 1922. He received his doctorate in psychology from Columbia University in 1932. During the next decade his work involved the interpretation of surveys and the handling of questionnaires. He joined the staff of the University of Michigan in 1946, becoming the founding Director of the Institute for Social Research in 1949 and holding this position until his retirement in 1970.

Likert scale A scale, usually of approval or agreement, used in *questionnaires. The *respondent is asked to say whether, for example, they 'Strongly agree', 'Agree', 'Disagree', or 'Strongly disagree' with some statement.

Lilliefors test A modification of the *Kolmogorov–Smirnov test for testing whether a *population has a *normal distribution.

linear combination A *variable y is a linear combination of the variables $x_1, x_2, \ldots, x_m$ if

$$y = \sum_{j=1}^{m} a_j x_j,$$

where $a_1, a_2, \ldots, a_m$ are constants. *See also* MATRIX.

linear constraints *See* LINEAR PROGRAMMING.

linear dependence The *variables y and x display linear dependence if, for some constants a and $b(\neq 0)$,

$$y = a + bx.$$

linear interpolation *See* INTERPOLATION.

linear-logistic model *See* LOGISTIC REGRESSION MODEL.

linearly dependent *See* MATRIX.

linearly independent *See* MATRIX.

linear model A *regression model that represents the expectation of a *random variable, Y, as a linear combination of functions of the variables $(x_1, x_2, \ldots)$. An example is the model

$$\mathrm{E}(Y) = \alpha + \beta x_1 + \gamma x_2 + \delta x_2^2 + \eta \sin(x_3),$$

where $\alpha, \beta, \gamma, \delta$ and η are unknown *parameters. *See also* GENERALIZED LINEAR MODEL; LINEAR REGRESSION; MULTIPLE REGRESSION MODEL.

linear piecewise (break point) regression model A model that describes the situation where the graph relating the expectation of the *random variable Y and the variable x is linear for $x \leq x_0$ and is also linear for $x \geq x_0$ but with a different slope. For example,

$$\mathrm{E}(Y) = \alpha + \beta_1 x, \qquad\qquad x \leq x_0,$$
$$\mathrm{E}(Y) = \{\alpha + (\beta_1 - \beta_2)x_0\} + \beta_2 x, \quad x \geq x_0.$$

linear programming A method for determining the optimal use of limited resources. The phrase 'linear programming' was used by *Dantzig in 1949. A typical problem requires the maximization (or minimization) of a linear function of the non-negative variables $x_1, x_2, \ldots, x_n$:

$$c_1 x_1 + c_2 x_2 + \cdots + c_n x_n,$$

where the coefficients $c_1, c_2, \ldots, c_n$ have known values. The function is known as the **objective function**. The maximization is subject to a set of **linear constraints** such as

$$a_{11}x_1 + a_{12}x_2 + \cdots + a_{1n}x_n \leq a_{10},$$
$$a_{21}x_1 + a_{22}x_2 + \cdots + a_{2n}x_n \geq a_{20},$$

$$\vdots \quad \vdots \quad \vdots$$

$$a_{m1}x_1 + a_{m2}x_2 + \cdots + a_{mn}x_n = a_{m0},$$

where the coefficients $\{a_{ij}\}$ have known values.

In order to solve the problem, inequalities are turned into equalities by introducing non-negative **slack variables** so that, for example, the first inequality would become

$$a_{11}x_1 + a_{12}x_2 + \cdots + a_{1n}x_n + x_{n+1} = a_{10}.$$

Any solution satisfying all the constraints (with the addition of slack variables as required) is called a **feasible solution**. A feasible solution involving exactly m non-zero x-variables (including any slack variables) is called a **basic feasible solution**.

One standard approach to the general problem is to use the *simplex method (introduced by Dantzig in 1947). Special algorithms are required for **integer programming** (where the x-values are required to be integers) and for *assignment problems, *network flow problems, and *transportation problems.

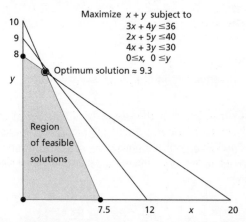

Linear programming. The solution must lie at one of the inner vertices marked with dots: these are the basic feasible solutions. The values of $x + y$ at the other two vertices are 7.5 and 8.

linear regression The simplest and most used of all statistical regression *models. The model states that the *random variable Y is related to the variable x by

$$Y = \alpha + \beta x + \varepsilon,$$

where the *parameters α and β correspond to, respectively, the intercept and the slope of the line, and ε denotes a *random error. With observations $(x_1, y_1), (x_2, y_2), \ldots, (x_n, y_n)$ the usual assumption is that the random errors are *independent observations from a *normal distribution with *mean 0 and *variance σ^2. In this case the parameters are usually estimated using *ordinary least squares. The estimates, denoted by $\hat{\alpha}$ and $\hat{\beta}$, are given by

$$\hat{\alpha} = \bar{y} - \hat{\beta}\bar{x} \quad \text{and} \quad \hat{\beta} = \frac{S_{xy}}{S_{xx}},$$

where $\bar{x}$ and $\bar{y}$ are the means of $x_1, x_2, \ldots, x_n$ and $y_1, y_2, \ldots, y_n$, respectively, and where

$$S_{xy} = \sum_{j=1}^{n} x_j y_j - n\bar{x}\bar{y} \quad \text{and} \quad S_{xx} = \sum_{j=1}^{n} x_j^2 - n\bar{x}^2.$$

The variance σ^2 is estimated by

$$\hat{\sigma}^2 = \frac{1}{n-2} \sum_{j=1}^{n} \left\{ y_j - \left(\hat{\alpha} + \hat{\beta}x_j \right) \right\}^2$$

$$= \frac{S_{xx}S_{yy} - S_{xy}^2}{(n-2)S_{xx}},$$

where $S_{yy} = \sum_{j=1}^{n} y_j^2 - n\bar{y}^2$.

A $100(1-\theta)\%$ confidence interval for β is provided by

$$\hat{\beta} \pm t_{n-2}(\theta)\sqrt{\frac{\hat{\sigma}^2}{S_{xx}}},$$

where $t_v(\theta)$ is the upper θ *percentage point of a *t-distribution with v *degrees of freedom. A $100(1-\theta)\%$ *confidence interval for the expected value of Y when $x = x_0$ is

$$\left(\hat{\alpha} + \hat{\beta}x_0 \right) \pm t_{n-2}(\theta)\sqrt{\hat{\sigma}^2 \left\{ \frac{1}{n} + \frac{(x_0 - \bar{x})^2}{S_{xx}} \right\}}.$$

A $100(1-\theta)\%$ *prediction interval for the value y_0 of Y when $x = x_0$ is

$$\left(\hat{\alpha} + \hat{\beta}x_0\right) \pm t_{n-2}(\theta)\sqrt{\hat{\sigma}^2\left\{1 + \frac{1}{n} + \frac{(x_0 - \bar{x})^2}{S_{xx}}\right\}}.$$

line graph *See* BAR CHART.

line transect A path, usually straight, across a region of interest, that will be used for sampling. If the path has constant width then the sampling procedure may be to count the organisms in predefined regions of the path (for example, in a sequence of *quadrats). Alternatively, the sampler may walk along the line transect either counting visible organisms or recording their positions.

link function *See* GENERALIZED LINEAR MODEL.

LISREL (Linear Structural Relations) A computer program for analysing structural equation models.

ln *See* NATURAL LOGARITHM.

local maximum A high value surrounded only by lower values, but not necessarily the highest value possible (which is the **global maximum**). Thus the peak of Ben Nevis is a local maximum, whereas that of Mount Everest is both a local maximum and a global maximum.

loess (lowess) A computationally intensive method for fitting smooth curves or surfaces to a set of data. The procedure fits *polynomials (usually linear or quadratic) to local subsets of the data, using *weighted least squares so as to pay less attention to distant points. Loess requires no predetermined model for the entire data set, but therefore provides no explicit formula for the smoothed values.

logarithm The logarithm to base $a(>0)$ of a positive number b is denoted by $\log_a b = x$, where $a^x = b$. Logarithms have the fundamental property $\log_a (bc) = \log_a b + \log_a c$. A logarithm to base 10 is usually denoted by log and a *natural logarithm, i.e. a logarithm to base e, by ln. Before the advent of computers and calculators, logarithms were used in evaluating products.

logarithmic distribution (log-series distribution) A *discrete distribution with one *parameter θ $(0 < \theta < 1)$, for which

$$P(X = x) = -\frac{\theta^x}{x \ln (1 - \theta)}, \qquad x = 1, 2, \ldots,$$

where ln is the *natural logarithm. The distribution has *mean and *variance

$$-\frac{\theta}{(1-\theta)\ln(1-\theta)} \qquad \text{and} \qquad -\frac{\theta}{(1-\theta)^2\ln(1-\theta)}\left(1+\frac{\theta}{\ln(1-\theta)}\right),$$

respectively. A simple *recurrence relation links successive probabilities:

$$P(X=x) = \frac{(x-1)\theta}{x}P[X=(x-1)], \qquad x = 2, 3, \ldots.$$

The distribution is much in use in the context of species diversity. If, when sampling a mixed-species population, n species occur just once, then it is commonly found that, for some θ, the numbers of species occurring twice, thrice, ... are approximately $\frac{1}{2}n\theta$, $\frac{1}{3}n\theta^2$, ... The ratio n/θ is the **index of diversity**. The distribution is also suitable for modelling the numbers of items of a product bought by a buyer in a given time period.

logarithmic link *See* GENERALIZED LINEAR MODEL.

logistic curve *See* GROWTH CURVE.

logistic distribution A continuous *distribution with *probability density function f given by

$$f(x) = \frac{\exp\left\{\frac{-(x-\alpha)}{\beta}\right\}}{\beta\left[1+\exp\left\{-\frac{(x-\alpha)}{\beta}\right\}\right]^2}, \qquad -\infty < x < \infty,$$

where α is the *mean of the distribution and β is a positive *parameter. The distribution has *variance $\frac{1}{3}\beta^2\pi^2$.

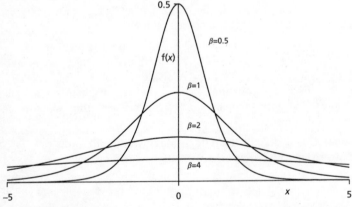

Logistic distribution. The distributions shown are for $\alpha = 0$. The variance of these distributions is $\frac{1}{3}\beta^2\pi^2$.

logistic link *See* GENERALIZED LINEAR MODEL.

logistic regression model *Linear models in which the dependent variable is a *logit and at least one explanatory variable is *continuous. The case where all the explanatory variables are *categorical is called a *logit model.

logistic transformation The use of the *logit, which takes values in the interval $(-\infty, \infty)$, in place of a *probability with values in the finite interval $(0, 1)$.

logit (log-odds) The quantity

$$\ln\left(\frac{p}{1-p}\right),$$

where p is a *proportion. The term 'logit' was introduced by *Berkson in 1944. Modelling variations in proportions directly is hampered by the need to ensure that estimated probabilities lie in the interval $(0, 1)$. Since corresponding values for the logit lie in the unrestricted interval $(-\infty, \infty)$, models for proportions are usually constructed in terms of logits. *See also* PROBIT.

logit models *Linear models in which the dependent variable is a *logit and the explanatory variables are *categorical. They can often be re-expressed as *log-linear models. If one or more of the explanatory variables is not categorical then the model is called a *logistic regression model.

log-likelihood The *natural logarithm of the *likelihood of a set of *data.

log-linear models *Linear models that describe the variations in the expectation of the *logarithm of the response variable. They arise naturally in the context of modelling the variations in the cell frequencies of a *contingency table.

In the commonly used **hierarchical log-linear models**, the inclusion of a term corresponding to a multi-variable *interaction necessitates the inclusion of terms corresponding to all simpler interactions.

For all log-linear models the *expected frequencies are determined by the *marginal totals of the table. In some cases there is a simple algebraic formula (as in the case of the *independence model)—models for which this is the case are called *direct models. *See also* LOGIT MODELS.

lognormal distribution The *distribution of a *random variable X

such that $\ln(X - \theta)$ has a *normal distribution, where θ is a parameter. There are two other parameters, $\delta\,(>0)$ and γ. The *probability density function f is given by

$$f(x) = \frac{\delta}{(x - \theta)\sqrt{2\pi}} \exp\left[-\frac{1}{2}\{\gamma + \delta\ln(x - \theta)\}^2 \right], \qquad x > \theta.$$

If we write

$$\alpha = \exp\left(-\frac{\gamma}{\delta}\right) \qquad \text{and} \qquad \beta = \exp\left(\frac{1}{\delta^2}\right),$$

the distribution has *mean $(\theta + \alpha\sqrt{\beta})$ and *variance $\alpha^2\beta(\beta - 1)$. The distribution has *mode at $\theta + \alpha/\beta$. *See* JOHNSON DISTRIBUTIONS.

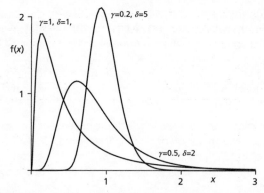

Lognormal distribution. The distributions illustrated are for $\theta = 0$. They have mode at $\exp\left(-(1 + \gamma\delta)/\delta^2\right)$.

log-odds *See* LOGIT.

log-rank test A test for comparing two (or more) sets of *survival times. The data consist of the completed lifetimes of individuals who have died and the current ages of those still living. The *null hypothesis is that the sets come from the same *population. Suppose there are m distinct survival times, $t_{(1)} < t_{(2)} < \cdots < t_{(m)}$. For the two-sample case the test statistic is g, given by

$$g = \sum_{j=1}^{m} \frac{n_{2j}d_{1j} - n_{1j}d_{2j}}{N_j},$$

where n_{kj} is the number of individuals in group k who are alive just before time $t_{(j)}$, d_{kj} is the number of these who die at that time, and $N_j = n_{1j} + n_{2j}$.

The distribution of g can be approximated by a *normal distribution with, under the null hypothesis, *mean 0 and *variance V, where

$$V = \sum_{j=1}^{m} \frac{n_{1j}n_{2j}D_j(N_j - D_j)}{N_j(N_j - 1)},$$

and $D_j = d_{1j} + d_{2j}$.

log-series distribution *See* LOGARITHMIC DISTRIBUTION.

longitudinal data *Data collected at a sequence of time points for each of a *sample of individuals. Because each individual contributes several *observations, these observations are usually not *independent and this has to be taken into account in the analysis.

longitudinal study A study, usually in the context of medicine or the social sciences, conducted over time. The *data that result are called *longitudinal data.

Lorenz curve A graphical representation of social inequality introduced by the American economist Max O. Lorenz in 1905 in

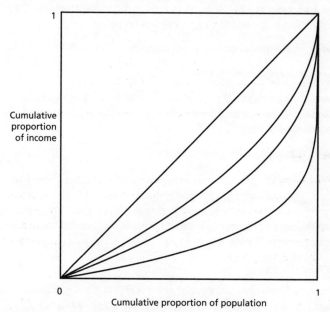

Lorenz curve. The curves shown result from *Pareto distributions with parameters 1, 1.2, 1.6, and 2.

connection with the distribution of wealth in a population.
Suppose a sample of n individuals have incomes $x_{(1)} \leq x_{(2)} \leq \cdots \leq x_{(n)}$.
The Lorenz curve is approximated by the polygon joining the origin to
the successive points with coordinates

$$\left(\frac{r}{n}, \quad \frac{\sum_{j=1}^{r} x_{(j)}}{\sum_{j=1}^{n} x_{(j)}} \right), \qquad r = 1, 2, \ldots, n.$$

Thus the x-coordinate represents a cumulative proportion of the
sample, while the y-coordinate represents the corresponding
cumulative proportion of the total wealth of the sample. If f is
the *probability density function of income, then the Lorenz curve is
given by the plot of

$$\frac{\int_0^y x f(x) dx}{\int_0^\infty x f(x) dx} \quad \text{against} \quad \int_0^y f(x) dx.$$

loss function The function that is minimized in the process of fitting a
*model (for example, the sum of squared residuals).

lower percentage point *See* PERCENTAGE POINT.

lower quartile *See* QUARTILE.

lower-tail probability The probability of a *random variable X
taking the value x or less, where, usually, x will be appreciably less than
the *mean of X.

lower triangular matrix *See* MATRIX.

lowess *See* LOESS.

low-pass filter *See* FILTER.

lowspread The difference between the sample *median and the
minimum sample value.

loyalty model *See* SQUARE TABLE.

l_p-metric A measure of distance between points in n-dimensional
space. Let $(x_1, x_2, \ldots, x_n)$ and $(y_1, y_2, \ldots, y_n)$ be two points in the
n-dimensional space $\mathbb{R}^n$. For the l_p-metric, the distance between these
points is

$$\left(\sum_{j=1}^{n} |x_j - y_j|^p \right)^{\frac{1}{p}},$$

where $p \geq 1$. In the case $p = 2$ the distance is the ordinary Euclidean distance. In the case $p = 1$ the distance is $\sum_{j=1}^{n} |x_j - y_j|$, which is the city-block metric. In the case $p = \infty$ the distance is taken to be $\max_j |x_j - y_j|$. *See* DISTANCE MEASURES.

LSD Abbreviation for 'least significant difference' (*see* MULTIPLE COMPARISON TESTS).

MA models *See* TIME SERIES.

Maclaurin, Colin (1698–1746; b. Kilmodan, Scotland; d. Edinburgh, Scotland) A Scottish mathematician. Orphaned by the age of nine, Maclaurin was a student at Glasgow University when aged eleven. At the age of nineteen he was appointed Professor of Mathematics at Aberdeen University. In 1719 he was elected a Fellow of the Royal Society. In 1725 he was appointed Professor at Edinburgh University, where he remained for the rest of his life. A lunar crater is named after him.

Maclaurin expansion *See* MACLAURIN SERIES.

Maclaurin series (Maclaurin expansion) A series representation for a function f having continuous derivatives of all orders. The series is

$$f(0) + f'(0)\frac{x}{1!} + f''(0)\frac{x^2}{2!} + \cdots = \sum_{r=0}^{\infty} f^{(r)}(0)\frac{x^r}{r!},$$

where $f^{(r)}(0)$ means $d^r f(x)/dx^r$ evaluated at $x = 0$. The series may converge for all values of x, or for $|x| < R$, for some positive R, or it may only converge when $x = 0$. If $f(x)$ is a *polynomial then the series is finite and the sum is $f(x)$. For most practical cases, if the series converges then its sum is $f(x)$—this is true for e^x and $\sin x$, when the series is convergent for all x, and for $(1 + x)^{1/2}$ and $\ln(1 + x)$, when the series is convergent for $|x| < 1$. In such cases the first $(n + 1)$ terms give a polynomial approximation $g(x)$ to $f(x)$, which has the property that at $x = 0$, the first n derivatives of $g(x)$ and $f(x)$ are equal. There are however non-zero functions f such that $f(x)$ and all its derivatives are zero at $x = 0$, and in this case the Maclaurin series vanishes—for example, $f(x) = \exp(-1/x^2)$ for $x \neq 0$, with $f(0) = 0$.

The Maclaurin series is of use in the theoretical development of *moment-generating functions and *probability-generating functions. It is the particular case of a *Taylor series when $a = 0$.

MAD *See* MEAN ABSOLUTE DEVIATION; MEDIAN ABSOLUTE DEVIATION

Mahalanobis, Prasanta Chandra (1893–1972; b. Calcutta, India; d. Calcutta, India) An Indian statistician who was the founder (in 1931) of the renowned *Indian Statistical Institute. Mahalanobis studied physics at Calcutta and Cambridge Universities. In the library of King's College, Cambridge, Mahalanobis was asked by a fellow student to express his opinion about an article in *Biometrika*. It was this encounter that stimulated his subsequent research in *Statistics. Practical problems led to theoretical advances in many areas of Statistics—it was an anthropometric problem that led to his distance statistic, D^2. As statistical adviser to the Indian government, Mahalanobis established, in 1950, the first national survey of India's population. He was elected a Fellow of the Royal Society in 1945.

Mahalanobis's D^2 The two-sample version of *Hotelling's T^2. The quantity D^2, introduced by *Mahalanobis in a 1930 paper, may be considered to be a measure of the distance between two groups of p-dimensional observations. Let $\bar{\mathbf{x}}_j$ be the $p \times 1$ *vector of *means for group j, with $\mathbf{S}_j$ the corresponding *variance–covariance matrix. The distance is given by

$$D^2 = (\bar{\mathbf{x}}_1 - \bar{\mathbf{x}}_2)'\mathbf{S}^{-1}(\bar{\mathbf{x}}_1 - \bar{\mathbf{x}}_2),$$

where $\mathbf{S}$ is given by

$$\mathbf{S} = \frac{n_1\mathbf{S}_1 + n_2\mathbf{S}_2}{n_1 + n_2}$$

and n_1 and n_2 are the sample sizes.

main diagonal *See* MATRIX.

main effect *See* FACTORIAL EXPERIMENT.

Mallows, Colin Lingwood (1930–; b. Great Sampford, England) An English statistician who has spent most of his career in the United States. Mallows was educated at University College, London (UCL), obtaining his PhD in 1953 and joining the UCL staff in 1955. In 1960 he moved to the United States to join the staff at Princeton University at the invitation of *Tukey. Since 1987 he has been a member of the research staff of AT&T Labs.

Mallows's C_p A statistic, introduced by *Mallows in 1964, that is used as an aid in choosing between competing *multiple regression models. With n observations and k explanatory variables, define s^2 as the estimate of the *experimental error *variance. Then, for a model using just p of the k variables, Mallows's C_p is given by

$$C_p = \frac{1}{s^2} \sum_{j=1}^{n} (y_j - \hat{y}_j)^2 - n + 2p,$$

where $y_1, y_2, \ldots, y_n$ are the observed values and $\hat{y}_1, \hat{y}_2, \ldots, \hat{y}_n$ are the corresponding *fitted values. A model that fits well should have a C_p value close to p. An acceptable fit is provided by a model for which

$$C_p < (2p - k - 1)kF_{k, n-k-1}(\alpha),$$

where $F_{a, b}(\alpha)$ is the value exceeded by chance on $100\alpha\%$ of occasions by a *random variable having an *F-distribution with a and b *degrees of freedom. Typically, $\alpha = 0.05$ or 0.01. For alternative approaches to model selection, *see* AIC; STEPWISE PROCEDURES.

Mandelbrot, Benoit B. (1924–; b. Warsaw, Poland) A French mathematician and engineer, famed for his work on *fractals. Mandelbrot graduated from the École Polytechnique in Paris in 1947 with a diploma in engineering. He obtained his MSc from California Institute of Technology in 1948 and his doctorate from the Faculté des Sciences de Paris in 1952. In 1958 he joined the IBM research centre in the state of New York. He was elected to Fellowship of the American Academy of Arts and Sciences in 1982 and was made a Chevalier, L'Ordre National de la Légion d'Honneur in 1989.

Manhattan distance *See* DISTANCE MEASURES.

manifest variable A *variable that is directly measurable. *Compare* LATENT VARIABLE.

Mann, Henry B. (1905–2000; b. Vienna, Austria; d. Tucson, Arizona) An Austrian mathematician who spent most of his career in the United States. Mann obtained his PhD in number theory at Vienna University. He emigrated to New York in 1938 and, after a variety of temporary positions, arrived at Ohio State University in 1946, where he remained until 1964. One of his first research students was D. Ransom *Whitney, with whom his name is now linked as joint originator of the *Mann–Whitney test. In 1964 he moved from Ohio State University to the University of Wisconsin, and subsequently to the University of Arizona where he was Professor Emeritus. His many publications show clearly his love of analytic number theory and its applications to *experimental design.

Mann–Whitney test A two-sample *non-parametric test, equivalent to the **Wilcoxon rank-sum test**, introduced in 1947 by *Mann and *Whitney. It is assumed that the *samples are random and come from *populations (X and Y) that have the same *distribution after a translation of size k:

$$P(X < x) = P(Y < x + k), \qquad \text{for all values of } x.$$

The *null hypothesis is that the populations have the same distribution (i.e. $k = 0$).

With samples of sizes m and n ($m \leq n$), the first stage is the replacement of the $(m + n)$ observed values by their *ranks in the combined sample. For the (smaller) sample of size m, denote the sum of the ranks by R. The distribution of R is approximately

$$N\left(\frac{1}{2}m(m + n + 1), \frac{1}{12}mn(m + n + 1)\right),$$

so that the test statistic, z, is given by

$$z = \frac{R - \frac{1}{2}m(m + n + 1) \pm \frac{1}{2}}{\sqrt{\frac{1}{12}mn(m + n + 1)}},$$

where the $\pm\frac{1}{2}$ is a *continuity correction with sign chosen so as to reduce the absolute magnitude of the numerator.

For example, suppose that the marks obtained by a small random sample of statistics students were as follows:

| Boys | 10, 22, 42, 59, 61, 63, 65, 83, 85, 90, 93 |
| Girls | 36, 53, 54, 56, 69, 84, 88 |

The question of interest is whether the data support the null hypothesis of a common mark distribution. The ranks are

| Boys | 1, 2, 4, 8, 9, 10, 11, 13, 15, 17, 18 |
| Girls | 3, 5, 6, 7, 12, 14, 16 |

Working with the (smaller) set of girls, m is 7 and R is $3 + 5 + \cdots + 16 = 63$. Using $n = 11$, the test statistic, z, is given by

$$z = \frac{(63 - 66.5 + 0.5)}{\sqrt{121.9}} = -0.272.$$

Since $|z| < 1.96$, we accept, at the 5% significance level, the hypothesis that the two sets of marks have come from the same distribution.

MANOVA *See* MULTIVARIATE ANALYSIS OF VARIANCE.

Mantel, Nathan (1919–2002; b. New York City; d. Potomac, Maryland) An American statistician. Mantel graduated from the City College (now City University) of New York in 1939. From 1947 to 1974 he worked at the National Cancer Institute, where Haenszel was a colleague. Mantel was co-author (with *Haenszel) of the 1959 paper which introduced the eponymous *Mantel–Haenszel test of *independence. Subsequently he was Research Professor first at George Washington University and then at the American University in Washington. He was made an Honorary Fellow of the *Royal Statistical Society in 1992.

Mantel–Haenszel statistic *See* MANTEL–HAENSZEL TEST.

Mantel–Haenszel test A test suitable for testing the *null hypothesis of *independence between two *dichotomous variables using *data from a *population subdivided into L classes: it is, therefore, a test for use with a $2 \times 2 \times L$ *contingency table. The test, which was introduced in 1959 by *Mantel and *Haenszel, is most used in medical contexts where, for example, one variable is outcome ('success' or 'failure'), one variable is treatment ('control' or 'new'), and the L classes correspond to different patient categories. The test assumes that any *association between the dichotomous variables is unaffected by the third variable.

The test statistic, M, is computed as follows. Denote by f_{jkl} the number of patients in class l ($= 1, 2, \ldots, L$) who experience outcome j ($= 1$ or 2) when given treatment k ($= 1$ or 2), and write $f_{0kl} = f_{1kl} + f_{2kl}$. Then

$$M = \left\{ \left| \sum_{l=1}^{L} \left(f_{11l} - \frac{f_{10l} f_{01l}}{f_{00l}} \right) \right| - \frac{1}{2} \right\}^2 \bigg/ \sum_{l=1}^{L} \frac{f_{10l} f_{20l} f_{01l} f_{02l}}{f_{00l}^2 (f_{00l} - 1)}.$$

Under the null hypothesis, the distribution of M is approximated by a *chi-squared distribution with one *degree of freedom. When $L = 1$ the test is equivalent to the *Yates-corrected chi-squared test. The $\frac{1}{2}$ term is a *continuity correction.

The test combines information from each of the L classes. In a similar way, the **Mantel–Haenszel statistic**, ψ, combines information about the strength of the relationship between the dichotomous variables. This statistic, given by

$$\psi = \sum_{l=1}^{L} \frac{f_{11l}\, f_{22l}}{f_{00l}} \Bigg/ \sum_{l=1}^{L} \frac{f_{12l}\, f_{21l}}{f_{00l}},$$

is an aggregate estimate of the *odds ratio for the two variables.

MAPE *See* MEAN ABSOLUTE PERCENTAGE ERROR.

marginal distribution *See* BIVARIATE DISTRIBUTION.

marginal homogeneity The property possessed by a square
*contingency table, in which, for all j, the total for row j is equal to
the total for column j. For example, the rows might refer to a set of
possible birthplaces and the columns to a set of current locations, using
the same location categories. Marginal homogeneity would imply that,
for all j, the *proportion born in location j is the same as the proportion
living in location j.

marginal totals If the cell frequencies of a (multidimensional)
*contingency table are totalled over one or more of the
categorizing variables the result is a set of marginal totals. For a
two-dimensional table, the marginal totals are the row and column
totals.

Markov, Andrei Andreevich (1856–1922; b. Ryazan, Russia; d. St
Petersburg, Russia) A Russian mathematician. Markov was educated at
St Petersburg University (where one teacher was *Chebyshev). On
graduation, he joined the staff and specialized in *probability. Markov's
outstanding contribution was the idea underlying *Markov processes.
He is also remembered for his contribution to the mathematics of linear
models through the *Gauss–Markov theorem. A lunar crater is named
after him.

Markov chain *See* MARKOV PROCESS.

Markov chain Monte Carlo (MCMC methods) Methods, making
heavy use of computers, that are very useful in the analysis of large
arrays of *correlated observations, such as satellite images of the Earth.
Suppose we wish to calculate θ, the *expected value of a function h of r
correlated *random variables. Let **X** be the $r \times 1$ *vector of random
variables. A standard *Monte Carlo method would generate a sequence
of numerical values, $\mathbf{X}_1, \mathbf{X}_2, \ldots, \mathbf{X}_n$, from which θ could be estimated by
$\hat{\theta}$, given by

$$\hat{\theta} = \frac{1}{n} \sum_{j=1}^{n} h(\mathbf{X}_j).$$

However, it can be difficult to generate a vector $\mathbf{X}_j$ when the joint distribution of the r variables involves a complex correlation structure—for example, when the variables refer to the values displayed in an array of neighbouring pixels in a satellite image of the Earth's surface.

Let X_{jk} be the kth random variable in $\mathbf{X}_j$. MCMC methods aim to generate r *Markov chains, where the kth chain is the sequence of values $x_{1k}, x_{2k}, \ldots$. An arbitrary set of values is chosen for the first chain, subsequent chains being generated from their predecessors by Monte Carlo methods. In calculating $\hat{\theta}$ it is necessary to ignore the early chains in the sequence, since these will be affected by the initial choice of values. This is called the **burn-in period**.

A popular method for obtaining the required Markov chains, which need to be *stationary, is the **Metropolis–Hastings algorithm**. Let the stationary probability for the random variable X_j being in state m be p_m and let $\mathbf{Q}$ be a known matrix with non-negative elements. The transition probability matrix governing the sequence $X_1, X_2, \ldots$ is defined by

$$p_{jk} = c_j q_{jk} a_{jk}, \qquad \text{if } j \neq k$$
$$p_{jj} = c_j q_{jj} + c_j \sum_{m \neq j} q_{jm}(1 - a_{jm}),$$

where a_{jk} is given by

$$a_{jk} = \min\left(\frac{p_k q_{kj}}{p_j q_{jk}}, 1\right),$$

and each c_j is chosen so that $\sum_k p_{jk} = 1$.

The most used version of the Metropolis–Hastings algorithm incorporates the **Gibbs sampler**. The aim is to generate a random vector $\mathbf{X}$ with elements satisfying a specified relationship. Denote the initial values in $\mathbf{X}$ by $x_1^{(0)}, x_2^{(0)}, \ldots$. A single element of the vector (element j, say) is chosen at random, and a potential new value, x, is generated by random selection from the *conditional distribution of X_j given the values of the remaining variables, $\{X_k, k \neq j\}$. If the new value is in accord with the specified relationship, then it replaces the previous value so that $x_j^{(1)} = x$; otherwise the previous value is retained: $x_j^{(1)} = x_j^{(0)}$. This process is repeated until approximate *equilibrium has been reached.

Markov process A Markov process is a *stochastic process, $X_1, X_2, \ldots$, for which the value taken by the *random variable X_m is, for all $m > 2$, *independent of the values taken by $X_1, X_2, \ldots, X_{m-2}$. The variable X_m may be described as exhibiting the **Markov property**. If there is a common finite (or *countable) number of possible outcomes for all the X-values then the process is a **Markov chain**.

In a **discrete-time Markov chain** the random variables $X_1, X_2, \ldots$ are ordered in time. If $X_m = r$ then the process is said to be in **state** r at time point m. The Markov property implies that the state occupied at time point m is dependent on the state occupied at time point $(m-1)$ but not on the state occupied at any previous time point.

The probability, p_{jk}, that a chain in state j at any time point will be in state k at the next time point is called the **transition probability** and is assumed to be unchanging over time. The square *matrix, $\mathbf{P}$, in which p_{jk} is the element in the jth row and kth column, is the **transition matrix**. The elements on each row of $\mathbf{P}$ sum to 1.

If there is a state that, once entered, can never be left, then this is called an **absorbing state** (or **absorbing barrier** or **trapping state**). Such a state will be indicated by a 1 on the leading diagonal of $\mathbf{P}$.

Denote by $p_{jk}^{(n)}$ the probability that a chain in state j at any time point will be in state k at n time points later (so that $p_{jk} = p_{jk}^{(1)}$). If $p_{jk}^{(n)} > 0$, for some n, j, and k, then state k is **accessible** from state j and the two states are **communicating states**. A Markov chain is **irreducible** if all states communicate with each other.

A state is **transient** if, starting from that state, the probability of ever returning to it is < 1; if the probability is equal to 1 then the state is **recurrent** (or **persistent**). A recurrent state with a finite expected time until return is called an **ergodic state**.

In a **stationary chain** $\mathrm{P}(X_m = k)$ is independent of m for all k. Denoting by p_k the probability of being in state k in a stationary chain, this implies that

$$p_k = \sum_j p_j p_{jk}$$

for all k.

In a **time-reversible stationary chain**, the probability of moving from state j at time m to state k at time $(m+1)$ is equal to the probability of moving from state k at time m to state j at time $(m+1)$:

$$p_j p_{jk} = p_k p_{kj},$$

for all j and k.

(*continued overleaf*)

If s refers to a past time, t to the present time, and u to a future time, then, in a **continuous-time Markov chain**, the probability of being in state l at time u, given that the process is in state k at time t, is independent of the state occupied at time s. If the transition probabilities depend only on the time lag $(u - t)$ rather than on u or t, then the process is a **time-homogeneous Markov chain**. Examples of continuous-time Markov chains include *birth-and-death processes, *Brownian motion, *queues, *random walks, and *renewal processes.

Let $P_{jk}(t)$ denote the probability that a time-homogeneous Markov chain in state j at time 0 will be in state k at time t. The **Chapman–Kolmogorov equation** states that

$$P_{jk}(t) = \sum_l P_{jl}(s)P_{lk}(t - s)$$

for all j and k and any s in the interval $(0, t)$.

Markov property *See* MARKOV PROCESS.

Markov random field A characterization of a *spatial process in which the value observed at a location depends only on the values at the immediately neighbouring locations. The form of the dependence is the same for all locations. The *probability distributions for which a Markov random field can exist are specified by the *Hammersley–Clifford theorem. A Markov random field is an undirected **graphical model**.

Markov's inequality Let X be a *random variable for which $|X|$ has expectation v. Then, for any positive constant a,

$$P(|X| \geq a) \leq \frac{v}{a}.$$

See also CHEBYSHEV'S INEQUALITY; HÖLDER'S INEQUALITY; MINKOWSKI'S INEQUALITY.

martingale The sequence of *random variables $X_1, X_2, \ldots$ (with finite expectations) is a martingale if the *conditional expectation of X_{n+1}, given the values of $X_1, X_2, \ldots, X_n$, is given by

$$E(X_{n+1}|X_1 = x_1, X_2 = x_2, \ldots, X_n = x_n) = x_n.$$

An example is provided by letting X_n be the amount of a gambler's winnings after the nth of a sequence of fair games.

matched pairs Pairs of experimental units chosen to resemble each other as closely as possible. Alternatively, the phrase may refer to single experimental units measured at two points in time, or in two different situations. *See also* EXPERIMENTAL DESIGN; HYPOTHESIS TEST; WILCOXON SIGNED-RANK TESTS.

matching coefficient A measure of the similarity between two individuals in a *population. Each individual is assessed with respect to q characteristics that are either present or absent. For each characteristic there are three possible outcomes: 'Absent for both individuals', 'Present for just one of the individuals', 'Present for both individuals'. If the corresponding *counts are denoted by n_0, n_1, and n_2, respectively, with $n_0 + n_1 + n_2 = q$, the **matching coefficient** is the proportion of variables in which the classifications of the two variables agree:

$$\frac{n_0 + n_2}{q}.$$

See also JACCARD'S COEFFICIENT.

Matheron, Georges François Paul Marie (1930–2000) A French geologist and statistician. Matheron was a graduate of the École Polytechnique and the École des Mines de Paris. From 1954 to 1963 he worked for the French Geological Survey in Algeria and France, and then, after a period at the Nancy School of Mines, he returned in 1968 to the École des Mines de Paris. By 1986 this had been reorganized and Matheron was Director of the Centre de Géostatistique. In 1962, in his *Traité de Géostatistique Appliquée* Matheron set out the fundamentals of *geostatistics, including the optimal spatial interpolation procedure that he called *kriging. He was made an Honorary Fellow of the *Royal Statistical Society in 1988.

matrices *See* MATRIX.

matrix An $r \times c$ **matrix** consists of a rectangular array with r rows and c columns, in which the elements are either numbers or algebraic expressions. Example **matrices** (the plural form) are:

$$\begin{pmatrix} 5 & 3 & 2 \\ 1 & 0 & 4 \\ 2 & -1 & 0 \\ 3 & 1 & 1 \end{pmatrix} \qquad \begin{pmatrix} x_{11} & x_{12} & x_{13} & x_{14} \\ x_{21} & x_{22} & x_{23} & x_{24} \end{pmatrix}.$$

When the array is not written out in full, a matrix is usually denoted by a bold-face capital letter, *e.g.* **X**, or by a typical **element** (or **entry**) from the array, shown in curly brackets, e.g. $\{x_{jk}\}$, where x_{jk} is the element in the jth row and kth column of the matrix. If $r = c$ the matrix is **square**.

If a matrix $\{x_{jk}\}$ is multiplied by the real number s, then the result is a matrix in which the element in the jth row and kth column is sx_{jk}. Such a real number s is often referred to as a **scalar**.

Two matrices, **A** and **B**, can be multiplied together only if the number of columns of one matrix is equal to the number of rows of the other matrix. If **A** is an $m \times n$ matrix and **B** is an $n \times p$ matrix then the product **AB** is an $m \times p$ matrix. However, if $p \neq m$ then the product **BA** does not exist. The rule for the construction of the product is as follows. Let e_{jk} denote the element in the jth row and kth column of the product **AB**, with a_{jk} and b_{jk} denoting typical elements in **A** and **B**. Then e_{jk} is given by

$$e_{jk} = \sum_{l=1}^{n} a_{jl}b_{lk}.$$

If the matrices, **A** and **B**, have the same values of r and c and if $a_{jk} = b_{jk}$ for all j and k, then $\mathbf{A} = \mathbf{B}$.

A **diagonal matrix** is a square matrix with all elements equal to 0, except for those on the **leading diagonal** (which runs from top-left to bottom-right). This diagonal is also called the **main diagonal**.

An **identity matrix**, usually denoted by **I**, is a diagonal matrix with on-diagonal elements all equal to 1. The size of an identity matrix may be indicated using a suffix: $\mathbf{I}_n$ is an $n \times n$ identity matrix.

The **transpose** of an $m \times n$ matrix **M** is the $n \times m$ matrix formed by interchanging the elements of the rows and columns of **M**. It is denoted by $\mathbf{M}'$. The jth row of $\mathbf{M}'$ is the transpose of the jth column of **M** and vice versa:

$$\mathbf{M} = \begin{pmatrix} 5 & 3 & 2 \\ 1 & 0 & 4 \end{pmatrix} \qquad \mathbf{M}' = \begin{pmatrix} 5 & 1 \\ 3 & 0 \\ 2 & 4 \end{pmatrix}.$$

If a square matrix **S**, with typical element s_{jk}, is equal to its transpose, $\mathbf{S}'$, then it is a **symmetric matrix** satisfying

$$s_{jk} = s_{kj}, \quad \text{for all } j, k.$$

A square matrix that is not symmetric is an **asymmetric matrix**. If a square matrix **S** satisfies the equation $\mathbf{SS} = \mathbf{S}$ then it is **idempotent**. The product **SS** may be written as $\mathbf{S}^2$. If it exists, the **inverse** of a square matrix, **S**, is denoted by $\mathbf{S}^{-1}$. It satisfies the relations that

$$SS^{-1} = S^{-1}S = I.$$

Only square matrices can have an inverse (but see 'generalized inverse' below). If S^{-1} exists then it will be the same size as S. A matrix that has an inverse is said to be **non-singular** (or **regular**, or **invertible**). A square matrix without an inverse is said to be **singular**.

A **generalized inverse** of the $m \times n$ matrix M is any $n \times m$ matrix M^- satisfying

$$MM^-M = M.$$

If a matrix M is multiplied by its transpose (to give either MM' or $M'M$) then the result is a symmetric matrix. If M is square and the product MM' is an identity matrix, then $M' = M^{-1}$ and M is said to be an **orthogonal matrix**.

A matrix with just one row is called a **row vector**. A matrix with just one column is called a **column vector**. Column vectors are usually denoted with a bold-face lower-case letter, e.g. x; row vectors are written as their transpose, e.g. x'. A vector with a single element (i.e. a 1×1 matrix) is a *scalar.

Vectors multiply together in the same way as matrices (see above). Thus, if v is an $n \times 1$ column vector, and v' is its transpose, then the product vv' is an $n \times n$ symmetric matrix, and the product $v'v$ is a scalar.

The set of $n \times 1$ vectors $v_1, v_2, \ldots v_m$ is **linearly independent** if the only values of the scalars $a_1, a_2, \ldots, a_m$ for which

$$\sum_{j=1}^{m} a_j v_j = 0,$$

where 0 is an $n \times 1$ vector with every element equal to 0, is $a_1 = a_2 = \cdots = a_m = 0$. If the set is not linearly independent then it is **linearly dependent**, in which case there are values for the scalars $a_1, a_2, \ldots, a_m$, not all equal to 0, such that $\sum_{j=1}^{m} a_j v_j = 0$. A linearly independent set with two or more vectors satisfies the requirement that at least one of the vectors, v_k, say, is a **linear combination** of the others, i.e.

$$v_k = \sum_{j \neq k} b_j v_j,$$

for some scalars $b_1, b_2, \ldots, b_k, b_{k+1}, \ldots, b_m$.

The rank of a matrix is the maximum number of linearly independent rows, which is the same as the maximum number of linearly independent columns. Thus the rank of a matrix is equal to that of its transpose. If a matrix has r rows and c columns, with $r \leq c$, then the rank is $\leq r$; if $r > c$ then the rank is $\leq c$. If the rank is equal to the smaller of r and c then the matrix is of **full rank**.

If A is a square matrix, x is a column vector not equal to 0, and λ is a scalar such that

$$Ax = \lambda x$$

then x is an **eigenvector** of A and λ is the corresponding **eigenvalue**. Eigenvectors and eigenvalues are also referred to as **characteristic vectors** and **characteristic values**. If x is the column vector $(x_1 \; x_2 \ldots x_n)'$ and A is an $n \times n$ symmetric matrix with typical element a_{jk}, then the product $x'Ax$, which is a scalar, is described as a **quadratic form** because it is equal to

$$\sum_{j=1}^{n} \sum_{k=1}^{n} a_{jk} x_j x_k,$$

which is a linear combination of all the squared terms (such as x_1^2) and cross-products (such as $x_1 x_2$).

A symmetric matrix A is a **positive definite matrix** if $x'Ax > 0$ for all non-zero x; it is a **positive semi-definite matrix** if $x'Ax \geq 0$ for all x and there is at least one non-zero x for which $x'Ax = 0$.

The **trace** of a square matrix is the sum of the terms on the leading diagonal. The **determinant** of a 2×2 square matrix, A, is written as $|A|$ or $\det(A)$, and is given by

$$|A| = a_{11}a_{22} - a_{12}a_{21}.$$

The determinant of a larger matrix is defined recursively in terms of **cofactors**. The cofactor of the entry a_{jk} is equal to the product of $(-1)^{j+k}$ and the determinant A_{jk} of the matrix obtained by eliminating the jth row and kth column of A. The recursive definition is $|A| = \sum_{j=1}^{n} a_{j1}A_{j1}$. In fact $\sum_{j=1}^{n} a_{jk}A_{jl} = |A|$ if $k = l$ (otherwise the sum is 0). Similarly, $\sum_{k=1}^{n} a_{jk}A_{lk} = |A|$ if $j = l$ and is otherwise 0. Thus, for a 3×3 matrix, A,

$$|A| = a_{11}(a_{22}a_{33} - a_{23}a_{32}) - a_{12}(a_{21}a_{33} - a_{23}a_{31}) + a_{13}(a_{21}a_{32} - a_{22}a_{31}).$$

The eigenvalues of a square matrix A are the roots of the **characteristic equation**

$$\det(A - \lambda I) = 0.$$

Mauchly's test *See* SPHERICITY TEST.

maximum *See* STATIONARY POINT.

maximum flow / minimum cut theorem *See* NETWORK FLOW
PROBLEMS.

maximum likelihood A commonly used method for obtaining an
estimate of an unknown *parameter of an assumed population
distribution. The *likelihood of a data set depends upon the
parameter(s) of the *distribution or *probability density function
from which the *observations have been taken. In cases where one
or more of these parameters are unknown, a shrewd choice as
an estimate would be the value that maximizes the likelihood. This is
the **maximum likelihood estimate (mle)**. Expressions for
maximum likelihood estimates are frequently obtained by
maximizing the *natural logarithm of the likelihood rather than the
likelihood itself (the result is the same). Sir Ronald *Fisher introduced
the method in 1912.

MCB *See* HSU'S MCB TEST.

MCMC *See* MARKOV CHAIN MONTE CARLO METHODS.

McNemar, Quinn (1900–86; b. West Virginia; d. Palo Alto, California)
An American psychologist educated at Stanford University. The author
of the influential text *Psychological Statistics*, he was President of the
Psychometric Society in 1951 and President of the American
Psychological Association in 1964.

McNemar's test A test, introduced by *McNemar in 1947, for use
with paired data when the observed variable is *dichotomous.
Suppose, for example, that competing drugs (A and B) are tested on
pairs of patients. The outcome is either success or failure.
Information about the relative merits of the drugs is provided only
by occasions where just one is successful. Let a denote the number
of pairs in which only drug A succeeds, and b denote the number of
cases in which only drug B succeeds. With a *continuity correction,
McNemar's test statistic is

$$\frac{(|a - b| - 1)^2}{a + b},$$

which, if the drugs are really equally effective, may be taken to be an
observation from a *chi-squared distribution with one *degree of

freedom. The generalization to more than two matched samples is provided by **Cochran's Q test**.

mean (data) The mean of a set of n items of *data $x_1, x_2, \ldots, x_n$ is $(\sum_{j=1}^{n} x_j)/n$, which is the arithmetic mean of the numbers $x_1, x_2, \ldots, x_n$. The mean is usually denoted by placing a bar over the symbol for the variable being measured. If the variable is x the mean is denoted by $\bar{x}$. If the data constitute a *sample from a *population, then $\bar{x}$ may be referred to as the **sample mean**; it is an unbiased estimate of the *population mean.

For example, the numbers of eruptions of the *Old Faithful geyser during the first eight days of August 1978 were 13, 13, 13, 14, 14, 14, 13, and 13. The mean is

$$(13 + 13 + 13 + 14 + 14 + 14 + 13 + 13)/8 = 13.375.$$

If the data are collected in *frequency form so that values $x_1, x_2, \ldots, x_n$ are obtained with frequencies $f_1, f_2, \ldots, f_n$ the mean is

$$\frac{\sum_{j=1}^{n} f_j x_j}{\sum_{j=1}^{n} f_j}.$$

For example, with the eruption data there are just two values, $x_1 = 13$ and $x_2 = 14$. Their respective frequencies are $f_1 = 5$ and $f_2 = 3$, so the mean is $\{(5 \times 13) + (3 \times 14)\}/(3 + 5) = 13.375$.

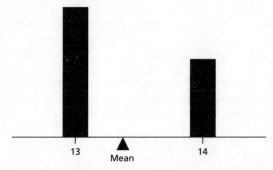

Mean. The data are the numbers of eruptions of the Old Faithful geyser during the first eight days of August 1978. The mean is seen to be the balance point of the observations.

The mean can be interpreted as the centre of gravity, or centre of mass (*see diagram*). If the data are grouped into classes with mid-values $x_1, x_2, \ldots, x_c$ and corresponding class frequencies $f_1, f_2, \ldots, f_c$, an approximate value for the mean of the original data is

$$\frac{\sum_{j=1}^{c} f_j x_j}{\sum_{j=1}^{c} f_j}.$$

mean (random variable) *See* EXPECTED VALUE.

mean absolute deviation (MAD) A *measure of spread. For *observations $x_1, x_2, \ldots, x_n$, with *mean $\bar{x}$ and *median m, the mean absolute deviation about the mean is given by

$$\frac{1}{n} \sum_{j=1}^{n} |x_j - \bar{x}|,$$

and the mean absolute deviation about the median is given by

$$\frac{1}{n} \sum_{j=1}^{n} |x_j - m|.$$

mean absolute percentage error (MAPE) A *statistic that measures the *goodness-of-fit of a *model to a set of *data. Let x_j and $\hat{x}_j$ be *observed and *fitted values; then, for a sample of n observations,

$$\text{MAPE} = \frac{100}{n} \sum_{j=1}^{n} \frac{|x_j - \hat{x}_j|}{|x_j|}.$$

mean deviation An alternative name for the *mean absolute deviation about the mean. For a *data set $x_1, x_2, \ldots, x_n$, with frequencies $f_1, f_2, \ldots, f_n$, and with *mean $\bar{x}$, the mean deviation is

$$\frac{\sum_{j=1}^{n} f_j |x_j - \bar{x}|}{\sum_{j=1}^{n} f_j}.$$

mean square *See* ANOVA.

mean squared error *See* ESTIMATOR.

measure of agreement A single statistic used to summarize the agreement between the rankings or classifications of objects made by two or more observers. Examples are the *coefficient of concordance, *Cohen's kappa, and *rank correlation coefficients.

measure of association A statistic used to summarize the extent to which two *variables are or are not *independent. Examples include the *correlation coefficient, *Cramér's V, *Goodman and Kruskal's lambda, *Yule's Q, and the *rank correlation coefficients. *See* ORDINAL VARIABLE.

measure of goodness-of-fit A statistic that compares the observed data with the *expected values estimated according to some proposed *model. The most common measures are the *chi-squared test statistic and the *likelihood-ratio goodness-of-fit statistic.

measure of location For a set of *data, or a *population, a single number, or data value, which is in some sense in the middle of the data, or the population. *See* MEAN; MEDIAN; MODE.

measure of spread A measures of the extent to which the values of a *variable, in either a *sample or a *population, are spread out. The most commonly used measures of spread are *variance, *standard deviation, *mean deviation, *median absolute deviation, *range, *interquartile range, and *semi-interquartile range. Of these only the variance is not measured in the same units as the observations.

median If a set of numerical *data has n elements and is arranged in order so that either

$$x_1 \leq x_2 \leq \cdots \leq x_n \qquad \text{or} \qquad x_1 \geq x_2 \geq \cdots \geq x_n,$$

then the median is $x_{\frac{1}{2}(n+1)}$ if n is odd, and $\frac{1}{2}(x_{\frac{1}{2}n} + x_{\frac{1}{2}n+1})$ if n is even.

For example, the time intervals (in minutes, to the nearest minute) between the eruptions of *Old Faithful on 1 August 1978 were

$$78, 74, 68, 76, 80, 84, 50, 93, 55, 76, 58, 74, 75.$$

Arranging these thirteen values in order, we get

$$50, 55, 58, 68, 74, 74, \mathbf{75}, 76, 76, 78, 80, 84, 93,$$

to give a median of 75 minutes. For 4 August 1978, there were fourteen inter-eruption times, which arranged in order, were

60, 66, 67, 68, 70, 72, **73**, **75**, 75, 75, 79, 84, 86, 86,

so that the median is $\frac{1}{2}(73 + 75) = 74$.

Alternatively, an approximate value for the median can be read from a
*cumulative frequency graph as the value of the variable corresponding
to a *cumulative relative frequency of 50%.

For a *continuous random variable X, the median m of the distribution
is such that $P(X \le m) = \frac{1}{2}$. For a discrete random variable taking values
$x_1 < x_2 < \cdots$, the median is x_i if $P(X < x_i) < \frac{1}{2}$ and $P(X > x_i) < \frac{1}{2}$, and it is
$\frac{1}{2}(x_i + x_{i+1})$ if $P(X \le x_i) = \frac{1}{2}$ and $P(X \ge x_{i+1}) = \frac{1}{2}$.

median absolute deviation (MAD) A *robust estimate of variability.
For *observations $x_1, x_2, \ldots, x_n$, with *median m, the median absolute
deviation is the median of the differences $|x_1 - m|, |x_2 - m|, \ldots, |x_n - m|$.

median polish A *robust method suggested by *Tukey for estimating
the *mean μ, row *parameters $r_1, r_2, \ldots, r_j$, column parameters
$c_1, c_2, \ldots, c_k$, and residuals ε_{jk} for the *model

$$y_{jk} = \mu + r_j + c_k + \varepsilon_{jk},$$

with $j = 1, 2, \ldots, J$ and $k = 1, 2, \ldots, K$. This *iterative algorithm alternates
row and column operations. Considering the rows first, for each row the
row *median is subtracted from every element in that row. For each
column, the median of the revised numbers is subtracted from every
element in that column. This continues until all medians are 0. The
outcome may vary slightly depending on whether rows or columns are
considered first. In the example, μ is estimated as 30, with $r_1 = -12$,
$c_4 = -7$, and $\varepsilon_{1,4} = 7$ (so that $30 - 12 - 7 + 7 = 18$, the original value).

Original table

13	17	26	18	29
42	48	57	41	59
34	31	36	22	41

$\rightarrow$

Parameter estimates

0	−1	0	7	0	−12
−1	0	1	0	0	18
9	1	−2	−1	0	0
−5	0	8	−7	11	30

Meier, Paul (1924–) An American statistician. Meier was co-author with
*Kaplan of the paper, published in 1958, that introduced the
*Kaplan–Meier estimate of the *survivor function. Meier was a graduate
of Oberlin College in Ohio in 1945, who gained his MA in logic from
Princeton University in 1947, with a PhD from Princeton in 1951. He

joined the staff of Johns Hopkins University in 1952, moving in 1957 to the University of Chicago where he was Professor of Statistics for more than thirty years. He is currently a Professor Emeritus at Columbia University. Meier was President of the *Institute of Mathematical Statistics in 1985.

mesokurtic *See* KURTOSIS.

***M*-estimates** *M*-estimates are *measures of location that are not as sensitive as the *mean to *outlier values. With *observations $x_1, x_2, \ldots, x_n$, the sample mean can be characterized as the value of θ that minimizes $\sum_{j=1}^{n} g(x_j - \theta)$, where $g(u) = u^2$. The sample *median can be characterized in a similar way, though now $g(u) = |u|$.

M-estimates can be characterized in this same way, but the functional forms for g are chosen to be less sensitive to *outlier values. One frequently used alternative as a *measure of location is the **Huber function**:

$$g(u) = \begin{cases} \frac{1}{2}u^2 & |u| \leq k \\ k|u| - \frac{1}{2}k^2 & |u| > k, \end{cases}$$

where k is a **tuning constant** (often set equal to twice the *median absolute deviation).

A second alternative is the **biweight function**:

$$g(u) = \begin{cases} \frac{1}{6}k^2 \left[1 - \left\{ 1 - (u/k)^2 \right\}^3 \right] & |u| \leq k \\ \frac{1}{6}k^2 & |u| > k, \end{cases}$$

where k is again a tuning constant and is here often set equal to seven times the median absolute deviation.

meta-analysis A statistical methodology in which *data from previous tests are considered and analysed together. For example, a series of small experiments may all show only slight signs of the same effect (e.g. that one medicine is better than another), whereas the aggregation of the experiments provides overwhelming evidence. A difficulty with this approach is that experimental conditions and experimental protocols may vary, so that the aggregate outcome may not be a fair reflection of the true situation.

method of maximum likelihood *See* MAXIMUM LIKELIHOOD.

method of moments An alternative to the method of *maximum likelihood as a method of *estimating the *parameters of a *distribution. Each *moment of a distribution can be expressed as a function of the parameters of the distribution, and often this implies that the parameters can be expressed as simple functions of the moments. In such cases, replacing the moments with their sample estimates provides estimates of the population parameters.

For example, the two-parameter *gamma distribution with *probability density function proportional to $x^{\alpha-1}e^{-x/\beta}$ has its first moment, μ_1' (equal to the mean, μ), given by $\mu = \alpha\beta$ and its second central moment, μ_2, given by $\mu_2 = \alpha\beta^2$. Solving these equations, we get $\alpha = \mu^2/\mu_2$ and $\beta = \mu_2/\mu$. The method of moments replaces the theoretical quantities μ and μ_2 with the corresponding sample quantities $\bar{x}$ ($= \frac{1}{n}\sum_j x_j$) and $\frac{1}{n}\sum_j (x_j - \bar{x})^2$ so that, for example, the estimator of α is $\tilde{\alpha}$ given by

$$\tilde{\alpha} = \frac{\left(\sum_j x_j\right)^2}{\left\{n\sum_j (x_j - \bar{x})^2\right\}}.$$

metric A generalized measure of distance. *See* l_p-METRIC.

Metropolis–Hastings algorithm *See* MARKOV CHAIN MONTE CARLO METHODS.

mgf *See* MOMENT-GENERATING FUNCTION.

mid-P With a *continuous random variable, X, it is evident that the sum of the usual *upper-tail probability, $P(X \geq x)$, and the *lower-tail probability, $P(X \leq x)$, is 1. However, with a *discrete random variable this is equal to $1 + P(X = x)$. This suggests that, for example, the calculation of the upper-tail probability should then use mid-P, defined as

$$P(X > x) + \frac{1}{2}P(X = x).$$

midrange For a set of numerical data the midrange is the arithmetic *mean of the greatest and least values in the set. It is sometimes used as a *measure of location, but it is unreliable, since, by definition, it depends only on extreme values. *See* OUTLIER.

midspread *See* INTERQUARTILE RANGE.

Mills's ratio For a *continuous random variable, the ratio, $R(x)$, of the *survivor function to the *probability density function:

$$R(x) = \frac{1 - F(x)}{f(x)}.$$

Investigations of the ratio have mostly been undertaken with respect to the *standard normal distribution, for which, for $x > 2$,

$$R(x) \approx \frac{4}{3x + \sqrt{x^2 + 8}} - \frac{2}{x^7}.$$

minimal cut vector; minimal path vector *See* NETWORK FLOW PROBLEMS; RELIABILITY THEORY.

minimax *See* STATIONARY POINT.

minimum *See* STATIONARY POINT.

minimum chi-squared A method of *estimation in which the value chosen for the *parameter estimate is the value that minimizes the value of the test statistic of the *chi-squared test for goodness-of-fit.

MINITAB A statistical package particularly designed for teaching purposes.

Minkowski, Hermann (1864–1909; b. Kaunas, Lithuania; d. Göttingen, Germany) A Lithuanian-born mathematician. Minkowski moved to Germany in 1872 and was educated at the University of Königsberg, gaining his PhD in 1885. His academic career took him successively to posts at the Universities of Bonn (1885), Zurich (1896), and Göttingen (1902). Minkowski had a passion for pure mathematics and his work underpinned that of Einstein on relativity.

Minkowski's inequality For *random variables X and Y,

$$\{E(|X + Y|^a)\}^{\frac{1}{a}} \le \{E(|X^a|)\}^{\frac{1}{a}} + \{E(|Y^a|)\}^{\frac{1}{a}}$$

for any positive constant a. *See also* CHEBYSHEV'S INEQUALITY; HÖLDER'S INEQUALITY; MARKOV'S INEQUALITY.

mixed effects design *See* EXPERIMENTAL DESIGN.

mixture distribution A *distribution made up of two or more component distributions. For example, suppose that light bulbs of type A have an *exponential lifetime with *parameter α and light bulbs of type B have an exponential lifetime with parameter β. A box contains a mixture of the two types of bulb, with a proportion p being of type A. Let

X be the lifetime of a randomly selected bulb. The *probability density function f of X is given by

$$f(x) = p\frac{1}{\alpha}e^{-\alpha x} + (1-p)\frac{1}{\beta}e^{-\beta x}, \qquad x > 0.$$

mle *See* MAXIMUM LIKELIHOOD.

mobility table A square *contingency table in which the rows and columns have equivalent classifications but the columns refer to a later time point than the rows. A **social mobility table** usually refers to the social classes (or occupations) of successive generations. **Voter mobility tables** cross-classify the votes of individuals in successive elections.

modal class *See* MODE.

modal frequency *See* MODE.

mode A data value, in a set of *categorical data, whose *frequency is not less than that of any other data value. For a set of numerical data, a mode is a data value whose frequency is not less than the frequency of neighbouring values. For a set of grouped numerical data, a **modal class** is a class whose frequency is not less than the frequency in neighbouring classes. The **modal frequency** is the frequency with which the mode occurs, or the frequency in the modal class.

For example, the time intervals (in minutes, to the nearest minute) between the eruptions of *Old Faithful on 1 August 1978 were

78, 74, 68, 76, 80, 84, 50, 93, 55, 76, 58, 74, 75.

The values 74 and 76 both occur twice, and the remaining values occur just once. The values 74 and 75 are the two modes, both having a modal frequency of 2. Combining the data from 1–8 August, we have the following summary table:

Time interval	40–49	50–59	60–69	70–79	80–89	90–99
Frequency	5	21	10	37	30	4

The modal classes are 50–59 (with modal frequency 21) and 70–79 (with modal frequency 37).

For a *discrete random variable a mode is a value whose *probability is not less than that of its neighbours. For a *continuous random variable

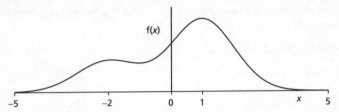

Mode. The multimodal probability density function illustrated is for a mixture distribution of two normal distributions, each with unit variance. One is centred on $x = 1$ and the other on $x = -2$.

a mode is a value such that the *probability density function has a local maximum. If there is only one mode the distribution is **unimodal**. If there is more than one mode the distribution is **multimodal**. If there are two modes it is **bimodal**. The word 'mode' was coined by Karl *Pearson in 1895.

model A simple description of a probabilistic process that may have given rise to observed *data. For example, if the data consist of the numbers shown by a fair die during a game of Snakes and Ladders, then a simple model would state that for each roll, and *independent of the outcomes of other rolls, the *distribution of the number shown is a *discrete uniform distribution, on $1, 2, \ldots, 6$.

Models form the bedrock of *Statistics. Specific distributions are often invoked. Many types of models are mentioned in this dictionary.

modulus *See* ABSOLUTE VALUE.

moment (**uncorrected moment**) For a *random variable X the rth moment (about the origin) is defined to be the expectation of X^r, where r is a non-negative integer. It is usually denoted by μ'_r. So $\mu'_0 = 1$ and $\mu'_1 = \mu$, the mean of the *distribution of X.

The rth **moment about the mean** (or **central moment** or **corrected moment**) is defined to be the expectation of $(X - \mu)^r$ and is usually denoted by μ_r. Thus $\mu_1 = 0$ and $\mu_2 = \sigma^2$, the *variance of X. The moments about the mean can be expressed as linear combinations of the uncorrected moments, for example:

$$\mu_2 = \mu'_2 - \mu^2,$$
$$\mu_3 = \mu'_3 - 3\mu'_2\mu + 2\mu^3,$$
$$\mu_4 = \mu'_4 - 4\mu'_3\mu + 6\mu'_2\mu^2 - 3\mu^4.$$

Either set of moments can also be expressed in terms of linear combinations of simple functions of the *cumulants. It should be noted that, for some distributions, μ_r' and μ_r may exist only for small values of r.

moment about the mean *See* MOMENT.

moment coefficient of skewness If the second and third
*moments about the mean of a *distribution are μ_2 and μ_3, respectively,
the moment coefficient of skewness is $\mu_3/\sqrt{\mu_2^3}$. It is invariant under
change of scale and change of origin.

moment-generating function (mgf) The moment-generating
function of a *random variable X involves a variable, usually denoted by
t, and is defined to be the expectation of e^{tX}. It is usually denoted by $M_X(t)$,
or $M(X, t)$. Expanding $M_X(t)$ in powers of t gives

$$M_X(t) = \sum \mu_r' \frac{t^r}{r!},$$

where μ_r' is the rth *moment of X about the origin. In virtue of the
*Maclaurin series, μ_r' is the value, when $t = 0$, of the rth derivative with
respect to t of $M_X(t)$. *See also* PROBABILITY-GENERATING FUNCTION.

Monte Carlo methods Methods that use *pseudo-random numbers
in order to determine the properties of some function or set of functions.
For example, *acceptance–rejection methods use pseudo-random
numbers to evaluate integrals, *simulation makes use of pseudo-random
numbers to study the outcome of some complex system, and *Monte
Carlo tests use pseudo-random numbers to decide on the outcome of
*hypothesis tests. The accuracy of the methods generally improves with
an increase in the number of pseudo-random numbers used.

Monte Carlo test A procedure suggested by *Barnard in 1963, for
testing whether a set of *data is consistent with a *null hypothesis. It is
appropriate for a situation where the theoretical *distribution of the
test statistic is unknown, although the distribution of the individual
observations is known. The test procedure is to use *Monte Carlo
methods to generate 99 (say) further data sets of the same size as the true
data, under the conditions defined by the null hypothesis. The value for
the test statistic is calculated for each data set and the distribution of
these values is examined. If the value of the test statistic for the actual
data is similar to the values obtained from the artificial data sets, then the
null hypothesis is accepted, whereas if it is extreme the hypothesis is
rejected.

Mood, Alexander McFarlane (1913–) An American mathematical statistician. As an undergraduate at the University of Texas, Mood had studied physics. On moving to Princeton University, he obtained (in 1940) his doctorate in *Statistics under the supervision of *Wilks. After a period on the faculty at the University of Texas, he rejoined Princeton University, where he is now Emeritus Professor of Management. Mood was co-author with *Graybill of the 1963 textbook *Introduction to the Theory of Statistics* which for many years was the recommended introductory textbook. He was President of the *Institute of Mathematical Statistics in 1957 and was awarded the *Wilks Medal of the *American Statistical Association in 1979.

Mood's dispersion test A two-sample *non-parametric test of the *null hypothesis of a common *variance, introduced by *Mood in 1954. The *population *distribution is assumed to be *symmetric, but need not be *normal. The test is based on replacing the original values with their *ranks in the combined sample. Suppose that the samples have sizes m and n, with $m + n = N$. The test statistic is M, given by

$$M = \sum_{j=1}^{m} \left\{ r_j - \frac{1}{2}(N+1) \right\}^2,$$

where r_j is the overall rank in the combined sample of the jth observation in the sample of size m. If m and n are reasonably large, then the transformed statistic z is an observation from an approximate *standard normal distribution, where

$$z = \frac{M - \frac{1}{12}m(N^2 - 1)}{\sqrt{\frac{1}{180} mn(N+1)(N^2 - 4)}}.$$

If there are tied ranks then the variance needs a correction.

Moran, Patrick Alfred Pierce (1917–88; b. Sydney, Australia) An Australian statistician. Moran was educated at the Universities of Sydney and Cambridge and was a researcher at Oxford University from 1946 to 1951, before returning to Australia. He was the founding Professor of Statistics at the Australian National University, from which he retired in 1982. He was elected to the Australian Academy of Science in 1962 and was elected a Fellow of the Royal Society in 1975. He was the first President of the Australian Statistical Society (now the *Statistical Society of Australia Inc.) in 1963 and was awarded the Society's Pitman

Medal in 1982. He was President of the Australian Mathematical Society in 1976.

Moran's *I* A *statistic, introduced by *Moran in 1950, that measures spatial *autocorrelation using information only from specified pairs of spatial *observations. Suppose there are n locations, and the observed value at location j is x_j and the overall mean value is $\bar{x}$. Let w_{jk} be 1 if the comparison of the value at location j to the value at location k is of interest, and let $w_{jk} = 0$ otherwise. By definition, $w_{jj} = 0$, for all j. Usually, the only comparisons of interest are those between immediate neighbours. The statistic is given by

$$I = n \sum_j \sum_k w_{jk}(x_j - \bar{x})(x_k - \bar{x}) \bigg/ \left\{ \sum_j \sum_k w_{jk} \right\} \left\{ \sum_j (x_j - \bar{x})^2 \right\}.$$

morbidity rate The *incidence rate of persons in a *population who become clinically ill during the period of time stated.

mortality rate (death rate) The number of deaths occurring in a *population during a given period of time, usually a year, as a *proportion of the number in the population. Usually the mortality rate includes deaths from all causes and is expressed as deaths per 1000. A disease-specific (or age-specific or sex-specific) mortality rate includes only deaths associated with one disease (or age or sex) and is usually reported as deaths per 100 000 people. The mortality rate may be standardized when comparing mortality rates over time, or between countries, to take account of differences in the population.

mosaic display A display (*see following diagram*) that highlights departures from *independence in a two-variable cross-classification. The display, which is asymmetric, emphasizes the variations in the *conditional probability of the categories of one variable, given the category of the second variable. For an alternative display that treats the variables in a symmetrical fashion, *see* COBWEB DIAGRAM.

most powerful test A *test of a *null hypothesis which has greater power than any other test for a given *alternative hypothesis. *See also* UNIFORMLY MOST POWERFUL TEST.

mother wavelet *See* WAVELET.

mover–stayer model A model for a *square table which classifies individuals into a group whose new location is *independent of their old

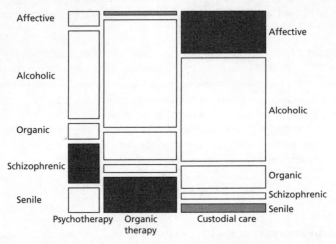

Mosaic display. The areas of the tiles are proportional to the cell frequencies. Dark shading indicates a large positive residual, pale shading a large negative residual.

location and a group who never move. The 'locations' may be geographical or they may refer to social class, occupation, or views on some subject. The model can provide a good fit to the data without actually making any sense when its implications are examined more closely.

moving average A method of smoothing a *time series to reduce the effects of *random variation and reveal any underlying *trend or *seasonality. For the time series $x_1, x_2, \ldots, x_t$ the simple three-point moving average would replace the value of $x_k, k = 2, 3, \ldots, t - 1$, with

$$\frac{1}{3}(x_{k-1} + x_k + x_{k+1}).$$

Often, different weights are used, as in this five-point moving average which could be used for $k = 3, 4, \ldots, t - 2$:

$$\frac{1}{12}(x_{k-2} + 3x_{k-1} + 4x_k + 3x_{k+1} + x_{k+2}).$$

The four-point moving averages (appropriate for quarterly data) are

$$\frac{1}{4}(x_1 + x_2 + x_3 + x_4), \qquad \frac{1}{4}(x_2 + x_3 + x_4 + x_5), \ldots.$$

Twelve-point moving averages are similarly defined and are appropriate for monthly data. For a cycle with an even period, e.g. quarterly or

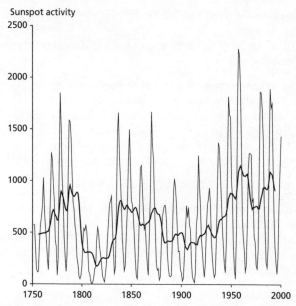

Moving average. The graph shows annual sunspot activity (in standardized units) from 1750 to 2000. There is a strong cycle with a period of around 11 years. The solid line is the 11-year moving average. This removes the obvious cycle but reveals longer-scale fluctuations.

monthly data, the **centred moving averages** are the arithmetic means of the successive moving averages as defined above. For example, in the case of quarterly data the first centred moving average is

$$\frac{1}{8}(x_1 + 2x_2 + 2x_3 + 2x_4 + x_5).$$

The advantage of these centred moving averages is that the resulting values are associated with a time point rather than the midpoint of the interval between two successive time points.

A graph of moving averages against time may show changes against time which are obscured by cyclical effects. A line of best fit to the moving averages is a **trend line**, and its slope is the **trend**. The trend line may be used to forecast future values (in the short term). For example, for monthly data the average deviation of the January data from the trend line can be used as an estimate of the future deviation of the January deviation from the trend line. The deviation can be measured as either a difference or a ratio.

Note that the use of moving averages can introduce spurious *cycles (*see* SLUTZKY–YULE EFFECT).

moving average models (moving average process) *Models for a *time series with a constant *mean (taken as 0). Let $x_1, x_2, \ldots$ be successive values of the *random variable X, measured at regular intervals of time and let $\varepsilon_1, \varepsilon_2, \ldots$ denote the corresponding *random errors. A pth-order moving average model with *parameters $\alpha_1, \alpha_2, \ldots, \alpha_p$ relates the value at time $j\, (\geq p + 1)$ to the preceding p error values by

$$x_j = \alpha_p \varepsilon_{j-p} + \alpha_{p-1}\varepsilon_{j-p+1} + \cdots + \alpha_1 \varepsilon_{j-1} + \varepsilon_j.$$

Such a model is written in brief as MA(p). The errors are presumed to be *independent and to have *mean 0 and hence the X-variables also have mean 0. Moving average models can also be expressed as *autoregressive models. Models combining both type of process include *ARMA models and *ARIMA models.

MSE *See* ESTIMATOR.

mu (μ) The Greek letter used to denote the *population mean.

multicollinearity *See* MULTIPLE REGRESSION MODEL.

multidimensional contingency table *See* CONTINGENCY TABLE.

multidimensional scaling A *multivariate method, resembling *principal components analysis, that gives a graphical representation of the main characteristics of a *proximity matrix using only a few dimensions. The success of the method is assessed by comparing the 'distances' between individuals as recorded by the proximity matrix with the (suitably scaled) *Euclidean distances as suggested by the method; the resulting *measure of goodness-of-fit is called the **stress** of the solution—a low value suggests a good solution.

As an example, the following proximity matrix shows, for seven European countries, the numbers of food products (out of twenty) that were found to similar extents (i.e. common in both countries or scarce in both countries) in each of the pairs of countries.

	GREAT BRITAIN	IRELAND	NETHER- LANDS	BELGIUM	FRANCE	ITALY	SPAIN
Great Britain	20	11	11	6	7	2	6
Ireland	11	20	11	10	11	10	10
Netherlands	11	11	20	12	11	6	6
Belgium	6	10	12	20	12	11	9
France	7	11	11	12	20	10	12
Italy	2	10	6	11	10	20	13
Spain	6	10	6	9	12	13	20

Multidimensional scaling of countries and food products. The results may appear surprising, but they reflect the (very crude) information from which they were derived. The (unlabelled) axes reflect different aspects of eating behaviour.

The biggest difference is between Great Britain and Italy: this is reflected in a two-dimensional graph of the positions of the countries.

multilevel models A multilevel model attempts to model all the components in a hierarchy of nested effects (*see* NESTED DESIGN). For example, the attitudes of schoolchildren, or the recovery rates of hospital patients, will be influenced at a number of levels. All children in the same class will have undergone almost identical teaching experiences. Their class will have had similar experiences to other classes in that school. The school may be similar to other schools in the neighbourhood. The neighbourhood may be ... (and so on). A multilevel model attempts to model all the components in the hierarchy of effects.

multimodal *See* MODE.

multinomial distribution The extension of the binomial distribution to the case of more than two classes. For example, suppose that the *probabilities of classes $1, 2, \ldots, m$ are $p_1, p_2, \ldots, p_m$ with $\sum_{j=1}^{m} p_j = 1$. Let X_j denote the number, in a sample of size n, that are in class j, for $j = 1, 2, \ldots, m$. The *random variables $X_1, X_2, \ldots, X_m$ have a *multivariate distribution given by

$$P(X_1 = n_1, X_2 = n_2, \ldots, X_m = n_m) = \frac{n!}{n_1! n_2! \ldots n_m!} p_1^{n_1} p_2^{n_2} \cdots p_m^{n_m},$$
$$0 \le n_1, n_2, \ldots, n_m \le n,$$

with $\sum_{j=1}^{m} n_j = n$. The random variable X_j has expectation np_j and *variance $np_j(1 - p_j)$ and the *covariance of X_j and X_k $(j \ne k)$ is $-np_j p_k$.

multiple bar chart A *bar chart for comparing the *frequencies of a *categorical variable for two or more *populations.

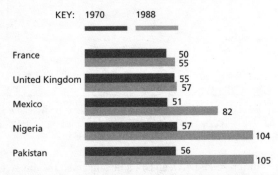

Multiple bar chart. This chart compares the population sizes in 1970 and 1988 (figures are in millions) for five countries. The small increases in the two European countries contrast sharply with the increases in countries on other continents.

multiple-choice question *See* CLOSED QUESTION.

multiple comparison tests Tests suitable for the simultaneous testing of several hypotheses concerning the equality of three or more population means. When *samples have been taken from several *populations, as a preliminary to the more general question of whether the populations differ, there is the simpler question of whether they have different *means. With m populations, and the mean of population j denoted by μ_j, the *null hypothesis is

$$H_0: \mu_1 = \mu_2 = \cdots = \mu_m.$$

If H_0 is rejected, then we wish to know where the differences lie. We could rewrite the null hypothesis in the equivalent form:

$$H_0: \mu_1 = \mu_2, \quad \mu_1 = \mu_3, \ldots, \mu_{m-1} = \mu_m.$$

In this form it can be seen that there are $\frac{1}{2}m(m-1)$ pairs of populations that could be compared. Some tests, such as *Tukey's test (the honestly significant difference test), preserve an overall significance level for all these possible tests. *Duncan's test is the related *multiple stage test based on testing for homogeneity successively smaller subsets of the original set of m populations. The *Newman–Keuls test is a variant of Duncan's test that provides a specified significance level for each separate test.

If one population corresponds to a *control treatment, then interest focuses on the $(m-1)$ comparisons involving this population, and *Dunnett's test is appropriate. *Hsu's MCB test restricts attention to comparisons involving the best treatment. If the comparisons of interest are contrasts of more than two population means, then *Scheffé's test is appropriate.

multiple range test *See* MULTIPLE STAGE TEST.

multiple regression model The extension of the *linear regression model to the case where there is more than one explanatory variable. For the case of p X-variables and n *observations the model is

$$E(Y_j) = \beta_0 + \beta_1 x_{1j} + \beta_2 x_{2j} + \cdots + \beta_p x_{pj}, \quad j = 1, 2, \ldots, n,$$

where $\beta_0, \beta_1, \ldots, \beta_p$ are unknown *parameters. An equivalent presentation is

$$Y_j = \beta_0 + \beta_1 x_{1j} + \beta_2 x_{2j} + \cdots + \beta_p x_{pj} + \varepsilon_j, \quad j = 1, 2, \ldots, n,$$

where $\varepsilon_1, \varepsilon_2, \ldots, \varepsilon_n$ are *random errors.

In *matrix terms the model is written as

$$\mathrm{E}(\mathbf{Y}) = \mathbf{X}\boldsymbol{\beta},$$

where $\mathbf{Y}$ is the $n \times 1$ column *vector of *random variables, $\boldsymbol{\beta}$ is the $(p+1) \times 1$ column vector of unknown parameters, and $\mathbf{X}$ is the $n \times (p+1)$ **design matrix** given by

$$\mathbf{X} = \begin{pmatrix} 1 & x_{11} & x_{21} & \dots & x_{p1} \\ 1 & x_{12} & x_{22} & \dots & x_{p2} \\ \vdots & \vdots & & \vdots & \\ 1 & x_{1n} & x_{2n} & \dots & x_{pn} \end{pmatrix}.$$

Equivalently,

$$\mathbf{Y} = \mathbf{X}\boldsymbol{\beta} + \boldsymbol{\varepsilon},$$

where $\boldsymbol{\varepsilon}$ is an $n \times 1$ column vector of random errors.

Usually it is assumed that the random errors, and hence the Y-variables, are *independent and have common *variance σ^2. In this case the *ordinary least squares estimates of the β-parameters are obtained by solving the set of simultaneous equations (the **normal equations**) which, in matrix form, are written as

$$\mathbf{X}'\mathbf{X}\hat{\boldsymbol{\beta}} = \mathbf{X}'\mathbf{y},$$

where X' is the *transpose of X and $\mathbf{y}$ is the $n \times 1$ column vector of observations $(y_1 \ y_2 \ \dots \ y_n)'$. The matrix $\mathbf{X}'\mathbf{X}$ is a symmetric matrix. If it has an inverse then the solution is

$$\hat{\boldsymbol{\beta}} = (\mathbf{X}'\mathbf{X})^{-1}\mathbf{X}'\mathbf{y}.$$

The *variance–covariance matrix of the *estimators of the β-parameters is $\sigma^2(\mathbf{X}'\mathbf{X})^{-1}$. The **Gauss–Markov theorem** shows that $\hat{\boldsymbol{\beta}}$ is the minimum variance linear unbiased estimate of $\boldsymbol{\beta}$.

If the random errors are not independent or have unequal variances then ordinary least squares is inappropriate and *weighted least squares may be appropriate. If the random errors are influenced by the X-variables, then a possible approach is to identify W-variables that are highly correlated with the X-variables but not with the errors. These W-variables are called **instrumental variables**. The subsequent analysis, commonly used in *econometrics, uses **two-stage least squares**, in which the first stage involves the regression of the X-variables on the W-variables, to obtain fitted $\hat{X}$ values:

$$\hat{X} = W(W'W)^{-1}W'X.$$

These values are then used in place of the X-values in the regression for Y to give the estimates $\boldsymbol{\beta}_2$:

$$\boldsymbol{\beta}_2 = (\hat{X}'\hat{X})^{-1}\hat{X}'y.$$

A perennial problem with multiple regression models is deciding which of the X-variables are really needed and which are redundant. For a discussion of this problem, *see* MODEL SELECTION PROCEDURES.

A related problem is that of **collinearity** (or **multicollinearity**). Suppose, for example, that the two explanatory variables X_j and X_k approximately satisfy the relation $X_j = a + bX_k$. In this case a multiple regression model that involves both X_j and X_k will run into problems, since either variable would be nearly as effective on its own. The situation can arise with other numbers of X-variables: in all cases the result is that the matrix $X'X$ is nearly singular. The equation $X'X\hat{\boldsymbol{\beta}} = X'y$ is then said to be **ill-conditioned**. One suggestion is to replace the usual parameter estimate by

$$\hat{\boldsymbol{\beta}}^{\star} = (X'X + kI)^{-1}X'y,$$

where k is a constant and I is an identity matrix. This technique is called **ridge regression**. A plot of an element of $\hat{\boldsymbol{\beta}}^{\star}$ against k is called a **ridge trace** and can be used to determine the appropriate value for k. The estimates of the β-parameters obtained using this method are biased but have smaller variance than the ordinary least squares estimates.

For methods of evaluating the fit of a multiple regression model, *see* REGRESSION DIAGNOSTICS. *See also* AIC; MALLOWS'S C_p; STEPWISE PROCEDURES.

multiple stage test A *multiple comparison test in which the comparisons are made in stages and testing stops when a non-significant result is obtained. This procedure results in increased power. Examples are *Duncan's test and the *Newman–Keuls test. These two tests are also called **multiple range tests** because they use the *Studentized range distribution.

multiplication law for probabilities Law for probabilities stating that if A and B are *independent events then

$$P(A \cap B) = P(A) \times P(B),$$

and, in the case of n independent events, $A_1, A_2, \ldots, A_n$,

$$P(A_1 \cap A_2 \cap \cdots \cap A_n) = P(A_1) \times P(A_2) \times \cdots \times P(A_n).$$

This is a special case of the more general **law of compound probability**, which holds for events that may not be independent. In the case of two events, A and B, this law states that

$$P(A \cap B) = P(A) \times P(B|A) = P(B) \times P(A|B).$$

For three events, A, B, and C, this becomes

$$P(A \cap B \cap C) = P(A) \times P(B|A) \times P(C|A \cap B).$$

There are six ($= 3!$) alternative right-hand sides, for example $P(C) \times P(A|C) \times P(B|C \cap A)$. The generalization to more than three events can be inferred. For definitions of symbols, *see* INTERSECTION; CONDITIONAL PROBABILITY.

multivariate analysis of variance (MANOVA) *ANOVA with a *multivariate distribution for the response variable. With g groups and n_j *observations in the jth group, the three basic *matrix quantities involve various sums of squares and cross-products. As with ANOVA the idea is to divide the total variation into a contribution explained by differences between the groups and a residual contribution resulting from variation within groups. The corresponding matrices are $\mathbf{T}$, $\mathbf{B}$, and $\mathbf{W}$, defined by

$$\mathbf{T} = \sum_{j=1}^{g} \sum_{k=1}^{n_j} \left(\mathbf{x}_{jk} - \bar{\mathbf{x}}\right)\left(\mathbf{x}_{jk} - \bar{\mathbf{x}}\right)'$$

$$\mathbf{B} = \sum_{j=1}^{g} n_j\left(\bar{\mathbf{x}}_j - \bar{\mathbf{x}}\right)\left(\bar{\mathbf{x}}_j - \bar{\mathbf{x}}\right)'$$

$$\mathbf{W} = \sum_{j=1}^{g} \sum_{k=1}^{n_j} \left(\mathbf{x}_{jk} - \bar{\mathbf{x}}_j\right)\left(\mathbf{x}_{jk} - \bar{\mathbf{x}}_j\right)',$$

where $\bar{\mathbf{x}}_j$ is the column vector of means for the jth group, and $\bar{\mathbf{x}}$ is the column vector of overall means.

The most usual test of the hypothesis that the groups come from a single population is **Wilks's lambda (Λ)**, given by

$$\Lambda = \frac{\det(\mathbf{W})}{\det(\mathbf{T})},$$

where $\det(\mathbf{M})$ denotes the determinant of the matrix $\mathbf{M}$. Alternative tests are provided by the **Hotelling–Lawley trace** (the sum of the eigenvalues of $\mathbf{BW}^{-1}$), the **Pillai–Bartlett trace** (the sum of the eigenvalues of $\mathbf{BT}^{-1}$), and **Roy's maximum root** (the largest eigenvalue of $\mathbf{BW}^{-1}$).

multivariate data Data collected on several variables for each sampling unit. For example, if we collect information on weight (w), height (h), and shoe size (s) from each of a random sample of individuals, then we would refer to the triples (w_1, h_1, s_1), (w_2, h_2, s_2), ... as a set of multivariate data.

multivariate distribution A description of the possible values, and corresponding *probabilities, of two or more (usually three or more) associated *random variables. Examples include the *multinomial and *multivariate normal distributions. The special case of two variables is described as a *bivariate distribution.

multivariate distribution function For *random variables $X_1, X_2, \ldots, X_n$, the multivariate distribution function $F(x_1, x_2, \ldots, x_n)$ is the *joint probability that $X_1 \leq x_1$ and $X_2 \leq x_2$ and ... and $X_n \leq x_n$.

multivariate methods Methods for analysing *multivariate data. Probably the most used procedure is *multiple regression. Other methods include *cluster analysis, *discriminant analysis, *factor analysis, *multidimensional scaling, and *principal components analysis.

multivariate normal distribution *Random variables $X_1, X_2, \ldots, X_n$ (each with ranges from $-\infty$ to ∞) have a multivariate normal distribution if their *joint probability density function f is given by

$$f(\mathbf{x}) = \frac{\exp\left\{-\frac{1}{2}(\mathbf{x} - \boldsymbol{\mu})'\boldsymbol{\Sigma}^{-1}(\mathbf{x} - \boldsymbol{\mu})\right\}}{\sqrt{(2\pi)^n \det(\boldsymbol{\Sigma})}},$$

where $\mathbf{x}$ is the $n \times 1$ *vector of values, $\boldsymbol{\mu}$ is the $n \times 1$ vector of *means, $\boldsymbol{\Sigma}$ is the $n \times n$ *variance–covariance matrix, and $\det(\boldsymbol{\Sigma})$ is the determinant of $\boldsymbol{\Sigma}$.

For the special case of the **bivariate normal distribution**, with random variables X and Y, the joint probability density function f is given by,

$$f(x, y) = \frac{1}{2\pi\sigma_x\sigma_y\sqrt{1 - \rho^2}} \exp\left[-\frac{1}{2(1 - \rho^2)}\left\{\frac{(x - \mu_x)^2}{\sigma_x^2} - 2\rho\frac{(x - \mu_x)(y - \mu_y)}{\sigma_x\sigma_y} + \frac{(y - \mu_y)^2}{\sigma_y^2}\right\}\right],$$

where the mean and variance of X are μ_x and σ_x^2, the mean and variance of Y are μ_y and σ_y^2 and ρ is the *correlation coefficient between the two variables.

mutually exclusive events Two events (*see* SAMPLE SPACE), A and B, are mutually exclusive if they cannot both occur simultaneously, i.e. if $A \cap B = \phi$, the empty set. In this case $P(A \cap B) = 0$. The n events $A_1, A_2, \ldots, A_n$ are mutually exclusive if, for all $j \neq k$, $A_j \cap A_k = \phi$.

mutually independent events *See* INDEPENDENT EVENTS.

National Lottery A nationwide opportunity to gamble, in which the probabilities of wins of various types are known in advance. The current United Kingdom National Lottery offers prizes for correctly choosing three or more of the balls drawn from a pool of 49 individually numbered balls. Assuming that each ball is equally likely to be drawn—which appears to be the case at the time of writing—the *probability of winning the lowest prize (which requires three correctly identified balls) is

$$\frac{\binom{6}{3}\binom{43}{3}}{\binom{49}{6}} = \frac{8\ 815}{499\ 422} \approx \frac{1}{57}.$$

The probability of identifying all six balls is

$$\frac{\binom{6}{6}\binom{43}{0}}{\binom{49}{6}} = \frac{1}{13\ 983\ 816}.$$

If all six numbers are identified correctly, then the amount won depends on the number of other individuals who have selected the same six numbers. A good choice would therefore be an unpopular set of numbers: perhaps, 44, 45, 46, 47, 48, and 49.

natural logarithm The natural logarithm of a positive real number a is denoted by $\ln a$ and defined by

$$\ln a = \int_1^a \frac{1}{x}\,dx,$$

so it is the (signed) area enclosed by the curve $y = 1/x$, the x-axis, and the lines $x = 1$ and $x = a$. If $a > 1$ then $\ln a > 0$, and if $a < 1$ then $\ln a < 0$. Also $\ln 1 = 0$. The natural logarithm is related to the *exponential function by $\exp(\ln x) = x$ for all $x > 0$ and $\ln\{\exp(x)\} = x$ for all values of x. It is also referred to as the *logarithm to base e. For historical reasons a natural logarithm is sometimes referred to as a Napierian logarithm, after the Scottish mathematician John Napier (1550–1617).

natural parameter *See* EXPONENTIAL FAMILY.

nearest-neighbour methods 1. In the context of *point processes, methods underlying *tests of randomness. If locations are positioned randomly in the plane, at a density λ per unit area, then the squared distance from a location to its nearest neighbour will have an *exponential distribution with *mean $1/(\pi\lambda)$, and the distance to its kth nearest neighbour will have a *gamma distribution with mean $k/(\pi\lambda)$.

2. In the context of *experimental design, methods used to improve estimates of plot yields by using neighbouring values as an indicator of the fertility of a plot of interest.

3. In the context of *cluster analysis and *discriminant analysis, methods used to determine from which *population, of a given set of populations, an observation has arisen.

negative binomial distribution (Pascal distribution) For trials classified as 'success' or 'failure', the *distribution of X, the number of trials required in order to obtain n successes. The trials are presumed to be *independent and it is assumed that each trial has the same *probability of success, p ($\neq 0$ or 1). The *probability function is

$$P(X = r) = \binom{r - 1}{n - 1} p^n (1 - p)^{r-n}, \qquad r = n, n + 1, \ldots.$$

The *mean of this distribution is n/p and the *variance is $n(1 - p)/p^2$.

Special cases of the distribution were discussed by *Pascal in 1679. The name 'negative binomial' arises because the probabilities are successive terms in the *binomial expansion of $(P - Q)^{-n}$, where $P = 1/p$ and $Q = (1 - p)/p$. Writing $Y = X - n$, an equivalent form for the distribution is

$$P(Y = r) = \binom{n + r - 1}{n - 1} p^n (1 - p)^r, \qquad r = 0, 1, 2, \ldots.$$

The variable X may be regarded as the sum of n independent *geometric variables, each with *parameter p. The case $n = 1$ therefore corresponds to the *geometric distribution.

The **arc-sinh transformation**, suggested by *Anscombe in 1948,

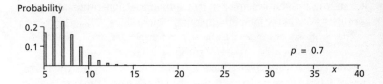

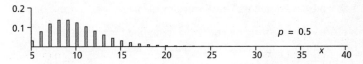

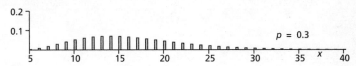

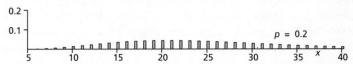

Negative binomial distribution. In each case $n = 5$; the shape is critically dependent upon the value of p.

$$S = \left(\sqrt{n - \frac{1}{2}} \right) \sinh^{-1} \sqrt{\left(X + \frac{3}{8} \right) \bigg/ \left(n - \frac{3}{4} \right)},$$

results in a random variable S having an approximate standard *normal distribution.

negative correlation *See* CORRELATION.

negatively skewed; negative skewness *See* SKEWNESS.

Nelder, John Ashworth (1924–) An English statistician. Nelder is a graduate of Cambridge University. In 1951 he joined the staff of the National Vegetable Research Station at Wellesbourne, becoming Head of

Statistics. In 1968 he joined the staff of Rothamsted Experimental Station, becoming Head of the Biomathematics Division. Since 1972 he has been a visiting Professor at Imperial College, London. Nelder, and his colleague *Wedderburn, were co-authors of the influential 1972 paper that introduced the *generalized linear model and led to the construction of the *GLIM statistical package. He was elected a Fellow of the Royal Society in 1981. He was President of the *International Biometric Society in 1978. He was awarded the *Royal Statistical Society's *Guy Medal in Silver in 1977 and was President of the Society in 1985.

Nelder–Mead simplex method A method for finding the maximum (or minimum) of a function. The method begins by evaluating the function at $(k + 1)$ locations, where k is the number of variables. Subsequent locations are selected automatically, using simple geometric patterns that allow for both the extension of the search area (when, apparently, far from the maximum) and contraction (when close to the maximum).

Nelson–Aalen estimate *See* KAPLAN–MEIER ESTIMATE.

nested design An *experimental design in which the *variables have an implicit hierarchy. For example, a hospital has two wings (I and II). Patients in wing I are randomly assigned to either consultant A or consultant B. Patients in wing II are randomly assigned to either consultant C or consultant D. Thus consultants A and B are nested within the wing I patients, and consultants C and D are nested within the wing II patients. If, instead, all the patients in the hospital had been randomly assigned to one of the four consultants then this would have been a *crossed design.

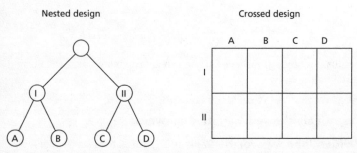

Nested design. The figure illustrates the difference between nested and crossed designs; in the latter, every combination of the crossed variables occurs.

Netherlands Society for Statistics and Operations Research
The Society that has published *Statistica Neerlandica* since 1946.

network A graphical representation of a problem by means of **nodes** connected by **arcs** of varying length or capacity. The arcs are usually directed. Examples include finding the shortest path between two nodes, finding the shortest set of arcs that connects all the nodes, the *Chinese postman problem, and the *travelling salesman problem. *See also* CRITICAL PATH ANALYSIS; NETWORK FLOW PROBLEMS; RELIABILITY THEORY.

network flow problems *Linear programming problems in which the objective is to maximize the overall flow from an initial **source** to a final **sink**. The *network consists of nodes connected by directed arcs, each arc having a given direction and a limited capacity. A **cut** is a line that breaks the network into two sections, one containing the source and the other containing the sink. The value of a cut is defined as the sum of the values on the broken arcs, ignoring arcs flowing from the sink section to the source section. The maximum flow can often be easily found using the **maximum flow / minimum cut theorem**, which states that these two quantities are equal.

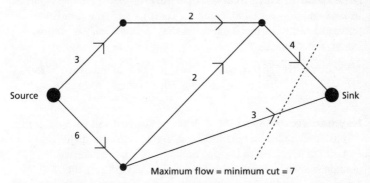

Maximum flow = minimum cut = 7

Network flow problems. This simple network illustrates the fact that the maximum flow equals the minimum cut.

neural net A mathematical structure that attempts to replicate the process by which the brain learns. The structure is conceived as consisting of a series of layers, the values of statistics computed in one layer being passed to processes ('neurons') in the next layer. The aim is that the output layer will provide accurate information about the process under investigation. The layers between the input and the output are called **hidden layers**. In order to be effective a neural network must be 'trained'—this is analogous to the *estimation of the parameters of a model. Much of the process of statistical modelling can be viewed as the construction of a neural net.

Newman–Keuls test *See* DUNCAN'S TEST.

Newton–Raphson method An algorithmic method for constructing a sequence of approximations to a root of an equation. Suppose it is desired to solve the equation f(x) = 0 and that x_1 is an approximate value for the root of the equation. If we write f′(x_n) for the value of the derivative of f(x) (with respect to x) evaluated at the point $x = x_n$, the sequence defined by

$$x_{n+1} = x_n - \frac{f(x_n)}{f'(x_n)}$$

usually converges to a root of the equation. It is based on constructing the tangent to the curve y = f(x) at the point $(x_n, f(x_n))$, and taking x_{n+1} to be the x-coordinate of the point where this tangent cuts the x-axis.

New Zealand Statistical Association This association was founded in 1948 and has about 400 members. Together with the *Statistical Society of Australia, Inc. it publishes the *Australia and New Zealand Journal of Statistics*.

New Zealand Statistician The original journal published by the *New Zealand Statistical Association. Superseded by *Australia and New Zealand Journal of Statistics*.

Neyman, Jerzy (1894–1981; b. Bendery, Russia; d. Oakland, California) A Polish probabilist and mathematical statistician who spent the majority of his career in the United States. Neyman's paternal grandfather was a Polish nobleman, burned alive during the 1863 uprising against the Russians. His father was exiled as a twelve-year-old to the Crimea, where he trained as a lawyer. Jerzy, who was born Jerzy Splawa-Neyman, studied physics and mathematics at the University of Kharkov, where *Bernstein was an influential teacher. On the outbreak of war between Poland and Russia, Neyman was arrested as an enemy alien. On his release he got a job as a statistician although claiming to know no statistics! To train, he went first to Berlin and then to study under Karl *Pearson at University College, London, where he met *Gosset and, crucially, Egon *Pearson. His work with Pearson laid down the Neyman–Pearson theory of *hypothesis testing with the introduction of the ideas of an *alternative hypothesis and of power. At the onset of another war, fearing a further spell of imprisonment as an enemy alien, Neyman moved to the USA, where he became the director of the Statistical Laboratory at the University of California at Berkeley. Neyman was President of the *Institute of Mathematical Statistics in 1949 and was elected to the National Academy of Sciences in 1963. He was awarded the *Royal Statistical Society's *Guy Medal in Gold in 1966 and the *Wilks Medal of the *American Statistical Association in 1968.

Neyman allocation *See* STRATIFIED SAMPLING.

Neyman–Pearson lemma A lemma, introduced in 1933 by *Neyman and Egon *Pearson, that gives a sufficient condition, in a *hypothesis test with *null hypothesis $\theta = \theta_0$ and *alternative hypothesis $\theta = \theta_1$, for choosing a *critical region, with given significance level, that maximizes the power of the test.

NH *See* NULL HYPOTHESIS; HYPOTHESIS TEST.

Nightingale, Florence (1820–1910; b. Florence, Italy; d. East Wellow, England) A pioneer of data analysis and graphical presentation. Nightingale is best known for her work as a nurse at Scutari during the Crimean War, where the soldiers called her the Lady with the Lamp. She was an efficient hospital administrator and compiled quantities of statistics in her drive for hospital reform: she standardized the reporting of deaths using *Miss Nightingale's Scheme for Uniform Hospital Statistics*. She has also been described as the 'the Passionate Statistician' and she wrote that Statistics is 'the most important science in the whole world'.

N (μ, σ^2) The *normal distribution with *mean μ and *variance σ^2.

NNT *See* NUMBER NEEDED TO TREAT.

node *See* NETWORK FLOW PROBLEMS.

noise A term used by analysts of *time series *data to describe random fluctuations that may obscure the true *signal. A sequence of errors for successive observations that consists of *independent random values from a *normal distribution with zero *mean is termed **white noise**.

nominal variable *See* CATEGORICAL VARIABLE.

nomogram A graphical method for calculating the value of one variable, given the values of other variables. A curve, or straight line, is drawn for each variable, and scales, usually non-linear, are marked on each curve in such a way that the points corresponding to any set of related values of the variables are collinear. Thus a straight line joining the points corresponding to the values of two of the variables will cross the curve corresponding to another variable at the point corresponding to the value of this variable.

non-ageing *See* FORGETFULNESS PROPERTY.

non-central chi-squared distribution If $Z_1, Z_2, \ldots, Z_v$ are *independent *standard normal variables and $\delta_1, \delta_2, \ldots, \delta_v$ are constants then X, given by

$$X = \sum_{k=1}^{v} (Z_k + \delta_k)^2,$$

is said to have a non-central chi-squared distribution with v *degrees of freedom and **non-centrality parameter** λ, where

$$\lambda = \sum_{k=1}^{v} \delta_k^2.$$

The *probability density function f of X is given by

$$f(x) = \frac{x^{(\frac{1}{2}v-1)}e^{-\frac{1}{2}(x+\lambda)}}{2^{\frac{1}{2}v}} \sum_{j=0}^{\infty} \frac{x^j \lambda^j}{\Gamma\left(\frac{1}{2}v+j\right)2^{2j}j!}, \qquad x > 0,$$

where Γ is the *gamma function. The distribution, which is *unimodal, has *mean $(v + \lambda)$ and *variance $2(v + 2\lambda)$. If $\lambda = 0$ then the distribution becomes the *chi-squared distribution with v degrees of freedom.

non-central *F*-distribution If Y_1 has a *non-central chi-squared distribution with v_1 *degrees of freedom and *non-centrality parameter λ, and if Y_2 is *independent of Y_1 and has a *chi-squared distribution with v_2 degrees of freedom, then the ratio X, given by

$$X = \frac{Y_1}{v_1} \bigg/ \frac{Y_2}{v_2},$$

is said to have a non-central *F*-distribution with non-centrality parameter λ and with v_1 and v_2 degrees of freedom. The *probability density function f of X is given by

$$f(x) = \frac{e^{-\frac{1}{2}\lambda} v_1^{\frac{1}{2}v_1} v_2^{\frac{1}{2}v_2} x^{(\frac{1}{2}v_1 - 1)}}{B\left(\frac{1}{2}v_1, \frac{1}{2}v_2\right)(v_2 + v_1 x)^{\frac{1}{2}(v_1 + v_2)}}$$
$$\times \sum_{j=0}^{\infty}\left[\left(\frac{\frac{1}{2}\lambda v_1 x}{v_2 + v_1 x}\right)^j \frac{(v_1 + v_2)(v_1 + v_2 + 2)\cdots(v_1 + v_2 + 2\{j - 1\})}{j! v_1 (v_1 + 2)\cdots(v_1 + 2\{j - 1\})}\right], \quad x > 0,$$

where B is the *beta function. The distribution has *mean and *variance given by

$$\frac{v_2(v_1 + \lambda)}{v_1(v_2 - 2)} \quad \text{and} \quad \frac{2v_2^2\{(v_1 + \lambda)^2 + (v_1 + 2\lambda)(v_2 - 2)\}}{v_1^2(v_2 - 2)^2(v_2 - 4)},$$

respectively (with $v_2 > 4$). The distribution is *unimodal.

non-centrality parameter *See* NON-CENTRAL CHI-SQUARED DISTRIBUTION; NON-CENTRAL F-DISTRIBUTION; NON-CENTRAL t-DISTRIBUTION.

non-central t-distribution If Y_1 has a standard *normal distribution, if Y_2 is *independent of Y_1 and has a *chi-squared distribution with v *degrees of freedom, and if λ is a non-zero constant, then X, given by

$$X = \frac{(Y_1 + \lambda)}{\sqrt{Y_2/v}},$$

is said to have a non-central t-distribution with v degrees of freedom and *non-centrality parameter λ. The *probability density function f of X is given by

$$f(x) = \frac{e^{-\frac{1}{2}\lambda^2}}{\sqrt{\pi v}\,\Gamma\left(\frac{1}{2}v\right)}\left(\frac{v}{v + x^2}\right)^{\frac{1}{2}(v+1)}\sum_{j=1}^{\infty}\frac{\Gamma\left(\frac{1}{2}\{v + j + 1\}\right)}{j!}\left(\frac{x\lambda\sqrt{2}}{\sqrt{v + x^2}}\right)^j, \quad -\infty < x < \infty,$$

where Γ is the *gamma function. The *mean and *variance of X are

$$\left(\frac{1}{2}v\right)^{\frac{1}{2}}\frac{\Gamma\left(\frac{1}{2}\{v - 1\}\right)}{\Gamma\left(\frac{1}{2}v\right)}\lambda \quad \text{and} \quad \frac{v(1 + \lambda^2)}{(v - 2)} - \frac{1}{2}v\left[\frac{\Gamma\left(\frac{1}{2}\{v - 1\}\right)}{\Gamma\left(\frac{1}{2}v\right)}\right]^2\lambda^2,$$

respectively (with $v > 2$). For large values of v the distribution of X is approximately a normal distribution with mean λ and variance 1.

non-informative prior *See* BAYESIAN INFERENCE.

non-linear model A model that involves a non-linear combination of its *parameters. For example,

$$\mathrm{E}(Y) = \alpha + \beta e^{\gamma x},$$

where α, β, and γ are parameters.

non-parametric tests (distribution-free tests) Tests that make no distributional assumptions about the *populations under investigation. The adjective 'non-parametric' was introduced by *Wolfowitz in 1942. Non-parametric tests are usually very simple to perform (e.g. the *runs test and the *sign test) and often make use of *ranks (examples include the *Mann–Whitney and *Wilcoxon signed-rank tests). *See also* KENDALL'S τ; KOLMOGOROV–SMIRNOV TEST; SPEARMAN'S ρ.

non-response A common problem for analysts of survey *data. There are two types: complete non-response in which the individual selected for interview is absent, dead, or uncooperative; and **item non-response** where the interviewee fails to answer some (but not all) questions. Provided the lack of response is not a consequence of the questions being asked, neither causes bias in the subsequent analysis. *Imputation provides a partial solution to item non-response.

nonsense correlation A term used to describe a situation where two variables (X and Y, say) are *correlated without being causally related to one another. The usual explanation is that they are both related to a third variable, Z. Often the third variable is time. For example, if we compare the price of a detached house in Edinburgh in 1920, 1930, ... with the size of the population of India at those dates, a 'significant' positive correlation will be found, since both variables have increased markedly with time. The first comprehensive study of nonsense correlation was undertaken in 1926 by *Yule, who considered the apparent connection between the fall in Church of England marriages and the concurrent increase in life expectancy.

non-singular matrix *See* MATRIX.

normal approximation to the binomial distribution *See* BINOMIAL DISTRIBUTION.

normal approximation to the Poisson distribution *See* POISSON DISTRIBUTION.

normal distribution (Gaussian distribution) The distribution of a *random variable X for which the *probability density function f is given by

$$f(x) = \frac{1}{\sigma\sqrt{2\pi}} \exp\left(\frac{(x-\mu)^2}{2\sigma^2}\right), \quad -\infty < x < \infty.$$

The *parameters μ and σ^2 are, respectively, the *mean and *variance of the distribution. The distribution is denoted by $N(\mu, \sigma^2)$. If the random variable X has such a distribution, then this is denoted by $X \sim N(\mu, \sigma^2)$. The graph of f(x) approaches the x-axis extremely quickly, and is effectively zero if $|x - \mu| > 3\sigma$ (hence the *three-sigma rule). In fact, $P(|X - \mu| < 2\sigma) \approx 95.5\%$ and $P(|X - \mu| < 3\sigma) \approx 99.7\%$. The first derivation of the form of f is believed to be that of *de Moivre in 1733. The description 'normal distribution' was used by *Galton in 1889, whereas 'Gaussian distribution' was used by Karl *Pearson in 1905.

The normal distribution is the basis of a large proportion of statistical analysis. Its importance and ubiquity are largely a consequence of the *Central Limit Theorem, which implies that averaging almost always leads to a bell-shaped distribution (hence the name 'normal'). *See* BELL-CURVE.

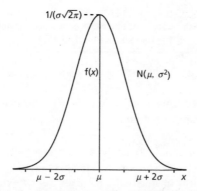

Normal distribution. The diagram illustrates the probability density function of a normal random variable X having expectation μ and variance σ^2. The distribution has mean, median, and mode at $x = \mu$, where the density function has value $1/(\sigma\sqrt{2\pi})$. Note that almost all the distribution (99.7%) lies within 3σ of the central value.

The **standard normal distribution** has mean 0 and variance 1. A random variable with this distribution is often denoted by Z and we write $Z \sim N(0, 1)$. Its probability density function is usually denoted by ϕ and is given by

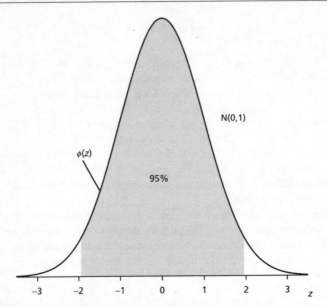

Standard normal distribution. The distribution is centred on 0, with 99.7% falling between -3 and 3 and 95% falling between -1.96 and 1.96.

$$\phi(z) = \frac{1}{\sqrt{2\pi}} e^{-\frac{1}{2}z^2}, \qquad -\infty < z < \infty.$$

If X has a general normal distribution $N(\mu, \sigma^2)$ then Z, defined by the **standardizing transformation**

$$Z = \frac{X - \mu}{\sigma},$$

has a standard normal distribution. It follows that the graph of the probability density function of X is obtained from the corresponding graph for Z by a stretch parallel to the z-axis, with centre at the origin and scale-factor σ, followed by a translation along the z-axis by μ.

The *cumulative probability function of Z is usually denoted by Φ and tables of values of $\Phi(z)$ are commonly available (*see* APPENDIX VII; *see also* APPENDIX VI). These tables usually give $\Phi(z)$ only for $z > 0$, since values for negative values of z can be found using

$$\Phi(z) = 1 - \Phi(-z).$$

The tables can be used to find cumulative probabilities for X via the standardizing transformation given above, since, for example,

$$P(X < x) = \Phi\left(\frac{x - \mu}{\sigma}\right).$$

As an example, if $X \sim N(7, 25)$ then the probability of X taking a value between 3 and 10 is given by

$$P\left(\frac{3 - 7}{5} < Z < \frac{10 - 7}{5}\right) = \Phi(0.6) - \Phi(-0.4) = \Phi(0.6) - \{1 - \Phi(0.4)\}$$

$$\approx 0.7257 + 0.6554 - 1 \approx 0.38.$$

The normal distribution plays a central part in the **theory of errors** that was developed by *Gauss. In the theory of errors, the **error function (erf)** is defined by

$$\mathrm{erf}(x) = 2\Phi(x\sqrt{2}) - 1.$$

An important property of the normal distribution is that any *linear combination of *independent *normal variables is normal: if $X_1 \sim N(\mu_1, \sigma_1^2)$ and $X_2 \sim N(\mu_2, \sigma_2^2)$ are independent, and a and b are constants, then

$$aX_1 + bX_2 \sim N(a\mu_1 + b\mu_2, a^2\sigma_1^2 + b^2\sigma_2^2),$$

with the obvious generalization to n independent normal variables. Many distributions can be approximated by a normal distribution for suitably large values of the relevant parameters. *See also* BINOMIAL DISTRIBUTION; POISSON DISTRIBUTION; CHI-SQUARED DISTRIBUTION; t-DISTRIBUTION.

normal equations *See* MULTIPLE REGRESSION MODEL.

normal inspection *See* QUALITY CONTROL.

normality The property of a *random variable or *population having a *normal distribution.

normal population A *population of values having a *normal distribution.

normal probability paper Specialized graph paper that can be used to help decide whether a *sample of n *observations is consistent with a *normal distribution. The observations are ranked in order: $x_{(1)} \leq x_{(2)} \leq \cdots \leq x_{(n)}$ and p_j, the proportion of values in the sample that are less than or equal to $x_{(j)}$, is plotted against $x_{(j)}$ on the graph paper. A sample from a normal distribution will appear as a series of

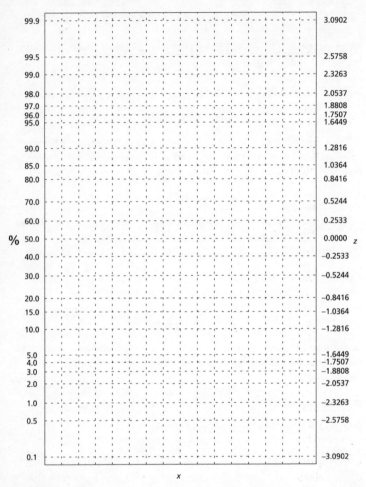

Normal probability paper. Plotting the sample cumulative proportions on this paper will result in an approximate straight line if the data have arisen from a normal distribution.

approximately collinear points. A strong departure from linearity in the central region of the graph will provide an indication that the distribution is not normal.

In order to get a better idea of the behaviour in the extremes of the distribution, the observations may be replaced by their absolute values before being arranged in order. The resulting plot, with the theoretical probabilities appropriately adjusted, is a **half-normal plot**. *See also* Q-Q PLOT.

normal random variable A *random variable having a *normal distribution.

normal scores The normal score corresponding to the kth largest of n *observations is the *expected value of the kth largest of n *independent observations from a standard *normal distribution. Several *non-parametric tests have been devised in which the ordered observations have their values replaced by the corresponding normal scores. These tests usually have greater power than the corresponding tests based on *ranks.

normal test *See* Z-TEST.

normal variable *See* NORMAL DISTRIBUTION.

not significant (n.s.) *See* HYPOTHESIS TEST.

nugget model A model of a *time series or *spatial process that allows for a non-zero *autocorrelation at lag 0. This acknowledges the fact that, with real data, there are measurement errors such that repeat measurements at the 'same' time or place may actually vary.

nuisance parameter A *parameter that must be estimated, even though it is of no immediate interest. The most familiar example occurs in the construction of a *confidence interval for an estimate of a *population mean. The sample *mean is the obvious estimate, but the population *variance is a nuisance parameter that must also be estimated in order to determine the size of the interval.

null hypothesis (NH) The hypothesis, in a *hypothesis test, which is used to obtain the *probability distribution, and hence the *critical region, of the *statistic used in the test. The phrase 'null hypothesis' was introduced by Sir Ronald *Fisher in 1935.

number needed to treat (NNT) A phrase used in medical contexts. Suppose past results show that treatment j ($j = 1, 2$) is successful for a proportion p_j of patients. Suppose $p_1 > p_2$. On average, with N patients, treatment 1 will cure $N(p_1 - p_2)$ more patients than treatment 2. Thus the number needed to treat so that one extra patient is successfully treated is M, given by

$$M = \frac{1}{p_1 - p_2}.$$

If treatment 1 costs an amount k per patient more than treatment 2, then one interpretation of the above is that it will be worth using treatment 1 rather than treatment 2 only if treating a patient successfully has a value greater than kM.

numerical Usually referring to a *variable whose possible values are numbers (as opposed to, for example, categories).

Nyquist frequency The shortest detectable frequency in a *time series. All time series are, in practice, recorded at discrete time points (for example, one reading per second). Periodic variations that happen more rapidly than the shortest time interval cannot be detected. With time intervals of size t the Nyquist frequency is t^{-1} cycles per unit time.

objective function *See* LINEAR PROGRAMMING.

observation A result of an experiment or trial in which a variable, either *numerical or *categorical, is measured.

observed frequency *See* FREQUENCY.

OC-curve *See* ACCEPTANCE SAMPLING.

Ockham's razor Essentially the **principle of parsimony** which states that if one is provided with a variety of explanations (*e.g.* a variety of statistical models) one should prefer the simplest. William of Ockham (*c.* 1285–1349) was an English philosopher who held that a complicated explanation should not be accepted without good reason, and wrote '*Frustra fit per plura, quod fieri potest per pauciora.*' ('It is vain to do with more what can be done with less.')

odds If the *probability of an event happening is p, then the odds on that event happening (as opposed to it not happening) are

$$\frac{p}{1-p}.$$

If the odds are expressed as a simple fraction $\frac{m}{n}$, where m and n are positive integers, then the probability of the event happening is

$$\frac{m}{m+n}.$$

In the case $m > n$ the odds are said to be 'm to n on'. In the case $m < n$ the odds are said to be 'n to m against'. Thus, for example, if $p = \frac{4}{5}$ then the odds are '4 to 1 on', and if $p = \frac{1}{5}$ then the odds are '4 to 1 against'. Confusingly, in the case $p = \frac{1}{2}$ the odds are said to be 'evens'!

odds ratio The ratio of the *odds on something occurring in one situation to the odds of the same event occurring under a second situation. An odds ratio of 1 implies that the odds on an event occurring (and hence the probability of its occurrence) are unaffected by the change in situation: they are *independent of the situation. The *interaction parameters of *log-linear models can be interpreted in terms of odds ratios (or ratios of odds ratios). *See also* RELATIVE RISK.

ogive A diagram representing grouped numerical data in which *cumulative frequency is plotted against upper *class boundary, and a curve is fitted to the resulting points. *See* CUMULATIVE FREQUENCY POLYGON.

Old Faithful Geyser in the Yellowstone National Park which erupts in a fashion that has challenged analysts of *time series. The data in the table refer to the period 1–8 August in 1978. For each day, the table shows X, the duration (in minutes, to the nearest 0.1 minute) of the current eruption, and Y, the time (in minutes, to the nearest minute) until the next eruption. Observations refer to daylight hours only.

DAY		DURATION AND TIME INTERVAL FOR SUCCESSIVE ERUPTIONS												
1	X	4.4	3.9	4.0	4.0	3.5	4.1	2.3	4.7	1.7	4.9	1.7	4.6	3.4
	Y	78	74	68	76	80	84	50	93	55	76	58	74	75
2	X	4.3	1.7	3.9	3.7	3.1	4.0	1.8	4.1	1.8	3.2	1.9	4.6	2.0
	Y	80	56	80	69	57	90	42	91	51	79	53	82	51
3	X	4.5	3.9	4.3	2.3	3.8	1.9	4.6	1.8	4.7	1.8	4.6	1.9	3.5
	Y	76	82	84	53	86	51	85	45	88	51	80	49	82
4	X	4.0	3.7	3.7	4.3	3.6	3.8	3.8	3.8	2.5	4.5	4.1	3.7	3.8	3.4
	Y	75	73	67	68	86	72	75	75	66	84	70	79	60	86
5	X	4.0	2.3	4.4	4.1	4.3	3.3	2.0	4.3	2.9	4.6	1.9	3.6	3.7	3.7
	Y	71	67	81	76	83	76	55	73	56	83	57	71	72	77
6	X	1.8	4.6	3.5	4.0	3.7	1.7	4.6	1.7	4.0	1.8	4.4	1.9	4.6	2.9
	Y	55	75	73	70	83	50	95	51	82	54	83	51	80	78
7	X	3.5	2.0	4.3	1.8	4.1	1.8	4.7	4.2	3.9	4.3	1.8	4.5	2.0
	Y	81	53	89	44	78	61	73	75	73	76	55	86	48
8	X	4.2	4.4	4.1	4.1	4.0	4.1	2.7	4.6	1.9	4.5	4.5	4.8	4.1
	Y	77	73	70	88	75	83	61	78	61	81	81	80	79

Old Faithful. Long time intervals between eruptions are generally followed by long eruptions, but predicting the future eruption pattern is a considerable challenge to time series experts.

OLS *See* ORDINARY LEAST SQUARES.

one-sided confidence interval *See* CONFIDENCE INTERVAL.

one-sided test *See* HYPOTHESIS TEST.

one-tailed test *See* HYPOTHESIS TEST.

one-tail probability For the *random variable X, with *median m, the one-tail probability associated with the value x is $P(X \leq x)$ if $x < m$ and $P(X \geq x)$ if $x > m$. If $x < m$, then the corresponding **two-tail probability** is $P(X \leq x) + P\{X \geq (2m - x)\}$. For a *symmetric distribution this amounts to $2P(X \leq x)$.

open question *See* CLOSED QUESTION.

operating characteristic curve *See* ACCEPTANCE SAMPLING.

operational research; operations research The application of statistical and scientific methods to the management of commercial and industrial processes and of military and governmental activities.

opinion poll A survey of opinions, usually social or political. People in a *sample are interviewed or reply to a questionnaire. The questions asked usually have a list of possible responses, and the responses are analysed. The results are usually given as percentages of those expressing an opinion on the particular question.

optimal design An *experimental design in which the values chosen for the explanatory variables lead to certain desirable properties. In the *multiple regression model the *matrix of *variances and *covariances is proportional to $X'X$, where X is the design matrix. A **D-optimal** design is one that maximizes the determinant of $X'X$. This is equivalent to minimizing the volume of the corresponding *confidence ellipsoid. An **A-optimal** design is one that minimizes the *trace of $(X'X)^{-1}$. This minimizes the average variance of the parameter estimates.

optimization The minimization or maximization of some function, usually subject to restrictions (which are often on the values of the variables over which the optimization takes place).

optimum allocation *See* STRATIFIED SAMPLING.

optimum interpolation The term used by meteorologists for the procedure more generally known as *kriging.

order statistics Particular values in a data set that has been ordered in magnitude. The best-known statistics based on ordered values are the *median and the *range, which provide measures of location and spread. An example of naturally ordered data is provided by the lifetimes of failing light-bulbs, all initially new. *See also* NORMAL SCORES; TRIMMED MEAN; Q-Q PLOT.

ordinal variable A *categorical variable in which the categories have an obvious order (e.g. strongly disagree, disagree, neutral, agree, strongly agree). Specialist models are required for ordinal variables to take account of their ordered categories.

There are also specialized measures of *association. Let j and j' be two categories of one ordinal variable, A, and let k and k' be two categories of a second ordinal variable, B. The quantities S, D, and T_B are defined in terms of pairs of observations, one belonging to cell (j, k) and the other to cell (j', k'):

S = the total number of pairs for which, when $j > j', k > k'$;

D = the total number of pairs for which, when $j > j', k < k'$;

T_B = the total number of pairs for which, when $j > j', k = k'$.

These quantities are calculated using every pair of observations.

Goodman and Kruskal's gamma, proposed by *Goodman and *Kruskal in 1954, is given by

$$\gamma = \frac{S - D}{S + D}.$$

Somers's d_{BA}, proposed by the sociologist Robert H. Somers in 1962, is preferred when B is believed to depend on A. The formula is

$$d_{BA} = \frac{S - D}{S + D + T_B}.$$

ordinary least squares (OLS) The method of least squares applied in a context in which equal weight is given to each of the differences between the observed values $y_1, y_2, \ldots, y_n$ and the fitted values $\hat{y}_1, \hat{y}_2, \ldots, \hat{y}_n$. *See also* GENERALIZED LEAST SQUARES; WEIGHTED LEAST SQUARES.

ordinate *See* CARTESIAN COORDINATES.

origin *See* CARTESIAN COORDINATES.

Ornstein–Uhlenbeck process *See* BROWNIAN MOTION.

orthogonality *See* VARIANCE–COVARIANCE MATRIX.

orthogonal matrix *See* MATRIX.

orthogonal polynomials *Polynomials designed to simplify calculations when a response variable is believed to be related to a polynomial function of an explanatory variable. If $x_1, x_2, \ldots, x_n$ are n *observations on the *variable X, then the polynomials $\phi_1(x)$ and $\phi_2(x)$ are orthogonal if

$$\sum_{j=1}^{n} \phi_1(x_j)\phi_2(x_j) = 0.$$

If we wish to fit the model

$$\mathrm{E}(Y_j) = \beta_0 + \beta_1 x_j + \beta_2 x_j^2 + \cdots + \beta_m x_j^m, \qquad j = 1, 2, \ldots, n,$$

then we will need (*see* *multiple regression model) to invert the *matrix product $\mathbf{X'X}$, where

$$\mathbf{X} = \begin{pmatrix} 1 & x_1 & x_1^2 & \ldots & x_1^m \\ 1 & x_2 & x_2^2 & \ldots & x_2^m \\ \vdots & \vdots & \vdots & \ddots & \vdots \\ 1 & x_n & x_n^2 & \ldots & x_n^m \end{pmatrix}.$$

This may not be easy, since $\mathbf{X'X}$ will often be nearly non-singular. One solution is to rewrite the model in the form

$$\mathrm{E}(Y_j) = \gamma_0 \phi_0(x_j) + \gamma_1 \phi_1(x_j) + \cdots + \gamma_m \phi_m(x_j),$$

where $\phi_0(x) = 1$ and $\phi_k(x)$ is a polynomial of degree k. With n observations on x, denoted by $x_1, x_2, \ldots, x_n$, the polynomials are chosen to be orthogonal, i.e. to satisfy the requirements that

$$\sum_{j=1}^{n} \phi_k(x_j)\phi_l(x_j) = 0 \qquad \text{for all } k \text{ and } l, k \neq l.$$

The revised $\mathbf{X}$ matrix has a typical element $\phi_k(x_j)$ and is such that the product $\mathbf{X'X}$ is now a diagonal matrix and is therefore easily inverted.

Often the appropriate degree of dependence on x (in other words, the value of m) is unknown. The reformulation also makes it easy to select an appropriate model without the need for further iterations. In the *ANOVA table the contributions for each polynomial can be listed separately.

outcome *See* SAMPLE SPACE.

outcome variable *See* REGRESSION.

outlier An observation that is very different to other observations in a set of data. Since the most common cause is recording error, it is sensible to search for outliers (by means of *summary statistics and plots of the data) before conducting any detailed statistical modelling.

Various indicators are used to identify outliers. One is that an observation has a value that is more than 2.5 standard deviations from the mean. Another is that an observation has a value that lies more than $1.5I$ beyond the upper or the lower quartile, where I is the interquartile range (see BOXPLOT).

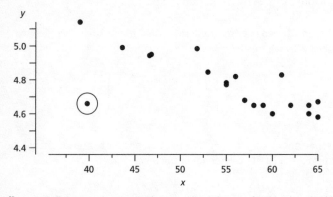

Outlier. A small data set showing an obvious outlier. Whenever feasible, data should be plotted, since including an outlier will usually make nonsense of any calculations.

overdispersed See INDEX OF DISPERSION.

overparameterized model A *model having more *parameters than can be estimated from the *data. For example, suppose that the yields of two types of tomatoes are to be compared using the data $\{y_{jk}\}$, where $j(=1,2)$ signifies the treatment and k is the number of the observation. Consider the model

$$y_{jk} = \mu + \tau_j + \varepsilon_{jk},$$

where the $\{\varepsilon_{jk}\}$ are *independent *random errors, each with *mean 0 and *variance σ^2. As it stands, this model is overparameterized, because for each value of j, only $(\mu + \tau_j)$ can be estimated. The value attributed to μ can be arbitrarily chosen. However, the problem can be solved either by rewriting the model

$$y_{jk} = \mu_j + \varepsilon_{jk},$$

or by retaining the original model and adding a constraint such as $\tau_1 + \tau_2 = 0$.

oversampling The deliberate selection of individuals of a rare type in order to obtain reasonably precise estimates of the properties of this type. In a *population which includes such a rare type, a random *sample of the entire population might result in very few (or none) of these individuals being selected. Oversampling implies the deliberate sampling of a much higher proportion of this type than of the rest of the population. This is a form of *stratified sampling.

Paasche, Hermann (1851–1925; b. Magdeburg, Germany; d. Detroit, Michigan) A German economist remembered for suggesting the eponymous *Paasche's price index, though he was not the first to suggest it.

Paasche's price index A measure of the value of money in which the cost of a standard collection, or basket, of goods and services at some particular date is compared with the cost of the same basket at a base date. The basket is chosen to represent the expenditure of a typical household at the date under consideration. The index, suggested by Hermann Paasche in 1874, is calculated as the ratio

$$\sum_j p_{nj}q_{nj} \Big/ \sum_j p_{0j}q_{nj},$$

where p_{0j} and p_{nj} are the prices at times 0 and n of the jth item in the basket, and q_{nj} is the quantity of that item at time n. *See also* FISHER'S INDEX; LASPEYRES'S PRICE INDEX; RETAIL PRICE INDEX.

PACF *See* PARTIAL AUTOCORRELATION FUNCTION.

paired comparisons *See* EXPERIMENTAL DESIGN.

paired-sample test *See* HYPOTHESIS TEST.

pairwise independent Events (*see* SAMPLE SPACE) $A_1, A_2, \ldots, A_n$ are pairwise independent if every possible pair of these events are independent, *i.e.*

$$P(A_j \cap A_k) = P(A_j) \times P(A_k) \qquad \text{for all } j, k \ (j \neq k).$$

It should be noted that pairwise independence is not sufficient to guarantee that

$$P(A_1 \cap A_2 \cap \cdots \cap A_n) = P(A_1) \times P(A_2) \times \cdots \times P(A_n).$$

See INDEPENDENT EVENTS; INTERSECTION.

panel study A *longitudinal study of the same group of people over

time. By contrast with a *cohort study the panel study gathers
information on the same people at each time point. In practice
the representation is not constant over time, because of *attrition.
A disadvantage compared with a *cross-sectional study is that the
panel ceases to be representative of the entire population
because of the absence of younger members in the later years of the
study.

parallel structure *See* RELIABILITY THEORY.

parameter A constant appearing in the *probability function or the
*probability density function of a family of *distributions. The shape of
the distribution depends on the value(s) given to the parameter(s). For
example, the parameters of a *binomial distribution are n and p, the
parameter of a *Poisson distribution is m (or λ), the parameters of a
normal distribution are μ and σ, and ν is the parameter of both a *chi-
squared distribution and a *t-distribution. The word 'parameter' was
used by Sir Ronald *Fisher in 1922.

Pareto, Marquis Wilfredo (1848–1923; b. Paris, France; d. Geneva,
Switzerland) An Italian economist famed for his studies of the
distribution of income. After training at the Polytechnic Institute,
Turin, Pareto began his career as a lecturer in mathematics and
engineering at the University of Florence. However, he became
increasingly concerned by what he saw as social injustices (*see*
EIGHTY/TWENTY RULE) and subsequently moved to Lausanne as
Professor of Economics. He is best remembered for his *Treatise on Sociology*,
completed in 1912 at his home, Angora Villa, which he shared with
eighteen Angora cats.

Pareto distribution A *random variable x with the simplest type of
Pareto distribution has *probability density function given by

$$f(x) = ak^a x^{-(a+1)}, \qquad x \geq k,$$

where a and k are positive constants. *See following diagram.*
  *Pareto proposed the distribution in terms of its *cumulative
distribution function, $\{1 - (k/x)^a\}$ which he believed gave a good
approximation to the *proportion of incomes that were less than x.
According to Pareto the shape of the income distribution was the same
for all countries, though the values of a and k varied from country to
country. *See* EIGHTY/TWENTY RULE.

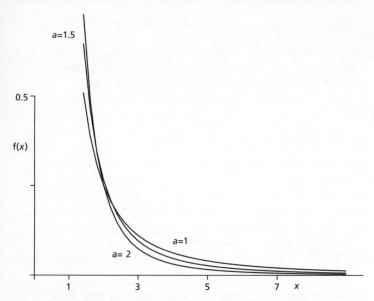

Pareto distribution. Pareto distributions were proposed to model the distribution of income. Those illustrated have $k = 1$.

Pareto plot A diagram similar to a *bar chart, with no gaps between the bars and with the bars ordered in decreasing order of the *frequencies they represent. The scale is usually *proportion rather than frequency. It is common to superimpose an indicator of cumulative proportion.

partial autocorrelation function (PACF) A function describing the relation between the lag k and the corresponding coefficient in an *autoregressive model. The plot of the PACF can be a useful guide to the appropriate order of an autoregressive *time series.

partial correlation The *correlation between two *variables after allowing for the effect of other variables. For variables A, B, and C, with, for example, r_{AB} denoting the correlation between A and B, the partial correlation between A and B allowing for C is $r_{AB|C}$, given by

$$r_{AB|C} = \frac{r_{AB} - r_{AC}r_{BC}}{\sqrt{\left(1 - r_{AC}^2\right)\left(1 - r_{BC}^2\right)}}.$$

partial least squares (PLS) A method for handling correlated

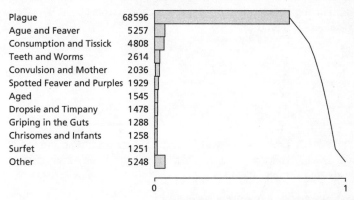

Plague	68596
Ague and Feaver	5257
Consumption and Tissick	4808
Teeth and Worms	2614
Convulsion and Mother	2036
Spotted Feaver and Purples	1929
Aged	1545
Dropsie and Timpany	1478
Griping in the Guts	1288
Chrisomes and Infants	1258
Surfet	1251
Other	5248

0   1

Proportions and cumulative proportion of total

Pareto plot. This plot illustrates the causes of death of Londoners in 1665. The data and spelling are taken from the *Annual Bill of Mortality for London*.

---

explanatory variables in the context of *multiple regression. In PLS the first stage is to determine k *uncorrelated variables that are linear combinations of the explanatory variables. The combinations are chosen for their predictive ability. **Principal components regression analysis** uses a different technique to achieve the same objective.

Parzen window *See* PERIODOGRAM.

Pascal, Blaise (1623–62; Clermont-Ferrand, France; d. Paris, France) A French mathematical prodigy, whose first paper was published when he was seventeen. His father, also a mathematician, was employed at one time as a tax collector and Blaise, at the age of nineteen, invented a mechanical calculator capable of working with a currency that used multiples of both 12 and 20. Together with *Fermat, Pascal laid the foundations for the modern treatment of *probability. He is quoted as saying, 'It is not certain that everything is uncertain.' A street in Paris and a lunar crater are named after him.

PASCAL A computer language named in honour of Blaise *Pascal.

Pascal distribution *See* NEGATIVE BINOMIAL DISTRIBUTION.

Pascal's triangle A triangle of numbers, named after *Pascal, in which the rth item on the nth row is the value of $^{n-1}C_{r-1}$, the number of different *combinations of $(r-1)$ objects chosen from $(n-1)$. The sum of the numbers on the nth row is 2^{n-1}.

Thus $^{6-1}C_{4-1} = {}^5C_3 = 10$. Apart from the 1 at the beginning and end of each row, each number is the sum of the two nearest numbers in the row above. For example $10 = 4 + 6$.

path vector *See* RELIABILITY THEORY.

PCA *See* PRINCIPAL COMPONENTS ANALYSIS.

pdf *See* PROBABILITY DENSITY FUNCTION.

Pearson, Egon Sharpe (1895–1980; b. London, England; d. Midhurst, Sussex) An English mathematical statistician and the only son of Karl *Pearson. Egon Pearson studied mathematics at Cambridge University, graduating in 1919. This was followed by graduate work in astronomy. In 1921 he joined his father on the staff at University College, London, where he remained (retiring as Head of Department in 1960). Egon's association with *Neyman started in 1926 and led to the *Neyman–Pearson lemma and the development of a standard approach to *hypothesis testing. He was Editor of *Biometrika* from 1936 to 1965. He was President of the *Royal Statistical Society from 1955 to 1957 and was awarded its *Guy Medal in Gold in 1955.

Pearson, Karl (1857–1936; b. London, England; d. London, England) An English mathematician, biometrician, and statistician. Pearson graduated from Cambridge University in 1879, joining the staff of University College, London, where he was appointed as Professor of Applied Mathematics and Mechanics in 1884. In 1890 he added the title of Gresham Lecturer in Geometry. It was not until 1893 that Pearson started publishing articles on statistics. By that time he already had a hundred publications to his name (including a number on German history and folklore). His first statistical work was entitled *The Chances of Death and Other Studies in Evolution*, and much of his subsequent work on statistical theory had a similar focus. During the period 1895–8 he presented a sequence of papers on *correlation and in 1900 he proposed the *chi-squared test. He founded the journal *Biometrika* in 1901 and was its Editor until his death, when his son (*see* PEARSON, EGON SHARPE) took over. In 1911 he was appointed Professor of Eugenics (the study of human evolution), a post he held until 1933. He was elected a Fellow of the Royal Society in 1896. He was also a Fellow of the Royal Society of Edinburgh.

Pearson family of distributions A family of *distributions that comprise distributions having *probability density functions of a variety of shapes. The family was proposed by Karl *Pearson in 1894 as a response to his recognition that not all *populations had *distributions that resembled the *normal distribution. He proposed twelve types of distribution which are variants of three basic distributions: Type I (the *beta distribution), Type VI (the *F-distribution), and Type IV (now little used).

Pearson goodness-of-fit test *See* CHI-SQUARED TEST.

Pearson residual *See* CHI-SQUARED TEST.

Pearson's coefficient of skewness A simple *statistic that uses the *mean, *mode, and *standard deviation:

$$\frac{\text{mean} - \text{mode}}{\text{standard deviation}}.$$

If the mode is unknown then the *median is used and the revised statistic is

$$3\frac{\text{mean} - \text{median}}{\text{standard deviation}}.$$

The coefficient is usually positive when the *distribution is positively *skewed, and negative when it is negatively skewed.

Pearson's correlation coefficient *See* CORRELATION.

penalty function A constraint specified in a *loss function which applies a penalty when an undesirable condition arises (e.g. an estimated value that is not feasible).

Penrose tiling *See* TESSELLATION.

percentage point (lower percentage point) For the *probability distribution of a *random variable X, the θ percentage point (or lower θ

percentage point) of the distribution is x_1, such that $P(X < x_1) = \theta/100$. The **upper θ percentage point** of the distribution is x_2, such that $P(X > x_2) = \theta/100$. For example, the upper 5% point of a *standard normal distribution is 1.645.

percentile An approximate value for the rth percentile of a *data set can be read from a *cumulative frequency graph as the value of the *variable corresponding to a cumulative relative frequency of $r\%$. So the lower *quartile is the 25th percentile and the *median is the 50th percentile. The term 'percentile' was introduced by *Galton in 1885.

periodogram A periodogram is a useful graphical tool in the analysis of *time series. Consider a time series consisting of n *observations $x_1, x_2, \ldots, x_n$, corresponding to the times $t = 1, 2, \ldots, n$. The *mean of the n observations is denoted by $\bar{x}$. The series may be thought of as resulting from a mixture of cyclic variations of different periodicities so that, for n odd,

$$x_t = \bar{x} + \sum_{j=1}^{h} \left\{ a_j \cos\left(\frac{2\pi jt}{n}\right) + b_j \sin\left(\frac{2\pi jt}{n}\right) \right\},$$

where

$$a_j = \frac{2}{n} \sum_{t=1}^{n} x_t \cos\left(\frac{2\pi jt}{n}\right), \qquad b_j = \frac{2}{n} \sum_{t=1}^{n} x_t \sin\left(\frac{2\pi jt}{n}\right),$$

and $h = \frac{1}{2}(n - 1)$. In the case where n is even the same expressions hold true, but now $h = \frac{1}{2}n - 1$ and there is one additional term, namely, $\frac{1}{2} a_{\frac{1}{2}n} \cos(\pi t)$.

The dominating contributions are indicated by large values for a_j or b_j. The periodogram is a plot of $I(j)$ against j, for $j \leq \frac{1}{2}(n - 1)$, where

$$I(j) = a_j^2 + b_j^2 = \frac{1}{2}c_0 + \sum_{k=1}^{n-1} c_k \cos\left(\frac{2\pi jk}{n}\right), \quad j = 1, 2, \ldots,$$

and

$$c_k = \frac{1}{n} \sum_{t=1}^{n-k} (x_t - \bar{x})(x_{t+k} - \bar{x}), \quad k = 1, 2, \ldots, n - 1.$$

Some authors define the periodogram using multiples (e.g. $2/\pi$) of the formula given here.

The *population counterpart of the time series is the *stationary process $X(t)$, and the counterpart of the periodogram is the **spectral density function** (or **power spectrum** or **spectrum**).

The periodogram is not a consistent *estimator of the power spectrum and it is usual to give more *weight to the first m terms of the summation in the periodogram, using

$$I_m(j) = \frac{1}{2}c_0 + \sum_{k=1}^{m} \lambda_k c_k \cos\left(\frac{2\pi jk}{n}\right),$$

where $m < n$ and the weight λ_k is the **lag window**. The best known lag windows are the **Tukey window**,

$$\lambda_k = \frac{1}{2}\left\{1 + \cos\left(\frac{\pi k}{m}\right)\right\}, \qquad k = 1, 2, \ldots, m,$$

the **Hamming window**,

$$\lambda_k = 0.54 + 0.46 \cos\left(\frac{\pi k}{m}\right), \qquad k = 1, 2, \ldots, m,$$

and the **Parzen window**,

$$\lambda_k = \begin{cases} 1 - 6\left(\frac{k}{m}\right)^2 + 6\left(\frac{k}{m}\right)^3, & 1 \le k \le \frac{1}{2}m, \\ 2\left(1 - \frac{k}{m}\right)^3, & \frac{1}{2}m \le k \le m. \end{cases}$$

A typical value for m is about $2\sqrt{n}$.

permutation An ordered arrangement of n different objects. The number of permutations (i.e. the number of different possible ordered arrangements) is $n!$.

For ordered selection of r objects from a set of n ($\ge r$) different objects, the number of permutations of r from n, i.e. the number of different possible ordered selections, is usually denoted by ^nP_r. In fact

$$^nP_r = n \times (n-1) \times (n-2) \cdots \times (n-r+1) = \frac{n!}{(n-r)!}.$$

Special values are $^nP_0 = 1$, $^nP_1 = n$, $^nP_n = n!$. For example, the 6 ($= 3!$) permutations of $\{A, B, C\}$ are $ABC, ACB, BAC, BCA, CAB, CBA$, and there are 24 ($= 4 \times 3 \times 2$) permutations of 3 letters from $\{A, B, C, D\}$.

If the n objects are not all different, and there are n_1 objects of type 1, n_2 objects of type 2, $\ldots$, n_k objects of type k, where $n_1 + n_2 + \cdots + n_k = n$, then the number of different ordered arrangements is

$$\frac{n!}{n_1!n_2!\dots n_k!}.$$

For unordered selection, *see* COMBINATION.

permutation matrix A square *matrix in which the rows are a permutation of the rows of an identity matrix.

Perron–Frobenius theorem The theorem stating that a square *matrix A with real non-negative elements has a positive real eigenvalue λ. The matrix is assumed to be irreducible, i.e. there is no *permutation matrix **P** such that **PAP**$'$ has a zero submatrix in the bottom left-hand corner. Furthermore, every eigenvalue of **A** has modulus not exceeding λ, the eigenvalue λ is simple (i.e. λ is a non-multiple root of the characteristic equation (*see* MATRIX)), and there is a corresponding eigenvector with positive elements. Perron originally established the theorem for matrices with positive elements, which are necessarily irreducible, and Frobenius extended the result.

persistent state *See* MARKOV PROCESS.

pgf *See* PROBABILITY-GENERATING FUNCTION.

pi (π) The ratio of the circumference of circle to its diameter; $\pi = 3.141\,592\,65\dots$.

pictogram A diagram in which *frequency or quantity is represented by symbols that are small images of the objects or material being counted. The diagram should indicate exactly how many, or how much, each symbol represents. *See following diagram.*

pie chart A diagram used when it is desired to emphasize the proportions of a set of *data when data items are grouped into classes according to the value of some *variable (usually *categorical). A circle is divided into sectors representing the classes. The area (or equivalently the angle) of a sector is proportional to the frequency of the corresponding class. If two pie charts are used to compare two populations, their areas can be made proportional to the sizes of the populations. *See following diagram.*

Pillai–Bartlett trace *See* MULTIVARIATE ANALYSIS OF VARIANCE.

pilot study A small survey taken in advance of a major investigation. The pilot study may show up problems in the organization of the

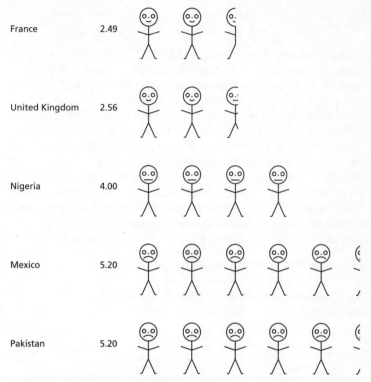

France	2.49
United Kingdom	2.56
Nigeria	4.00
Mexico	5.20
Pakistan	5.20

Pictogram. The diagram shows, for 1984, the average number of people per dwelling in five countries. In the diagram each extra stick person represents one extra person per dwelling. The two European countries are very similar and differ markedly from the other countries.

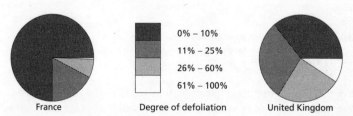

France   Degree of defoliation   United Kingdom

0% – 10%
11% – 25%
26% – 60%
61% – 100%

Pie chart. The charts contrast the amount of defoliation of conifers in France and the UK in 1989.

intended major study. It can also give information about response variability that will help determine the size of the major study.

Pitman, Edwin James George (1897–1993; b. Melbourne, Australia; d. Kingston, Australia) Australian mathematical statistician. Pitman's parents, emigrating from England, met on the ship to Australia. He was educated at Melbourne University, graduating with an MA in mathematics in 1923 after his studies had been interrupted by two years of war service. Pitman was Professor of Mathematics at the University of Tasmania from 1926 to 1962. An early task at the University of Tasmania was to lecture in *Statistics—a subject new to him. His subsequent research concerned the foundations of *statistical inference, including pioneering work on *non-parametric statistics. He was President of the Australian Mathematical Society in 1958. In 1978 the *Statistical Society of Australia instituted the award of the Pitman Medal for 'high distinction in Statistics': Pitman himself was the first recipient.

Pitman efficiency A method for comparing two *estimators of an unknown *population *parameter. Suppose that, with a *sample of n *observations, $\hat{\theta}$ and $\tilde{\theta}$ are two estimators of an unknown population parameter, θ. The Pitman efficiency of $\hat{\theta}$ relative to $\tilde{\theta}$ is

$$P\left(|\hat{\theta} - \theta| < |\tilde{\theta} - \theta|\right).$$

pivotal quantity A function of the *observations having a *distribution that does not depend on any parameter other than the parameter of interest. For example, if a random sample of n observations is taken from a *normal distribution with unknown *mean μ and *variance σ^2 then a pivotal quantity for the parameter μ is the *statistic t, given by

$$t = \frac{\bar{x} - \mu}{s/\sqrt{n}},$$

where $\bar{x}$ is the sample mean and s^2 is the sample variance (calculated using the $(n-1)$ divisor). The distribution of t (a *t-distribution) does not depend on the value of σ^2.

pixel An area, often rectangular, for which a value has been recorded. The term was coined in connection with the analysis of pictures, such as satellite images of ground conditions, or medical scans of parts of the human body. Each dot in such a picture is a pixel conveying information.

A **voxel** is the three-dimensional equivalent, representing a volume of space for which a value has been recorded.

placebo; placebo effect *See* BLINDING.

Plackett, Robert (Robin) Lewis (1920–) An English statistician. Plackett was a co-author of the 1946 paper that introduced Plackett and Burman designs. Formerly a lecturer at the University of Liverpool, in 1962 Plackett was appointed Professor of Statistics at the University of Newcastle upon Tyne, where he is now Professor Emeritus. He was awarded the *Royal Statistical Society's *Guy Medal in Bronze in 1968, in Silver in 1973, and in Gold in 1987.

Plackett and Burman designs *See* FACTORIAL EXPERIMENT.

platykurtic *See* KURTOSIS.

Playfair, William (1759–1823; b. Dundee, Scotland; d. Burntisland, Scotland) A Scottish political economist. At the age of thirteen he was apprenticed to a millwright. At twenty-one he became a draughtsman to James Watt (of steam-engine fame) in Birmingham. His elder brother was Professor of Mathematics at Edinburgh University, so mathematical ideas ran in the family. However, William Playfair became interested in politics and the economy, writing several books (in particular, in 1786, *The Commercial and Political Atlas*). His books contain many effective graphical devices and he is credited with the introduction of *pie charts and *time series graphs.

play-the-winner rule A rule for allocating treatments to patients in a clinical trial. It is supposed that patients join the trial one at a time, the outcome of the last treatment being known before the next allocation is made. The rule states that if the last treatment was successful then the next patient should also be given that treatment, whereas a failure implies that the next patient should be given the next treatment in the set of possible treatments. The procedure has ethical advantages but practical limitations.

plot *See* EXPERIMENTAL UNIT.

PLS *See* PARTIAL LEAST SQUARES.

point-biserial correlation *See* BISERIAL CORRELATION.

point estimate *See* ESTIMATOR.

point process A *stochastic process concerned with the random positions of locations in space or of points in time. *See* POISSON PROCESS.

Poisson, Siméon Denis (1781–1840; b. Pithiviers, France; d. Sceaux, France) A French mathematician. Poisson studied at the École Polytechnique in Paris. From 1802 to 1808 he taught at the École, attaining the chair in pure mathematics at the Faculté des Sciences. Poisson published in many branches of mathematics. His major work on probability, published in 1837, was *Recherches sur la Probabilité des Jugements en Matière Criminelle et en Matière Civile* (*Researches on the Probability of Criminal and Civil Verdicts*). In this long book (over 400 pages) only about one page is devoted to the derivation of the distribution that now bears his name. He is alleged to have said that 'Life is good for only two things: to study mathematics and to teach it.'

Poisson approximation to the binomial distribution *See* BINOMIAL DISTRIBUTION.

Poisson distribution A *random variable X, whose set of possible values consists of the non-negative integers, with *probability function given by

$$P(X = r) = \frac{e^{-\lambda}\lambda^r}{r!}, \qquad r = 0, 1, \ldots,$$

where λ is a positive constant, is said to have a Poisson distribution, or to be a **Poisson variable**, with parameter λ. (By convention, $\lambda^0 = 1$ and $0! = 1$.) The distribution was named after *Poisson, though the first derivation was by de Moivre in 1711. If we note that $P(X = 0) = e^{-\lambda}$, successive *probabilities can be calculated by using the *recurrence relation

$$P(X = r) = \frac{\lambda}{r}P(X = r - 1), \qquad r = 1, 2, \ldots.$$

The *mean and *variance of the *distribution are both λ. If λ is not an integer the *mode is the value of the integer r for which $r - 1 < \lambda < r$. If λ is an integer then $P(X = \lambda - 1) = P(X = \lambda)$ and both $(\lambda - 1)$ and λ are modes. If $\lambda < 1$ the graph of the probability function decreases steadily, whereas if $\lambda > 1$ the graph increases steadily to the value at the mode, then decreases steadily, tending to 0 as $r \to \infty$. A short table of cumulative probabilities for small values of λ is given in Appendix V.

For large values of λ a **normal approximation to the Poisson distribution** may be used:

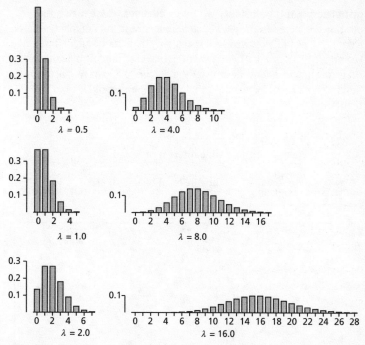

Poisson distribution. The outlines of Poisson distribution with parameter λ increasingly resemble that of a normal distribution as λ increases. Both the mean and the variance of a Poisson distribution with parameter λ are equal to λ.

$$P(X \leq r) \approx \Phi(z), \quad \text{where} \quad z = \frac{r + \frac{1}{2} - \lambda}{\sqrt{\lambda}}$$

and Φ is the cumulative distribution function for a standard *normal variable. The '$\frac{1}{2}$' is a *continuity correction. The approximation may be described as 'For large values of λ a Poisson variable with mean λ is approximately $N(\lambda, \lambda)$.'

In a *Poisson process the number of events in a given region, or a given time interval, has a Poisson distribution.

Poisson process Description of a situation where events occur randomly in time, or space, in such a way that for each small interval of time, or small region of space, the *probability that it contains exactly one event is proportional to the size of the interval or region. It is assumed that this probability is *independent of whether or not any other small intervals, or regions, contain an event, and it is also assumed that the probability of two events occurring in the same small interval or region is 0. This is a mathematical description of randomness. As the diagram shows, randomness usually gives rise to apparent clustering, despite the natural expectation that randomness would lead to regularity.

For events that occur at random instants in time, the formal defining properties of a Poisson process are

$$\frac{\text{P(one event in a time-interval of length } \delta t)}{\delta t} \to k \text{ as } \delta t \to 0,$$

$$\frac{\text{P(two or more events in a time-interval of length } \delta t)}{\delta t} \to 0 \text{ as } \delta t \to 0,$$

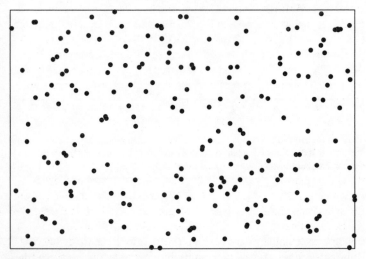

Poisson process. The diagram shows 200 points, randomly scattered such that each location was equally likely; the points were generated using pseudo-random numbers. The arrangement shows the typical characteristics of a spatial Poisson process with apparent clumps and blank spaces.

where k is a positive constant. The number of events in a time interval of length t has a *Poisson distribution with parameter kt.

The time intervals between events, and from the start of observations to the first event, have independent *exponential distributions with parameter k, *mean $1/k$, and *variance $1/k^2$. The time interval from the start of observations to the nth event has an Erlang distribution (*See* GAMMA DISTRIBUTION) with *probability density function f given by

$$f(x) = \frac{k^n x^{n-1} e^{-kx}}{(n-1)!}, \qquad x > 0.$$

The mean of this distribution is n/k and the variance is n/k^2.

An example of a naturally occurring Poisson process is the emission of α-particles from a radioactive source. A **compound Poisson process** is a process in which the events of interest occur in 'packets' (e.g. buses) that arrive at a steady rate. The number in each packet is an observation from the same distribution (e.g. a *uniform distribution between zero and the capacity of a bus).

Poisson variable *See* POISSON DISTRIBUTION.

polar coordinates A coordinate system in a plane in which the position of a point P is specified by the distance $r = OP$, where O is a fixed point (the pole), and the angle θ between OP and a fixed line through O (the initial line). The polar coordinates are given as (r, θ), where $r \geq 0$, and $-180° < \theta \leq 180°$ or $0° \leq \theta < 360°$. Referred to *Cartesian coordinates Oxy, with Ox along the initial line, Cartesian and polar coordinates are related by

$$x = r\cos\theta, \qquad y = r\sin\theta, \qquad r = \sqrt{x^2 + y^2}, \qquad \tan\theta = \frac{y}{x}.$$

Pólya, George (1887–1985; b. Budapest, Hungary; d. Palo Alto, California) A Hungarian mathematician who spent nearly half his life in the United States. Pólya obtained his PhD from the University of Budapest in 1912. From 1914 to 1940 he was a member of the faculty of the Swiss Federal Institute of Technology in Zürich. He then moved to the United States, occupying posts at Brown University and Smith College. His final post, from 1946 to 1953, was as Professor at Stanford University. He was elected to membership of the National Academy of Sciences in 1976. Pólya is quoted as saying that 'If you cannot solve a problem, then there is an easier problem you can solve: find it.'

Pólya distribution *See* BETA-BINOMIAL DISTRIBUTION.

Pólya's urn model A model that has been used to describe the spread of contagious diseases. In this *urn model the urn initially contains b black balls and w white balls. A ball is chosen at random and is then replaced by c (> 1) balls of the same colour. This process is repeated indefinitely.

polynomial A polynomial of **degree** n, is an expression of the form $a_0 + a_1 x + a_2 x^2 + \cdots + a_n x^n$, where x is a variable, $a_0, a_1, a_2, \ldots, a_n$ are constants, and $a_n \neq 0$.

polynomial time complexity *See* COMPLEXITY.

polytomous Adjective describing a *categorical variable with more than two categories. The corresponding noun is **polytomy**.

pooled estimate of common mean An estimate obtained by combining information from two or more *independent *samples taken from *populations believed to have the same mean. *Observations x_{11}, $x_{12}, \ldots, x_{1m}$ are randomly selected from a population. Their mean is $\bar{x}_1$, given by

$$\bar{x}_1 = \frac{1}{m}(x_{11} + x_{12} + \cdots + x_{1m}).$$

Random observations from a second population are denoted by x_{21}, x_{22}, $\ldots, x_{2n}$ and have mean $\bar{x}_2$. If the two populations are believed to have the same mean, then a pooled estimate of the common mean is $\bar{x}$, given by

$$\bar{x} = \frac{m\bar{x}_1 + n\bar{x}_2}{m + n}.$$

With k samples of sizes $n_1, n_2, \ldots, n_k$ and with means $\bar{x}_1, \bar{x}_2, \ldots, \bar{x}_k$, the pooled estimate is given by

$$\bar{x} = \frac{n_1\bar{x}_1 + n_2\bar{x}_2 + \cdots + n_k\bar{x}_k}{n_1 + n_2 + \cdots + n_k}.$$

The unbiased estimate of the variance of the first population is s_1^2, given by

$$s_1^2 = \frac{1}{m-1}\left\{ \sum_{j=1}^{m} x_{1j}^2 - \frac{1}{m}\left(\sum_{j=1}^{m} x_{1j} \right)^2 \right\}.$$

The corresponding estimate for the second population is s_2^2. If it is believed that the two populations have the same variance, but possibly different means, then the **pooled estimate of common variance** is s^2, given by

$$s^2 = \frac{(m-1)s_1^2 + (n-1)s_2^2}{m + n - 2}.$$

In the case of k samples, with the estimate from sample j being s_j^2, the pooled estimate is given by

$$s^2 = \frac{(n_1 - 1)s_1^2 + (n_2 - 1)s_2^2 + \cdots + (n_k - 1)s_k^2}{(n_1 + n_2 + \cdots + n_k) - k}.$$

See also HYPOTHESIS TEST.

pooled estimate of common variance See POOLED ESTIMATE OF COMMON MEAN.

population The complete set of all people in a country, or a town, or any region (or just the number of such people). By extension the term is used for the complete set of objects of interest; for example, all cars built by a particular company in the year 2001, all apples sold as Grade I by a particular supermarket, all students in a university, all smokers. These are all real populations and are finite, though they may be large. The term is also used for the infinite population of all possible results of a sequence of statistical trials; for example, tossing a coin.

population mean The mean value of some *variable that is measured for all members of a possibly infinite population. If the value of the variable for a randomly chosen member of the population is denoted by X, then the population mean is the expectation of X and is usually denoted by μ (the notation μ was introduced in 1936 by Sir Ronald *Fisher in the sixth edition of his *Statistical Methods for Research Workers*). Similarly, the **population variance**, usually denoted by σ^2, is the mean of the squared differences between the values of the members of the population and the population mean: this is the expectation of $(X - \mu)^2$.

population pyramid A diagram for representing the age distribution of a population. It is really a *histogram in which age is plotted vertically and *frequency, or relative frequency (i.e. *proportion), is plotted horizontally. Often drawn as a back-to-back pyramid with one side for males and the other side for females. Paired pyramids can be used to compare two populations. See following diagram.

population variance See POPULATION MEAN.

positive correlation See CORRELATION.

positive definite; positive semi-definite See MATRIX.

positively skewed; positive skewness See SKEWNESS.

posterior distribution; posterior probability See BAYESIAN INFERENCE.

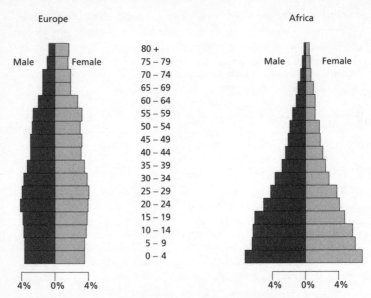

Population pyramid. The pyramids show the age distribution, by sex, for typical European and African nations. The pyramids have the same total area, so the wide base of the African pyramid reflects the high birth-rate, and its fast-tapering shape indicates the low expectation of life.

power; power curve *See* HYPOTHESIS TEST.

power-divergence statistics *Statistics, used in *goodness-of-fit tests, that have the form

$$\frac{2}{\lambda(\lambda+1)}\sum_{j=1}^{n}f_j\left\{\left(\frac{f_j}{e_j}\right)^{\lambda}-1\right\},$$

where f_j is an observed *frequency, e_j is the corresponding *expected frequency, and λ is a constant. The case $\lambda = 1$ gives the *chi-squared test statistic, the limiting case as λ approaches 0 gives the *likelihood-ratio goodness-of-fit statistic, and the case $\lambda = -\frac{1}{2}$ gives the *Freeman–Tukey test statistic.

power spectrum *See* PERIODOGRAM.

p-p plot *See* Q-Q PLOT.

PRE *See* PROPORTIONAL REDUCTION IN ERROR.

prediction interval A statement about the likely value of a future *observation. It has a similar interpretation to a *confidence interval but the interval is wider because it allows for the *random error associated with the future observation.

For example, if a random sample of n observations is taken from a population with known variance σ^2 and unknown mean μ, then the natural estimate of μ is $\bar{x}$, the sample mean. A confidence interval for μ is based on the uncertainty in the estimate, which has variance σ^2/n. However, associated with a future observation from this distribution is the future random sampling variation with variance σ^2. A prediction interval for a future observation is based on the sum of the two variances, $\sigma^2 + \sigma^2/n$.

predictor variable *See* REGRESSION.

PRESS statistic A *statistic that provides an indication of the extent to which a *multiple regression model can be generalized. This is achieved by fitting the model to every subset of $n-1$ of the n *observed values, y_1, $y_2, \ldots, y_n$ of the *response variable. Let $\hat{y}_{-j}$ be the *fitted value for y_j based on the model that uses all the observed values except y_j. The corresponding residual $\hat{\varepsilon}_{-j}$ is given by

$$\hat{\varepsilon}_{-j} = y_j - \hat{y}_{-j}.$$

The PRESS statistic is

$$1 - \frac{\sum_{j=1}^{n} \hat{\varepsilon}_{-j}^2}{\sum_{j=1}^{n} (y_j - \bar{y})^2},$$

where $\bar{y}$ is the *mean of the n observed values. The abbreviation PRESS is derived from predicted residual sum of squares. The statistic is an analogue of the R^2 statistic and values close to 1 are preferable to those close to 0.

primary sampling unit (psu) *See* CLUSTER SAMPLING.

principal components analysis (PCA) A technique for making *multivariate data easier to understand. Suppose that there is information on n *variables for each data item. These variables are unlikely to be *independent of one another; an increase in one is likely to be accompanied by a decrease in another. The idea of PCA is to replace the original n variables by $m(< n)$ *uncorrelated variables, each of which is a linear combination of the original variables, so that the bulk of the variation can be accounted for using just a few explanatory variables. In the diagram, the two original variables x_1 and x_2 can be replaced by the first **principal component**, y_1.

Explaining the variation
between observations
requires statements to be made
about both original variables

Explaining the variation
between observations
requires a statement to be made
about the principal component only

Principal components analysis. Principal components analysis aims to explain an n-dimensional situation using m ($< n$) uncorrelated variables. Here $n = 2$ and $m = 1$.

Let **R** denote the *correlation matrix for the case of p x-variables. The coefficients of the x-variables corresponding to the kth principal component are the elements of the eigenvector corresponding to the kth largest eigenvalue, λ_k, of **R**. All the eigenvalues are non-negative, since **R** is positive semi-definite. The proportion of the variation in the data explained by the kth principal component is $\lambda_k / \sum_{j=1}^{p} \lambda_j$.

The first few principal components should account for the majority of the observed variation. The hope is that the linear combinations thus identified will have some natural interpretation.

principal components regression analysis *See* PARTIAL LEAST SQUARES.

principle of parsimony *See* OCKHAM'S RAZOR.

prior distribution, prior probability *See* BAYESIAN INFERENCE.

probability The probability of an event (*see* SAMPLE SPACE) is a number lying in the interval $0 \leq p \leq 1$, with 0 corresponding to an event that never occurs and 1 to an event that is certain to occur. For an experiment with N equally likely outcomes the probability of an event A is n/N, where n is the number of outcomes in which the event A occurs. For some experiments, such as throwing a drawing pin and seeing whether it lands point up, there is no possible set of equally likely outcomes. In the 'frequentist' view of probability, the probability of getting 'point up' is the limit, in some sense, of the relative *frequency as the number of experiments tends to infinity. In the context of *Bayesian inference, each observer has his or her own *a priori* *distribution for the probability, which is then modified *a posteriori* in the light of whatever results have been obtained.

probability density; probability density function (pdf) For a *continuous random variable X the probability density function f is such that

$$P(x_1 < X < x_2) = \int_{x_1}^{x_2} f(x)dx,$$

for all $x_1 < x_2$. Because, for a continuous random variable, $P(X = x_j) = 0$ for any value x_j, either or both of the '$<$' signs in the left-hand side can be replaced with '$\leq$'. If the interval of possible values for X is (a, b), then

$$\int_a^b f(x)dx = 1.$$

It is often convenient to regard the function f as being defined for all real values of x. This can be achieved by taking $f(x) = 0$ for x outside the interval (a, b), so that

$$\int_{-\infty}^{\infty} f(x)dx = 1.$$

This property and the property that $f(x) \geq 0$ for all x are essential properties of a probability density function. It should be noted that, although f(x) is related to probability, f(x) can exceed 1.

The probability density function is often referred to simply as the probability density. It is related to the *cumulative distribution function F by

$$F(x) = \int_{-\infty}^{x} f(t)dt \quad \text{and} \quad f(x) = \frac{dF(x)}{dx}.$$

The probability density function is not necessarily continuous. At a point of discontinuity the value assigned to f(x) is immaterial.

probability distribution A description of the possible values of a
*random variable, and of the *probabilities of occurrence of these
values. For a *discrete random variable, *see* PROBABILITY FUNCTION. For a
*continuous random variable, *see* PROBABILITY DENSITY FUNCTION.

probability function For a *discrete random variable X, with possible
values $x_1, x_2, \ldots$, the function f defined by

$$f(x_j) = P(X = x_j), \qquad j = 1, 2, \ldots,$$

is the probability function of X. Of course, $\sum_j f(x_j) = 1$.

probability-generating function (pgf) For the *discrete random
variable X, with *probability distribution $P(X = x_j), j = 1, 2, 3, \ldots$, the
probability-generating function G is defined by

$$G(t) = \sum_j t^{x_j} P(X = x_j),$$

where t is an arbitrary variable. Note that $G(t)$ is the expectation of t^X and
$G(1) = 1$. If the set of possible values of x is infinite, $|t|$ needs to be small
enough for the series to converge.

If the first and second derivatives of $G(t)$ with respect to t are denoted
by $G'(t)$ and $G''(t)$, respectively, the expectation and *variance of X are
given by $G'(1)$ and $G''(1) + G'(1) - \{G'(1)\}^2$, where, for example, $G'(1)$
denotes the value of $G'(t)$ when $t = 1$.

Like the *moment-generating function the probability-generating
function can provide a useful alternative description of a probability
distribution. For example, if Y denotes the sum of n *independent
random variables, each having pgf $G(t)$, then $P(Y = y)$ is the coefficient of
t^y in $\{G(t)\}^n$.

Another useful property is that, if X and Y are independent random
variables with probability-generating functions $G_X(t)$ and $G_Y(t)$,
respectively, then the probability-generating function of $Z = X + Y$ is
$G_Z(t)$, given by

$$G_Z(t) = G_X(t)G_Y(t).$$

probability laws The basic properties of *probability. As well as the
*addition law and the *multiplication law, these comprise $0 \leq P(A) \leq 1$
for every event A, and $P(\phi) = 0$ and $P(S) = 1$, where ϕ and S represent the
empty set and the *sample space, respectively.

probability paper; probability plot *See* Q-Q PLOT.

probit A function of a *probability p defined by

$$y = \Phi^{-1}(p),$$

where Φ is the *cumulative distribution function of the *standard normal distribution. When it is believed that the probability of success, p, is dependent upon the values of some predictor variables, it is usual to work with some function of p that has an infinite range in order to avoid estimates of p outside the interval (0, 1). One such function is the *logit; another is the probit.

Procrustes transformation A procedure used in *factor analysis to transform a *matrix to represent a target matrix as closely as possible. The name derives from the Greek myth in which Procrustes was wont to trap travellers and transform their shapes (by the removal or stretching of extreme parts!) to match his bed.

producer's risk *See* ACCEPTANCE SAMPLING.

product-limit estimate *See* KAPLAN–MEIER ESTIMATE.

product-moment correlation coefficient *See* CORRELATION COEFFICIENT.

projection pursuit; exploratory projection pursuit (EPP) A simple alternative to *principal components analysis. To understand the underlying idea, visualize a data set as being a cloud of points in p-dimensional space. Now imagine shining a light on the cloud so that we see the projections of the points on some lower-dimensional 'wall'. If there is an obvious pattern that was not visible previously, then, indeed, light has been cast upon the problem! Typically, the patterns searched for are non-overlapping clusters.

proportion For a *population of size N, of which R have a particular characteristic, the population proportion is given by $P = R/N$. For a *sample of size n, of which r have the characteristic, the sample proportion is $p = r/n$.

proportional allocation *See* STRATIFIED SAMPLING.

proportional-hazards model A *model in which the *logarithm of the *hazard rate, h(t), for a component (or device), is related to one or more explanatory variables $x_1, x_2, \ldots$ by

$$\ln\{h(t)\} = \beta_0 + \beta_1 x_1 + \beta_2 x_2 + \cdots,$$

where $\beta_0, \beta_1, \beta_2, \ldots$ are *parameters.

proportional odds model *See* CUMULATIVE ODDS RATIO.

proportional reduction in error (PRE) A criterion underlying some *measures of association. The measures attempt to quantify the extent to which knowledge about one *variable helps with the prediction of another variable. Examples include R^2 and *Goodman and Kruskal's lambda.

prosecutor's fallacy A mis-statement of a *probability as a result of a misunderstanding of *conditional probability. *See also* DEFENDER'S FALLACY. As an example, suppose a blood type possessed by only 1% of the population is found at a crime scene. The accused has blood of this type. The prosecutor argues that there is only a 1% chance that the accused would have blood of this type (event A, say) if innocent (event B, say) and concludes that the accused is guilty. The prosecutor has quoted $P(A|B)$ when it is $P(B|A)$ that is relevant.

prospective study An alternative name for a *cohort study or *longitudinal study. The term is used in epidemiology.

protocol The step-by-step procedure for carrying out an *experimental design.

proximity matrix A square *matrix in which the entry in cell (j, k) is some measure of the similarity (or distance) between the items to which row j and column k correspond. A simple example would be a standard mileage chart—the smaller the entry, the closer together are the two items. Proximity matrices form the data for *multidimensional scaling. Asymmetric matrices can occur (for example, if the measurement is time taken, then the journey from top to bottom of a hill will be shorter than the journey from bottom to top).

proxy variable A measurable variable that is used in place of a variable that cannot be measured. For example, since husbands and wives usually have similar views, an interviewer might use the view expressed by a present wife in place of the view that could not be expressed by an absent husband. *See also* SURROGATE VARIABLE.

Prussian horse-kicks A set of *data introduced by *Bortkiewicz. It illustrates the fact that the *frequencies of occurrence of unlikely

CORPS	1875	1876	1877	1878	1879	1880	1881	1882	1883	1884	1885	1886	1887	1888	1889	1890	1891	1892	1893	1894
G	0	2	2	1	0	0	1	1	0	3	0	2	1	0	0	1	0	1	0	1
I	0	0	0	2	0	3	0	2	0	0	0	0	1	1	0	2	0	3	1	0
II	0	0	0	2	0	2	0	0	1	1	0	0	2	1	1	1	0	2	0	0
III	0	0	0	1	1	1	2	0	2	0	0	0	1	0	1	2	1	0	0	0
IV	0	1	0	1	1	1	1	0	0	0	0	1	0	0	0	0	1	1	0	0
V	0	0	0	0	2	1	0	0	1	0	0	1	0	1	1	1	1	1	1	0
VI	0	0	1	0	2	0	0	1	2	1	0	1	3	0	1	1	1	3	2	0
VII	1	0	1	0	1	1	0	0	0	0	0	1	0	1	1	0	0	1	0	0
VIII	1	0	0	0	0	0	0	1	0	1	0	0	0	0	1	0	1	1	0	0
IX	0	0	0	0	0	2	1	1	0	1	0	2	1	0	2	2	0	1	0	0
X	0	0	1	1	0	1	0	2	0	2	0	1	1	1	2	1	3	0	1	1
XI	0	0	0	0	2	4	0	1	3	0	1	1	1	1	2	1	3	1	3	1
XIV	1	1	2	0	1	3	0	4	0	1	0	3	2	1	0	2	1	1	0	0
XV	0	1	0	0	0	0	0	1	0	1	1	0	0	0	2	2	0	0	0	0

Prussian horse-kicks. Numbers of deaths due to horse-kicks in each corps of the Prussian Army during the period 1875–94.

*events follow a *Poisson distribution even when there may be variations in the *probabilities of the events. Bortkiewicz's data refer to the numbers of deaths in each corps of the Prussian Army during the period 1875–94. The data, given in the preceding table, show significant variations from year to year and from corps to corps, but have mean (0.70) and variance (0.76) approximately equal.

pseudolikelihood If the values at n locations of a *spatial process are denoted by $x_1, x_2, \ldots, x_n$, the pseudolikelihood is the product

$$\prod_{j=1}^{n} P(X = x_j | x_1, x_2, \ldots, x_n, \text{except for } x_j).$$

This product will depend on the parameters defining the spatial process. Usually, these are unknown and are estimated by maximizing the pseudolikelihood.

pseudo-random numbers Numbers generated by a mathematical formula which has the property that the generated numbers appear to be independent observations from a uniform distribution. Depending on the context, the distribution is usually a continuous uniform distribution on the interval (0, 1), or a discrete uniform distribution on the integers $0, 1, \ldots, 9$. Pseudo-random numbers are often referred to simply as **random numbers**. They form the basis for many statistical methods including the *bootstrap, *Markov Chain Monte Carlo methods and *simulation. A short table of pseudo-random numbers is given in Appendix XIV. It is conjectured that the decimal digits of π (3, 1, 4, 1, 5, 9, ...) may be a naturally occurring sequence of random digits.

pseudovalue *See* JACKKNIFE.

psu *See* CLUSTER SAMPLING.

psychometrics A branch of psychology in which a numerical measurement is made of psychological factors; for example, an individual's or a group's preference in foods, or assessment of artistic or personal qualities.

pure birth process *See* BIRTH-AND-DEATH PROCESSES.

Puri, Madan Lal (1929–; b. Sialkot, Pakistan) Indian born mathematician and statistician, now a US citizen. Puri arrived in the United States in 1957, studying mathematics at the University of

Colorado (Boulder). He obtained his PhD in statistics from the University of California (Berkeley) in 1962. He then joined the faculty at New York University before moving to Indiana University (Bloomington) in 1968, where he is Professor Emeritus. Puri has published extensively on *probability theory, extreme value theory, *time series, and *non-parametric tests. His 1971 book with *Sen on non-parametric tests in *multivariate analysis remains a standard reference. *See* SEN–PURI TEST.

p-value *See* HYPOTHESIS TEST.

q-q plot A plot for comparing two *probability distributions, usually the *sample distribution function and a theoretical *distribution function. The sample values are ranked in order: $x_{(1)} \leq x_{(2)} \leq \cdots \leq x_{(n)}$. Define the sample cumulative *proportion as p_j, typically calculated using $(j - \frac{1}{2})/n$, and denote the theoretical distribution function by F. In a q-q plot, $F^{-1}(p_j)$ is plotted against $x_{(j)}$, for all j. If the sample has come from the theoretical distribution, the plotted values will lie on an approximate straight line. Specialized graph paper, that has the values of F marked for interesting values of p is called **probability paper**.

The **p-p plot** is entirely equivalent: $F(x_{(j)})$ is plotted against p_j. Once again, if the sample has come from the theoretical distribution, the plotted values will lie on an approximate straight line. A plot of this type is called a **probability plot**. *See also* NORMAL PROBABILITY PAPER.

quadrant *See* CARTESIAN COORDINATES.

quadrat Originally, a square wooden frame thrown on to the ground by, for example, botanists who wished to count plant species at random locations. Now used more generally to refer to a sampled area of space.

quadratic form *See* MATRIX.

qualitative variable A nominal or *categorical variable.

quality control The analysis of a sequence of small *samples taken at regular intervals from the output of an industrial production process, with the aim of ensuring that the output meets its required specification.

Standard statistical procedures expect large samples whereas here the samples are small. The solution suggested by *Shewhart was the use of a **control chart (Shewhart chart)**. The basis of this chart is a simple plot of the successive sample means, though often a simultaneous parallel chart of some *measure of spread is maintained. The purpose of the chart is to give a quick visual indication of any trends in the production process (e.g. a drift in the mean), or an increase in variability (possibly a consequence of some problem in part of the process), without the need for advanced statistics. The diagram illustrates a combined **mean chart ($\bar{x}$-chart)** and **range chart (R-chart)**.

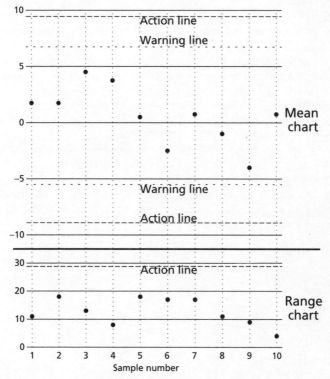

Quality control. This quality control chart plots the values of sample means (upper graph) and ranges (lower graph) as a function of time. If a value falls beyond an action line, or if a succession of sample means lie outside the warning lines, then production is stopped.

On the $\bar{x}$-chart are two pairs of lines: **warning lines** and **action lines**. If any sample mean lies outside an action line, or if a succession of sample means lie outside the warning lines then production is stopped. One set of rules to have found favour are the **Western Electric Rules** which state that the process should be judged out of control if any of the following occur:

1. The sample mean is more than three *standard errors from the notional mean (this is the *three-sigma rule);

2. The sample mean and one of its two predecessors are more than two standard errors from the notional mean (and on the same side of it);

3. The sample mean and three of its four predecessors are more than one standard error from the notional mean (and on the same side of it);

4. The sample mean and its seven predecessors all lie on the same side of the notional mean.

In some applications of quality control the discovery of a value outside the warning line will be a signal to inspect more often—this is **tightened inspection** (as opposed to **normal inspection**).

An alternative to the control chart is the **cumulative sum chart (cusum chart)**. If $\bar{x}_j$ denotes the jth sample mean and m denotes the expected mean of the production process, then, on the cusum chart, $\sum_{j=1}^{k} (\bar{x}_j - m)$ is plotted against k. If production is normal then the plot will be roughly horizontal, whereas a trend indicates a departure from the expected mean. *See* ACCEPTANCE SAMPLING.

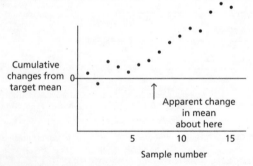

Cumulative sum chart. Small changes in mean show more clearly on a cusum chart than on a control chart.

quartile If a set of numerical *data has n elements and is arranged in increasing order,

$$x_1 \leq x_2 \leq \cdots \leq x_n,$$

the **lower quartile** may be taken to be the *median of the lower half of the data, *i.e.* of $x_1, x_2, \ldots, x_{\frac{1}{2}(n-1)}$ if n is odd, and the median of $x_1, x_2, \ldots, x_{\frac{1}{2}n}$ if n is even. The **upper quartile** may be taken to be the median of the upper half of the data, i.e. of $x_{\frac{1}{2}(n+1)}$, $x_{\frac{1}{2}(n+3)}, \ldots, x_n$ if n is odd, and the median of $x_{\frac{1}{2}(n+2)}, x_{\frac{1}{2}(n+4)}, \ldots, x_n$ if n is even.

Alternatively, approximate values for the lower and upper quartiles can be read from a *cumulative frequency graph as the values of the variable corresponding to cumulative relative frequencies of 25% and 75%, respectively.

For a *continuous random variable X, the lower quartile Q_1 of the distribution is such that $P(X < Q_1) = \frac{1}{4}$ and the upper quartile Q_3 is such that $P(X < Q_3) = \frac{3}{4}$. See INTERQUARTILE RANGE.

quartile coefficient of skewness If the lower and upper *quartiles of a set of *data are denoted by Q_1 and Q_3, respectively, and the *median is denoted by Q_2, the quartile coefficient of skewness is

$$\frac{Q_3 + Q_1 - 2Q_2}{Q_3 - Q_1}.$$

The coefficient takes values in the interval $(-1, 1)$.

quasi-independence The situation, in a *contingency table, in which the usual *independence model holds for a subset of the *cells in the table.

108	12	18	22	7
12	24	36	44	14
18	36	54	66	21

In the table illustrated, all the cells except that at the top left display perfect independence: the quasi-independence model fits the remaining cells perfectly. The *mover–stayer model is a special case of a quasi-independence model.

quasi-symmetry model *See* SQUARE TABLE.

QUEST (Quick, Unbiased, Efficient Statistical Trees) A program for constructing *classification trees.

questionnaire A list of questions—principally used to collect socio-economic or political data on individuals and households.

Quetelet, Lambert Adolphe Jacques (1796–1874; b. Ghent, Belgium; d. Brussels, Belgium) A Belgian mathematician, astronomer, and statistician. Quetelet obtained his PhD (on conic sections) from the University of Ghent in 1819. By 1833 he was working as an astronomer and meteorologist at the Royal Observatory in Brussels, but his international fame was due to his work as a statistician working on social science data. He spent much time constructing *tables and diagrams to show relationships between *variables. He was interested in the concept of an 'average man' as today we talk of the 'average family' and this was the subject of *Sur l'homme et le développement de ses facultés* (*On Man and the Development of his Faculties*), published in 1835. He was one of the founders of the *Royal Statistical Society and, in 1853, he organized the first international *Statistics conference. In 1839 he was elected a Fellow of the Royal Society. A lunar crater is named after him.

Quetelet index (body mass index) An index of obesity, given as the ratio (weight in kg)/(height in m)^2. A person with an index greater than 30 is officially obese.

queues Examples of continuous-time Markov chains. Their properties are much studied by analysts of *stochastic processes. Three components of a queuing system are the inter-arrival times (a *Markov process (M), a more general process (G), or a predetermined process (D)), the service times (also Markovian, general, or predetermined), and the number of servers (one or more). The standard nomenclature for queues describes them as M/M/1, M/D/1, M/G/1, or G/M/k queues, as appropriate.

The basic quantities of interest are the expected number in the queue, the expected number waiting (i.e. not being served) in the queue, the expected queueing time, and the expected waiting time (the sum of the queueing and service times).

quincunx A simple arrangement of pegs on a board that can be used to illustrate the *binomial and *normal distributions. A funnel allows a ball to roll down and strike the single peg on the top line. The ball rolls to left or right (ideally, with equal probability) and then falls to strike a peg on the next row and the process is repeated on each row. At the bottom the ball is held in one of a number of channels. When many balls are fed through the system it is found that the central channels will contain more balls than the extreme ones. Sir Francis Galton used a quincunx in his 1874 lecture on the normal distribution at the Royal Institution in London.

quota sampling A method of *sampling in a survey in which the interviewer is instructed to include certain prescribed percentages of people who come from various identifiable subpopulations, e.g. men over 60, women under 30, unemployed men.

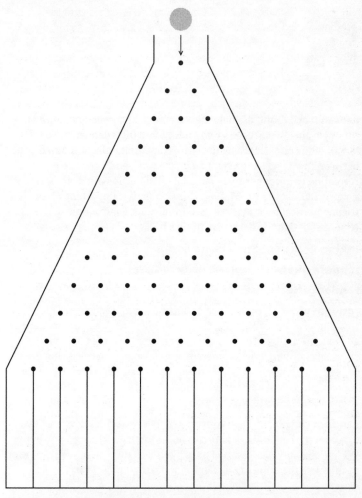

Quincunx. In the diagram each point represents a peg. A series of small balls is inserted at the top of the quincunx. Each ball hits a sequence of pegs before coming to rest in a channel at the bottom of the quincunx. The distribution of balls in channels will be a realization of a binomial distribution, with *n* being the number of rows of pegs.

R

radian (rad) A unit of angle measurement. The radian measure of the angle enclosed between two lines OP and OQ is defined to be the length of the circular arc, with centre O and unit radius, enclosed between the lines. Thus a right angle is $\frac{1}{2}\pi$ rad, and a complete revolution corresponds to 2π rad. The relationship with degrees is that π rad $= 180°$ or 1 rad $\approx 57.296°$. Use of radians is important when trigonometrical functions are used in calculus. For example in the relationship that the derivative of $\sin x$ is $\cos x$.

raking *See* DEMING–STEPHAN ALGORITHM.

random digits *See* PSEUDO-RANDOM NUMBERS.

random effects *See* EXPERIMENTAL DESIGN.

random error An error of measurement as a consequence of recording the value $x + \epsilon$ instead of the true value x, with ϵ being an *observation on a *random variable. The random error is often assumed to have a *normal distribution with *mean 0 and constant *variance (though this assumption should always be verified). In the case of errors introduced by rounding to a fixed number of decimal places, a *uniform distribution is appropriate. *See* SYSTEMATIC ERROR.

random graph A *graph constructed following rules governed by *probability. Let j and k denote two nodes (with $j = k$ being a possibility). With probability p_{jk}, construct an arc between these nodes. The value of p_{jk} might be the same for all pairs of nodes, or it might vary. *See following diagram.*

randomized blocks design An *experimental design in which the bt *experimental units are divided into b *blocks each of size t units. Within each block the t *treatments under comparison are allocated at random. In this example, four treatments (A–D) have been randomly allocated within each of three blocks:

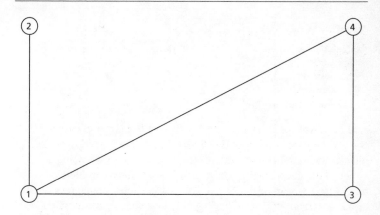

Random graph. The example shows a random graph that is not a connected graph, since node 5 is not connected to the other nodes.

---

random numbers *See* PSEUDO-RANDOM NUMBERS.

random sample *See* SAMPLE.

random variable (r.v.) When the value of a *variable is subject to *random variation, or when it is the value of a randomly chosen member of a population, it is described as a random variable—though the adjective 'random' may be omitted. *See* PROBABILITY DISTRIBUTION.

random variation A *variable is subject to random variation if its value is not predictable.

random walk A walk in which the walker's movements are a consequence of a sequence of *observations on one or more *random variables. For example, suppose that, at each time point an individual walks one step to the left (with *probability p) or one step to the right (with probability $1 - p$). This simple *Markov process is a one-dimensional

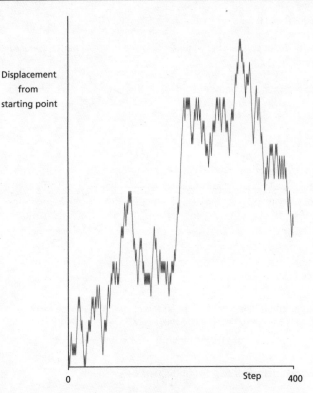

Random walk. The diagram shows the displacement from the starting point for the first 400 steps of a random walk in which each step has the same length, and moves up and down are equally likely. The x-axis represents the step number. The path is typical, in that it predominantly lies on one side of the starting line.

random walk. If the individual is allowed at each stage to move one step in any direction in two dimensions, then this is the **drunkard's walk**. The phrase 'random walk' was first used by Karl *Pearson in 1905. *See also* BROWNIAN MOTION.

   In the particular case where $p = 0.5$, it might be supposed that the proportion of time spent to the left of the starting point is likely to be close to a half. In fact the proportion is more likely to be near to 0 or 1. The probability of the time proportion being less than α ($0 \leq \alpha \leq 1$) is given by the approximate **arc-sine law**:

$$\frac{2}{\pi} \sin^{-1} (\sqrt{\alpha}).$$

random zero *See* STRUCTURAL ZERO.

range The difference between the largest and the smallest items in a set of numerical *data.

range chart *See* QUALITY CONTROL.

rank (of a data item) *See* RANKS.

rank (of a matrix) *See* MATRIX.

rank correlation coefficient A *non-parametric measure of the extent to which two *variables are related. The original *data may be *ranks, or measurements that are converted to ranks. For example, in a flower show, six sunflower exhibits (A–F) might be given the ranks 1, 3, 2, 5, 6, and 4. If the heights of these exhibits are 2.4, 1.9, 2.2, 1.8, 1.85, and 2.0 m then we might suppose that the heights affected the judge's ranking, since on converting the heights to ranks we get a similar pattern to the judge's ranks:

Exhibit	A	C	B	F	D	E
Judge's rank	1	2	3	4	5	6
Rank of height	1	2	4	3	6	5

An attraction of using a **rank correlation coefficient** is that the calculations are simple. Commonly used coefficients are *Spearman's rho and *Kendall's tau.

ranking The allocation of *ranks to data items.

ranks The numbers 1, 2, . . . assigned to a set of objects arranged in order according to some criterion such as increasing numerical value. Many *non-parametric tests are based on the use of ranks. If two or more observations are indistinguishable according to the chosen criterion then they are said to have **tied ranks**; they are usually assigned a rank equal to the average of the ranks that would have been assigned in the absence of ties. The following set of twelve values includes two examples of ties:

Value	11	14	15	15	16	17	17	17	20	25	40	113
Rank	1	2	3.5	3.5	5	7	7	7	9	10	11	12

Rao, Calyampudi Radhakrishnan (1920–; b. Huvina Hadagali, India) An Indian statistician who divides his time between India and the United States. Rao obtained his BSc in mathematics from Andhra University in 1940 and his MSc in statistics from Calcutta University in 1943. In 1944 he joined the *Indian Statistical Institute (ISI), working under *Mahalanobis. There he worked on the *Cramér–Rao inequality (allegedly proved overnight in response to a student inquiry) and on the theorem now known as the *Rao–Blackwell theorem (proved independently by *Blackwell two years later). Mahalanobis sent Rao to Cambridge University (where he is a Life Fellow of King's College) to analyse data under the guidance of Sir Ronald *Fisher—he was Fisher's research student. On obtaining his PhD in 1948, Rao returned to the ISI as head of the research section. By the time of his formal retirement in 1984 he was the Director of the ISI. He was the founding Director of the Center for Multivariate Analysis at Penn State University, where he is Professor Emeritus. He was President of the *International Biometric Society in 1974, President of the *Institute of Mathematical Statistics in 1977, and President of the *International Statistical Institute in 1982. He is a Fellow of the Royal Society and a Fellow of the American Academy of Arts and Science. He was awarded the *Wilks Medal of the *American Statistical Association in 1989 and the *Guy Medal in Silver of the *Royal Statistical Society in 1965. He was made an Honorary Fellow of the Society in 1969.

Rao–Blackwell theorem A theorem, proved independently by *Rao in 1945 and *Blackwell in 1947, that states, in effect, that any function of a sufficient statistic is the unique, minimum variance, unbiased *estimator of its *expected value.

Rasch, Georg (1901–80; b. Odense, Denmark; d. Laesoe, Denmark) A Danish statistician. Rasch was the son of a mathematics teacher who ran a mission high school for prospective seamen. He gained his first taste for mathematics through reading his father's trigonometry textbooks. In 1925 he graduated in mathematics at the University of Copenhagen in 1925, obtaining his PhD in 1930. His subsequent work as a consultant brought him into contact with *Statistics. In 1934 he joined Sir Ronald *Fisher in London. He returned to Denmark in 1935 to the Bio-statistical Department of the Danish National Serum Laboratory. In 1947 he was a founder member of the *International Biometric Society. In 1961 he returned to the University of Copenhagen as Professor of Statistics. His work on the *Rasch model was published in 1960. He was knighted in 1967.

Rasch model A simple model that describes the probability that, when a group of individuals are asked a number of questions, individual j makes a mistake in answering question k. The model, which was suggested by *Rasch in 1960, was originally applied to the answers given in an oral reading test and has subsequently been applied as a model of the responses to questions in many different tests (particularly psychological tests). The model is a particular type of *logit model and takes the form

$$\log\left\{\frac{\text{P(Individual } j \text{ answers question } k \text{ correctly)}}{\text{P(Individual } j \text{ answers question } k \text{ incorrectly)}}\right\} = \mu + \alpha_j + \beta_k,$$

where $\sum_j \alpha_j = 0$ and $\sum_k \beta_k = 0$.

ratio scale *See* INTERVAL SCALE.

Rayleigh, Lord (1842–1919; b. Maldon, England; d. Witham, England) An English mathematical physicist. The third Baron Rayleigh, John William Strutt graduated from Cambridge University as Senior Wrangler in 1865 and held a Fellowship there until 1871. After a period managing his 7000-acre estates he returned to Cambridge as Head of the Cavendish Laboratory (1879–84). From 1887 to 1905 he was Professor of Natural Philosophy at the Royal Institution. His early work in mathematical physics extended to cover a wide range of physical problems. He was President of the London Mathematical Society in 1876 and was awarded its de Morgan Medal in 1890. He was elected a Fellow of the Royal Society in 1873 and its President in 1905. He was awarded the Society's Copley Medal in 1899. He was awarded the Nobel Prize for Physics in 1904. Craters on the Moon and Mars are named after him.

Rayleigh distribution The distribution of the distance between a point and its nearest neighbour in a *spatial *Poisson process. Also the distribution of the distance from the origin in n-dimensional space to the point $(Y_1, Y_2, \ldots, Y_n)$, where $Y_1, Y_2, \ldots, Y_n$ are *independent *normal variables, each with expectation 0 and *variance σ^2. The *probability density function f is given by

$$f(x) = \frac{2x^{n-1}\exp\left\{-\frac{1}{2}\left(\frac{x}{\sigma}\right)^2\right\}}{(2\sigma^2)^{\frac{1}{2}n}\Gamma\left(\frac{1}{2}n\right)}, \qquad x > 0,$$

where $\sigma > 0$ and Γ is the *gamma function. The distribution has *mean and *variance given by

$$\frac{\sigma\sqrt{2}\,\Gamma\left\{\frac{1}{2}(n+1)\right\}}{\Gamma\left(\frac{1}{2}n\right)} \qquad \text{and} \qquad \sigma^2\left(n - 2\left[\frac{\Gamma\left\{\frac{1}{2}(n+1)\right\}}{\Gamma\left(\frac{1}{2}n\right)}\right]^2\right),$$

respectively. The distribution has *mode $\sigma\sqrt{n-1}$. In the case $n = 2$, the expressions for the mean and variance simplify to $\sigma\sqrt{\frac{1}{2}\pi}$ and $\frac{1}{2}\sigma^2(4-\pi)$, respectively.

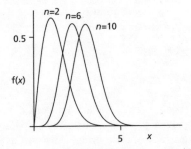

Rayleigh distribution. The shape of the distribution depends on the value of the parameters σ and n. The figure illustrates the dependence on n, with $\sigma = 1$.

Rayleigh test A *test that a set of *cyclic data are consistent with a *circular uniform distribution—in other words, that there is no preferred direction. The test was introduced by Lord Rayleigh in 1880. For a *data set of directions $\theta_1, \theta_2, \ldots, \theta_n$, the test statistic, R^2, is given by

$$R^2 = \left(\sum_{j=1}^{n}\sin\theta_j\right)^2 + \left(\sum_{j=1}^{n}\cos\theta_j\right)^2.$$

Under the *null hypothesis that the data have arisen from a circular uniform distribution, the approximate *probability of observing a value greater than or equal to R^2 is given by

$$\exp\left[\sqrt{1 + 4n + 4(n^2 - R^2)} - (2n+1)\right].$$

For very large values of n, $T = 2R^2/n$ is an *observation from an approximate *chi-squared distribution with two *degrees of freedom.

R-chart *See* QUALITY CONTROL.

recall data Data provided by an individual remembering events in the past. Recall data are notoriously unreliable. For example, if voters are asked after a general election whether or not they voted, the replies will usually overestimate the true proportion voting. Similarly, when voters are asked which party they voted for, support for the winning party will usually be overestimated.

reciprocal The reciprocal of x is $1/x$.

rectangular distribution An alternative name for a *uniform distribution for a *continuous variable.

recurrence relation An equation relating the nth term of a sequence to preceding terms. For example, the *probabilities in a *Poisson distribution with *parameter λ are related by

$$p_n = \frac{\lambda}{n} p_{n-1} \qquad \text{for } n = 1, 2, \dots.$$

recurrent state *See* MARKOV PROCESS.

recursion A method of generating successive terms of a sequence when there is a *recurrence relation expressing the nth term in terms of some or all of the preceding terms. For example, $a_n = na_{n-1}$, with $a_1 = 1$, gives a **recursive** definition of $n!$ (*factorial n).

refined boxplot *See* BOXPLOT.

regression A word, introduced by *Galton and deriving from the phrase '*regression towards the mean' that is often used as shorthand for *linear regression or *multiple regression models. In these models the *mean of one *variable Y is presumed to be dependent on one or more other variables $(x_1, x_2, \dots)$. The variable Y is variously known as the **response variable**, **dependent variable**, or **outcome variable**. The x-variables are known as **predictor variables**, **explanatory variables**, **controlled variables**, or, potentially confusingly, **independent variables**). In the context of a *factorial experiment the x-variables are the *factors.

regression diagnostics Various *statistics that give information about the reliability of the estimates of the regression model

$$\mathrm{E}(\mathbf{Y}) = \mathbf{X}\boldsymbol{\beta},$$

where $\mathbf{Y}$ is an $n \times 1$ *vector of *independent and identically distributed response variables, $\boldsymbol{\beta}$ is a $p \times 1$ vector of unknown *parameters, and $\mathbf{X}$ is an $n \times p$ matrix. If $\boldsymbol{\beta}$ is replaced by its least squares estimate, $\hat{\boldsymbol{\beta}}$, the estimated column vector of *fitted values, $\hat{\mathbf{y}}$, is given by

$$\hat{\mathbf{y}} = \mathbf{H}\mathbf{y},$$

where the $n \times n$ matrix $\mathbf{H}$, the **hat matrix**, is given by

$$\mathbf{H} = \mathbf{X}(\mathbf{X}'\mathbf{X})^{-1}\mathbf{X}',$$

$\mathbf{X}'$ is the transpose of $\mathbf{X}$, $(\mathbf{X}'\mathbf{X})^{-1}$ is the inverse of the matrix $\mathbf{X}'\mathbf{X}$, and $\mathbf{y}$ is the column vector of observed values. Denote the element in the jth row and kth column of $\mathbf{H}$ by h_{jk}. The fitted value, $\hat{y}_j$, for the jth observation, y_j, is given by

$$\hat{y}_j = h_{jj}y_j + \sum_{k \neq j} h_{jk}y_k.$$

Thus there is a direct link between the fitted and observed values in the form of h_{jj}. This is the **leverage**: a large value (e.g. $> 2/p$) indicates an observation having a large **influence** on the form of the fitted model.

The most obvious guide to the fit of a model are the **residuals**, $e_1, e_2, \ldots$, where e_j is given by

$$e_j = y_j - \hat{y}_j.$$

If the random variables have common *variance σ^2 and if s^2 is an unbiased estimate of σ^2, then the **standardized residual** is sometimes defined by e_j/s. However, an unbiased estimate of the variance of e_j is not s^2 but $s^2(1 - h_{jj})$ and a more appropriate residual (having unit variance if the model is correct) is given by r_j, where

$$r_j = \frac{e_j}{s\sqrt{1 - h_{jj}}}.$$

This is sometimes called the **standardized residual** and sometimes the **Studentized residual**.

The *deletion residual is given by

$$d_j = y_j - \hat{y}_{j,\,-j},$$

where $\hat{y}_{j,\,-j}$ is the fitted value for observation j based on the fit of the model to all the observations except the observation y_j. Dividing the deletion residual by its estimated *standard error, we get the **Studentized deletion residual** which can be written as

$$r_{-j} = \frac{e_j}{s_{-j}\sqrt{1 - h_{jj}}},$$

where s_{-j}^2 is the unbiased estimate of σ^2 obtained when observation j is omitted. Confusingly, this may also be called the **Studentized residual**. *See also* ANSCOMBE RESIDUALS; DEVIANCE RESIDUALS.

A related influence statistic is **DFFITS**, which is an abbreviation for difference in fits. For observation j, DFFITS$_j$ is given by

$$\text{DFFITS}_j = \frac{\hat{y}_j - \hat{y}_{j,-j}}{s_{-j}\sqrt{h_{jj}}}.$$

The influence statistic **DFBETA** (difference in beta values) applies the idea embodied in DFFITS to the parameter estimates rather than the fitted values. For β_k, DFBETA$_{k,-j}$ is given by

$$\text{DFBETA}_{k,-j} = \frac{\hat{\beta}_k - \hat{\beta}_{k,-j}}{s_{-j}\sqrt{m_{kk}}},$$

where $\hat{\beta}_k$ is the estimate of β_k from the complete data, $\hat{\beta}_{k,-j}$ is the estimate when observation j is omitted, and m_{kk} is the corresponding diagonal element of the $p \times p$ matrix $(\mathbf{X'X})^{-1}$.

A statistic that usefully combines information about leverage and influence is **Cook's statistic**, D_j, given by

$$D_j = \frac{h_{jj}r_j^2}{(p + 1)(1 - h_{jj})}.$$

This statistic can also be interpreted as measuring the effect on the parameter estimates of omitting the jth observation. Large values point to possible *outliers.

regression towards the mean An expression of the fact that, when we take pairs of related measurements, the most extreme values of one *variable will, on average, be paired with less extreme values of the other variable. In his early work on inheritance *Galton included a study of the heights of successive generations of people. Some of his data are summarized below.

| Mean height of parents (inches) | 72.5 | 70.5 | 68.5 | 66.5 | 64.5 |
| Mean height of their adult children (inches) | 72.2 | 69.5 | 68.2 | 67.2 | 65.8 |

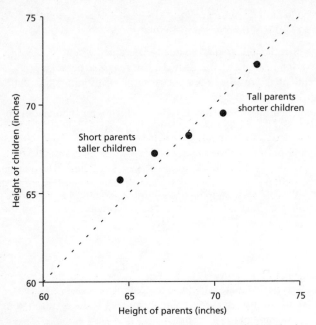

Regression towards the mean. The diagram illustrates Galton's data. The dotted line is the regression line.

Galton noted that, on average, the children of tall parents are shorter than their parents (72.2 < 72.5, etc.), whereas the children of short parents are taller than their parents (65.8 > 64.5, etc.): there is a regression towards the mean. These findings led Galton (in a talk entitled 'Regression towards Mediocrity in Hereditary Stature' given in 1885 to the British Association for the Advancement of Science) to refer to the summary line as a line of *regression.

regular *See* MATRIX.

regular counts *See* INDEX OF DISPERSION.

rejection method A method for *simulating a *random variable X. Suppose, for example, that X is *continuous, with *probability density function (pdf) f, and suppose that it is easy to simulate a random variable, Y, which has a pdf g that satisfies $f(y) \leq cg(y)$, for all y where c is a constant. The rejection method then has three stages: (i) generate y having pdf g; (ii) generate a *pseudo-random number u in the interval $0 < u < 1$; (iii) if $f(y) \geq ucg(y)$, set x equal to y; otherwise return to (i).

rejection region (critical region) The set of values of the *statistic, in a *hypothesis test, which lead to rejection of the *null hypothesis.

relative frequency *See* FREQUENCY.

relative risk A measure of the relative success of one treatment as opposed to another, that is applied in the context of a 2×2 *contingency table. Typically, the rows of the table might refer to two different medicines and the columns to their success or failure. The question is whether there is a difference between the medicines. If there are a failures out of m treated with one medicine, and b out of n for the other medicine, then the relative risk is the ratio

$$\frac{a}{m} \Big/ \frac{b}{n}.$$

reliability A measure of the confidence that we can have in the results obtained from a psychological test. A key question is whether the variability in the scores obtained by different individuals is due to real differences between the individuals or to chance variations resulting from inadequacies in the testing process. The ratio

$$\frac{\text{Var(true scores)}}{\text{Var(test scores)}}$$

is the **reliability index** and its square root is the **reliability coefficient**. The true scores are unknown, but the coefficient can be estimated by using repeat tests. Suppose n individuals are given k similar tests. Let x_m be the total score obtained by individual m over the k tests and let $\bar{x}$ and s^2 be the mean and variance, respectively, of $x_1, x_2, \ldots, x_n$. If each test consists of a single question for which an answer is either correct or incorrect, then p_j represents the proportion of correct answers to question j. Alternatively, if a variety of scores are possible on test j, let s_j^2 be the variance of those obtained. An approximation to the reliability coefficient is provided by **Cronbach's alpha**, given by

$$\alpha = \frac{k}{k-1} \left(1 - \frac{1}{s^2} \sum_{j=1}^{k} s_j^2 \right),$$

and other approximations are provided by the **Kuder–Richardson**

formulae KR_{20} and KR_{21} (named after the equation numbers in Kuder and Richardson's 1937 paper):

$$KR_{20} = \frac{k}{k-1}\left\{1 - \frac{1}{s^2}\sum_{j=1}^{k} p_j(1-p_j)\right\},$$

$$KR_{21} = \frac{k}{k-1}\left\{1 - \frac{1}{s^2}\left(\bar{x} - \frac{1}{k}\bar{x}^2\right)\right\}.$$

An alternative approach is the **split-half method**, in which each test provides two scores (for example, the score on the even questions and the score on the corresponding odd questions). Let r be the *correlation coefficient between the scores obtained on the odd questions and the scores obtained on the even questions. The **Spearman–Brown formula** measures the reliability as

$$\frac{kr}{1 + (k-1)r}.$$

reliability coefficient *See* RELIABILITY.

reliability function *See* SURVIVOR FUNCTION.

reliability index *See* RELIABILITY.

reliability theory Theory concerned with determining the *probability that a system (with n components) is working. Let $x_j = 1$ if the jth component is working and let $x_j = 0$ if it has failed. The *vector $\mathbf{x} = (x_1 \; x_2 \ldots x_n)$ is called the **state vector**. The function $\phi(\mathbf{x})$, which takes the value 1 when the system is working and 0 when it has failed, is called the **structure function**. For n components in **series**,

$$\phi(\mathbf{x}) = \min(x_1, x_2, \ldots, x_n) = x_1 \times x_2 \times \cdots \times x_n.$$

For n components in **parallel**

$$\phi(\mathbf{x}) = \max(x_1, x_2, \ldots, x_n) = 1 - (1 - x_1)(1 - x_2)\cdots(1 - x_n).$$

If $\phi(\mathbf{x}) = 1$ then $\mathbf{x}$ is a **path vector**: it traces a set of connected working components. If failure of any of its working components results in system failure, the vector is a **minimal path vector**. Correspondingly, if $\phi(\mathbf{x}) = 0$ then $\mathbf{x}$ is a **cut vector** and, if it is the case that repair of any of the failed components in $\mathbf{x}$ leads to the system working, then the vector is a **minimal cut vector**.

If p_j denotes the probability that the jth component continues to work during the next unit of time, then the probability that a structure consisting of n components in series continues to work is $p_1 \times p_2 \times \cdots \times p_n$ with the corresponding probability for n components in parallel being $1 - (1 - p_1)(1 - p_2) \cdots (1 - p_n)$.

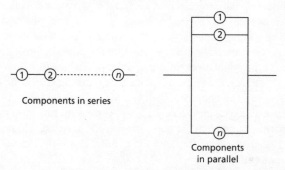

Components in series

Components in parallel

Reliability theory. Complicated systems can always be subdivided into combinations of components arranged either in series or in parallel.

renewal process A *stochastic process concerned with the times of replacement of components that are replaced as soon as they break down. Suppose a system has a single component, which is replaced immediately when it breaks down. Suppose the jth component has a lifetime X_j. The time, T_j, at which this component breaks down is given by

$$T_j = \sum_{k=1}^{j} X_k.$$

The sequence $T_1, T_2, \ldots$ is a renewal process. A useful result is that, if $N(t)$ is the number of breakdowns by time t, then

$$P(N(t) \geq n) = P(T_n \leq t).$$

renewal theory The study of *renewal processes.

repeated measures An *experimental design in which the measurements are taken at two or more points in time on the same set of *experimental units.

replication *See* EXPERIMENTAL DESIGN.

resampling The process of sampling from the *observations in a *sample, in order to obtain estimates and *confidence intervals for *population *parameters without making assumptions about the form of the population *distribution. Suppose that we have taken a random sample of n observations, and assume, for simplicity, that all the sample values $(x_1, x_2, \ldots, x_n)$ are different. If we have no other information about the population then the obvious estimate of the population *mean, μ, is the sample mean $\bar{x}$. This is not contentious. However, it is equally true that an unbiased estimate of the *probability of the value x_j is $\frac{1}{n}$. In a sense the sample is a surrogate for the population—if we want to know what other samples from the population might have looked like, we can find out by sampling from the sample. This is the process called resampling.

As an example, suppose that we wish to estimate the *median of a distribution. We take 10 observations and obtain the values

$$3.1, \quad 3.7, \quad 3.8, \quad 4.1, \quad 4.4, \quad 5.1, \quad 5.2, \quad 5.6, \quad 5.9, \quad 6.2.$$

A simple estimate of the median of the distribution is the median that we have observed, namely 4.75. Resampling enables us to derive an empirical *confidence interval for this estimate. Using the pseudo-random numbers in the first row of Appendix XIV, which begins 07552 37078 we generate a new sample

$$6.2, \quad 5.2, \quad 4.4, \quad 4.4, \quad 3.7, \quad 3.8, \quad 5.2, \quad 6.2, \quad 5.2, \quad 5.6,$$

which has median 5.2. Further resampling produces successive sets of ten 'observations' with medians 5.6, 4.4, 5.35, 4.4, 3.8, 4.25, 3.95, 4.75, 4.4, 4.75, 4.1. The fifteen resampled medians have mean 4.61 and *standard deviation 0.548 so that an approximate 95% confidence interval for the median is (3.5, 5.7). *See also* BOOTSTRAP; JACKKNIFE.

residual *See* REGRESSION DIAGNOSTICS.

residual sum of squares *See* ANOVA.

respondent A person who answers the questions posed by a *questionnaire.

response surface A surface in $(n + 1)$ dimensions that represents the variations in the *expected value of a response variable as the values of n explanatory variables are varied. Usually the interest is in finding the combination that gives a global maximum (or minimum). One interactive procedure is the **method of steepest ascent** (or descent), in which, in a sequence of experiments, the points corresponding to the successive values of the explanatory variables are collinear and lie on the estimated line of greatest (or least) slope that passes through the centre of the original *experimental design. *See* FACTORIAL EXPERIMENT.

response variable *See* REGRESSION.

resultant vector *See* CYCLIC DATA.

retail price index (RPI) A measure of the value of money in which the cost, C, of a standard collection, or basket, of goods and services at some particular date is compared with the cost, C_0, of an equivalent collection at a base date. The RPI is $100 \times \frac{C}{C_0}$, so that at the base date the RPI is 100. The basket is defined to represent the expenditure of a typical household, and the contents of the basket are changed periodically to accommodate changes in the pattern of expenditure. The RPI is a weighted mean of the prices of the goods and services in the basket. *See also* FISHER'S INDEX; LASPEYRES'S PRICE INDEX; PAASCHE'S PRICE INDEX.

retrospective studies Studies that make use of historical information. For example, if we are interested in the life expectancy of patients then it is convenient if all the patients have died, since only then will the lengths of their lives be known.

In a **case–control study** it is the outcomes (for example, death from a particular disease) that determine the sampling procedure. In effect, time is reversed and the early background of those with the outcome (the **cases**) is compared with those without the outcome (the **controls**). The approach allows one to *oversample the cases.

rho (ρ) The symbol customarily used for the *population *correlation coefficient.

ridge regression; ridge trace *See* MULTIPLE REGRESSION MODEL.

right-censored *See* CENSORED DATA.

$\mathbb{R}^n$ *See* CARTESIAN COORDINATES.

robust An adjective applied to a statistic or a statistical procedure which implies that the value of the statistic or the outcome of the procedure will be relatively unaffected by the presence of a small number of unusual or incorrect data values. Thus the *median is a robust *measure of location and the *interquartile range is a robust *measure of spread.

robust regression A method of *regression that is not greatly affected by discordant observations. The most usual method for obtaining regression estimates is *ordinary least squares (OLS), which is quite sensitive to the presence of *outliers. One example of a robust alternative to OLS is *iteratively reweighted least squares.

When there is a single x-variable a simple alternative robust method is as follows.

1. Divide the data into three approximately equal-sized groups according to the sizes of the x-values. Call the groups L (low), M (medium), and H (High). Find the *medians of the x-values and y-values in each group: (x_L, y_L), (x_M, y_M), and (x_H, y_H).
2. Estimate the slope of the line as

$$\tilde{\beta} = \frac{y_H - y_L}{x_H - x_L}.$$

3. Estimate the intercept as

$$\frac{1}{3}\left\{ \left(y_H - \tilde{\beta}x_H\right) + \left(y_M - \tilde{\beta}x_M\right) + \left(y_L - \tilde{\beta}x_L\right) \right\}.$$

root mean squared error *See* ESTIMATOR.

rose diagram An alternative to the *circular histogram as a method of displaying grouped *cyclic data. The areas of the sectors are proportional to the corresponding frequencies. *See following diagram.*

Rosenbrock pattern search A method for finding the maximum (or minimum) of a function. It is based on detecting and moving along ridges in the function.

rotatable design An *experimental design using n *continuous variables, in which all the observations are taken at points that are equidistant, in n dimensions, from the centre of the design.

rounding error (round-off error) Error that occurs when each number in a set of numbers is rounded according to some rule with the result that the total of the rounded numbers is not equal to the rounded

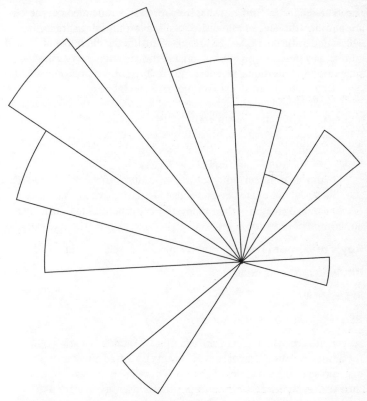

Rose diagram. A rose diagram illustrates the frequencies with which different directions occur. In this case the directions refer to the movements of Colorado beetles released in a wheat field.

version of the original total. For example, rounded to the nearest 10, the *counts 21, 22, 23, and 24 each become 20. However, $4 \times 20 = 80$ whereas the total of the unrounded numbers is 90.

Rounding errors are most often noticed when a set of percentages that should sum to 100% actually sums to something different (*e.g.* 99% or 101%).

round-off error *See* ROUNDING ERROR.

row vector *See* MATRIX.

Royal Statistical Society (RSS) The Royal Statistical Society, which was founded in 1834 and now has over seven thousand members across the world, is the earliest established of all the world's statistics societies. The stated purposes of the RSS are: (i) to nurture the discipline of statistics by publishing a journal, organizing meetings, setting and maintaining professional standards, accrediting university courses and operating examinations, (ii) to promote the discipline of statistics by disseminating and encouraging statistical knowledge and good practice among producers and consumers of statistics and in society at large, and (iii) to provide effective and efficient services to its members which will support their professional and academic interests and their endeavours to advance the other objectives of the Society.

From time to time (currently at three-year intervals) the Society awards the Guy Medal in Gold to persons 'judged to have merited a signal mark of distinction by reason of their innovative contributions to the theory or application of statistics'.

Roy's maximum root *See* MULTIVARIATE ANALYSIS OF VARIANCE.

RPI *See* RETAIL PRICE INDEX.

R^2 *See* ANOVA.

RSS *See* ROYAL STATISTICAL SOCIETY.

rum consumption A favourite example of *nonsense correlation. If the annual consumption of rum in Havana is plotted on a *scatter diagram against the salaries of ministers of religion in Massachusetts, then the graph will show a strong positive nonsense correlation that might suggest that it is the ministers who have been consuming the rum (and/or preaching in favour of its consumption).

runs A run of a certain type of element is an uninterrupted sequence of elements, all of the same type, bordered at each end either by the start or finish of the complete sequence or by an element of a different type.

runs test (Wald–Wolfowitz test) A versatile *non-parametric test. One application is as a test of the *null hypothesis that two *samples, of sizes m and n, have been taken from *populations with the same *distribution. Arrange the two sets of *observations in joint ascending order of size. Label the observations 1 or 2 according to their sample. This results in a sequence of 1s and 2s, *e.g.* 1–222–1–2–1111–2–1, containing a number of *runs (in this case, seven). If there are too few sequences (e.g. 1111111–22222), then this leads to rejection of the null hypothesis. The test statistic is the number of runs. Under the null hypothesis, for reasonably large m and n, this has an approximate *normal distribution, with mean and variance equal to

$$1 + \frac{2mn}{m+n} \quad \text{and} \quad \frac{2mn(2mn - m - n)}{(m+n-1)(m+n)^2},$$

respectively. Since the count is an integer, a *continuity correction of 0.5 will be needed.

As an example, suppose that the reaction times, in ms, of 20 girls and 25 boys were:

Girls   428, 444, 446, 479, 492, 513, 522, 533, 544, 545, 560, 566, 581, 582, 590, 595, 599, 612, 634, 655

Boys   415, 439, 442, 477, 500, 512, 523, 532, 577, 580, 613, 614, 622, 633, 670, 671, 680, 688, 701, 703, 722, 730, 744, 750, 777

The null hypothesis is that the reaction times are *independent samples from a common distribution, the alternative being that this is not the case. Denoting the boys by 2s, we get the sequence

2–1–22–11–2–11–22–11–22–11111–22–111111–2222–11–22222222222.

There are therefore just fifteen runs. The resulting *z-value is

$$\left\{ 15 + 0.5 - \left(1 + \frac{1000}{45} \right) \right\} \Big/ \sqrt{(1000 \times 955)/(44 \times 45^2)} = -2.36.$$

The corresponding *tail probability is about 0.009. Considering both tails, the probability is 0.018 and this provides strong evidence to reject the null hypothesis.

r.v. *See* RANDOM VARIABLE.

St Petersburg paradox A paradox that was the source of much correspondence among eighteenth-century mathematicians. Originally posed by Nicolaus *Bernoulli, it became well known as a result of the solution given by Daniel *Bernoulli in 1738 in the journal of the St Petersburg Academy.

Bernoulli's scenario was essentially as follows. Two players, A and B, play the following game. Player A repeatedly tosses a coin, stopping when a head is obtained. If A has to toss the coin k times, then A pays £2^k to B. Bernoulli's question is 'How much should B pay A in order to make the game fair?' The answer is that B must pay to A the average amount that A pays B. Half the time (assuming a fair coin) this will be £2. Half the remaining time it will be £4, and so on. However,

$$\left(\frac{1}{2} \times \text{£2}\right) + \left(\frac{1}{4} \times \text{£4}\right) + \cdots = \text{£1} + \text{£1} + \cdots.$$

Since the required number of tosses has no upper limit, this sum is infinite. Thus, for this game to be fair, B must pay A an infinite amount of money, even though B is certain to receive only a finite amount of money in exchange (and less than £10 on 87.5% of occasions).

sample A subset of a *population usually chosen in such way that it can be taken to represent the population with respect to some characteristic, for example, height, or cost, or gender, or make of car. A list of members of the population of interest is called the **sampling frame**. If each member of the sample is selected by the equivalent of drawing lots, the sample is a **simple random sample** or, commonly, a **random sample**. In this case each **sampling unit**, i.e. each member of the population, has the same probability of being in the sample, independently of whether any other object is in the sample, and so all possible samples are equally likely. When the sampling units are people, the sample is often referred to as a sample **survey**. Various modifications of the simple random sample are often used. *See* CLUSTER SAMPLING; STRATIFIED SAMPLING; SYSTEMATIC SAMPLING; QUOTA SAMPLING.

sample autocorrelation *See* AUTOCORRELATION.

sample distribution function The equivalent of the *distribution function for a *sample of *data. Let the ordered data be
$x_{(1)} \leq x_{(2)} \leq \cdots \leq x_{(n)}$; then the sample distribution function $F_n(x)$ is given by

$$F_n(x) = \begin{cases} 0 & x < x_{(1)}, \\ j/n & x_{(j)} \leq x < x_{(j+1)}, \quad 1 \leq j \leq (n-1), \\ 1 & x_{(n)} \leq x. \end{cases}$$

sample mean *See* MEAN (DATA); SAMPLING DISTRIBUTION.

sample size The number of *observations in a *sample.

sample space (universal set) A complete set of all possible results or **outcomes** for an experiment or observational procedure. The concept was introduced by *von Mises in 1931. The sample space is usually denoted by S or E.

An **event** is a particular collection of outcomes, and is a subset of the sample space. For example, when a die is thrown and the score observed, the sample space is 1, 2, 3, 4, 5, 6, a possible event is 'the score is even' (i.e. 2, 4, 6). If all the possible outcomes are equally likely, then the probability of an event A is given by

$$P(A) = \frac{\text{Number of events in subset of sample space corresponding to } A}{\text{Number of events in sample space}}.$$

The word event was used in this context by *de Moivre in 1718.

The subset of the sample space for the event 'the score is both even and odd' is the **empty set**, usually denoted by ϕ, and $P(\phi) = 0$.

The subset of the sample space for the event 'the score is less than 10', is the whole sample space, S, and $P(S) = 1$.

See also BOOLEAN ALGEBRA; COMPLEMENTARY EVENT; INTERSECTION; UNION.

sample variance *See* VARIANCE (DATA); SAMPLING DISTRIBUTION.

sampling distribution *Distribution that describes the variation in the values of a *statistic over all possible *samples. For example, if n values are sampled from a *population and if $X_1, X_2, \ldots, X_n$, are the *random variables representing the individual sample values, then the **sample mean** $\bar{X}$, given by

$$\bar{X} = \frac{1}{n}(X_1 + X_2 + \cdots + X_n).$$

is a random variable. The variability of the n values about their mean,

$$V^2 = \frac{1}{n}\left\{ (X_1 - \bar{X})^2 + (X_2 - \bar{X})^2 + \cdots + (X_n - \bar{X})^2 \right\},$$

is also a random variable. The form of the sampling distributions of $\bar{X}$ and V^2 will depend on the population, but statements can nevertheless be made about their *moments. If the population has *mean μ and *variance σ^2, then, for an infinite population, or for *sampling with replacement from a finite population, each of $X_1, X_2, \ldots$ has mean μ and variance σ^2. Consequently the expectation of $\bar{X}$ is μ and the expectation of V^2 is $\frac{n-1}{n}\sigma^2$. This shows that $\bar{X}$ and $S^2 = \frac{n}{n-1}V^2$ are unbiased estimators of μ and σ^2, respectively. The variance of the sample mean $\bar{X}$ is $\frac{1}{n}\sigma^2$.

The **sample variance** is usually taken to be the value of S^2, though the value of V^2 is sometimes used.

sampling fraction The *sample size, n, divided by N, the size of the finite *population from which it has been drawn. Its importance is that the *variance of the sample mean (*see* SAMPLING DISTRIBUTION) is not σ^2/n, but

$$\frac{\sigma^2}{n}\left(1 - \frac{n}{N}\right).$$

The term $(1 - \frac{n}{N})$ is the **finite population correction**.

sampling frame; sampling unit *See* SAMPLE.

sampling with replacement The case where *observations are taken one at a time from a *population. The sampled value is returned to the population before the next value is selected. For example, in a competition with several prizes it is possible for one lucky person to be selected as the winner of several prizes. By contrast, **sampling without replacement** implies that each member of the population can be chosen only once.

Sankhya The journal of the *Indian Statistical Institute, first published in 1933. There are two series; Series A concentrates onprobability and mathematical statistics, whereas the emphasis of Series B is on data analysis and statistical methodology.

SARIMA models *ARIMA models for *time series that display *seasonality.

SAS (Statistical Analysis System) A powerful statistical package permitting many types of analyses via user-written commands.

Satterthwaite's formula A formula for finding the approximate *distribution of a linear combination of *independent *chi-squared variables. Define the *random variable S^2 by

$$S^2 = \sum_j c_j S_j^2,$$

where $c_1, c_2, \ldots$ are known positive constants, and the chi-squared random variable S_j^2 has v_j *degrees of freedom. Satterthwaite's suggestion, made in 1946, was that the distribution of vS^2/σ^2 is approximately *chi-squared with v *degrees of freedom, where $\sigma^2 = \sum_j c_j v_j$, and v is given by

$$v = \frac{\left(\sum_j c_j s_j^2\right)^2}{\left\{\sum_j \frac{1}{v_j}\left(c_j s_j^2\right)^2\right\}},$$

where s_j^2 is the observed value of S_j^2. See also BEHRENS–FISHER PROBLEM.

saturated model A *model that perfectly fits the *data because it has as many *parameters as there are values to be fitted. It can provide useful information in the search for a simpler **unsaturated model** in which some of the parameters of the saturated model are set to 0.

scalar A number. The term is often used to distinguish a number from a *vector or *matrix.

scaling function See WAVELET.

Scandinavian Journal of Statistics A journal, published in English, concerned with advances in statistical theory and its applications. The first volume was published in 1974.

scatter diagram (scattergram; scatter plot) The simplest display when the *data consists of pairs of values. The data are plotted as a series of points using *Cartesian coordinates. If the data are ordered (for example, in time) then it may be sensible to join the successive points with a line. If there are other *categorical variables, their values can be indicated using different plotting symbols or different colours. A quantitative third variable can be indicated by varying the size of the plotting symbol.

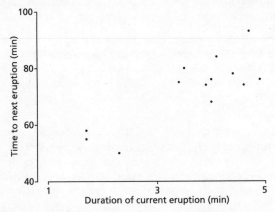

Scatter diagram. The diagram shows the Old Faithful data for 1 August 1974. Long eruptions are followed by long intervals before the next eruption.

Scheffé, Henry (1907–77; b. New York City; d. Berkeley, California) An American statistician. Scheffé worked as a technical assistant at the Bell Telephone Laboratories for four years after leaving school. He then studied at the University of Wisconsin, Madison, first as an undergraduate and finally as a postdoctoral instructor. Subsequently he held positions at many institutions, becoming *Neyman's Assistant Director at the University of California, Berkeley, in 1953. In 1959 his classic book *The Analysis of Variance* was published. He died in a traffic accident while cycling on the Berkeley campus. Scheffé was President of the *Institute of Mathematical Statistics in 1955.

Scheffé's test *See* MULTIPLE COMPARISON TESTS.

Schmidt net A method of illustrating three-dimensional *spherical data in two dimensions. The point with *spherical polar coordinates $(1, \theta, \phi)$ is represented by the point with *polar coordinates (d, ϕ), where, for $0° \leqslant \theta \leqslant 90°$,

$$d = 2\sin\left(\frac{1}{2}\theta\right).$$

The quantity d is the (chordal) distance from the pole at $\theta = 0°$. Data having $\theta > 90°$ are plotted in the same way, using either a different diagram, or a different plotting symbol, and replacing $\sin\left(\frac{1}{2}\theta\right)$ with $\cos\left(\frac{1}{2}\theta\right)$; the distance is now taken from the opposite pole at $\theta = 180°$.

Schwarz criterion *See* AIC.

scree plot A plot, in descending order of magnitude, of the eigenvalues of a *correlation matrix. In the context of *factor analysis or *principal components analysis a scree plot helps the analyst visualize the relative importance of the factors—a sharp drop in the plot signals that subsequent factors are ignorable.

seasonally adjusted A term applied to a *time series from which the *seasonality component has been removed.

seasonality A pattern in a *time series that repeats in a regular way—for example, daily average temperatures rise each summer.

second-order stationary *See* STATIONARY PROCESS.

secretary problem An interesting problem in *decision theory. There are n applicants for the post of secretary. The applicants are randomly ordered and each is interviewed in turn until an appointment is made. Before each interview, the employer must decide whether or not to appoint the previous applicant. The employer cannot subsequently decide to appoint a previously interviewed applicant. It turns out that the strategy that maximizes the probability of appointing the best secretary is to interview the first e^{-1} (~37%) of the applicants, and then to appoint the first subsequent applicant who is superior to all those in the first group. If there are none, then the employer is left having to appoint the nth applicant.

self-organizing feature maps *See* KOHONEN NETWORKS.

semi-interquartile range If the lower and upper **quartiles** of a *variable are denoted by Q_1 and Q_3, respectively, the semi-interquartile range is $\frac{1}{2}(Q_3 - Q_1)$. The term can be used for either a set of *data or a *probability distribution.

semi-Markov process A *renewal process in which the *random variable X can take one of N states. When it is in state j it stays there for a random length of time, with *mean μ_j, and then moves to state k with *probability p_{jk}.

semi-variogram *See* AUTOCORRELATION.

Sen, Pranab Kumar (1937–; b. Calcutta) An Indian statistician who has spent most of his career in the United States. Sen was educated at Calcutta University, where he gained his PhD in 1962. He spent three years on the faculty at Calcutta before moving to the University of California, Berkeley in 1964. In the following year he moved to the University of North Carolina at Chapel Hill, where he is now professor in the Departments of Biostatistics and Statistics. It was at Chapel Hill that he commenced his collaboration with *Puri that resulted in the publication of their influential 1971 book detailing *non-parametric methods for use with *multivariate data.

Sen–Puri test A *test, most used in medical contexts, in which individuals are re-examined at a sequence of t time points. The individuals belong to one of a set of groups (usually corresponding to different treatments). The Sen–Puri test is a *non-parametric test of the *null hypothesis that the sequences of *observations have arisen from a common *distribution. The test uses scores, based on *ranks, accumulated over the time points.

sensitivity A term used in connection with the diagnosis of diseases. Each patient in a *sample, who may or may not have a disease, is tested for that disease. The test gives either a positive result or a negative result. The four possible outcomes are shown in the table, together with the corresponding frequencies a, b, c, and d.

		Disease	
		yes	no
Test	positive	a	b
	negative	c	d

Sensitivity is the *conditional probability of the test correctly giving a positive result, given that the patient does have the disease. An estimate is $a/(a + c)$. **Specificity** is the conditional probability of the test correctly giving a negative result, given that the patient does not have the disease. An estimate is $d/(b + d)$.

sequential sampling Sampling in which *observations are taken one at a time, with an appraisal, after each observation, of whether there is any need for further sampling. This approach is useful when taking an observation is very expensive, or when there are ethical considerations. After each observation has been taken, the *data so far available are re-analysed and one of three possible decisions is taken: accept the *null hypothesis, reject the null hypothesis, or take another observation. *See also* DOUBLE SAMPLING.

serial correlation *Autocorrelation between values that are at a constant distance apart in time or space. Suppose $x_1, x_2, \ldots, x_n$ is an ordered sequence of observations. The serial correlation at lag k is r_k, given by

$$r_k = \frac{\sum_{j=1}^{n-k} (x_j - \bar{x}_1)(x_{j+k} - \bar{x}_n)}{\sqrt{\left\{\sum_{j=1}^{n-k}(x_j - \bar{x}_1)^2\right\}\left\{\sum_{j=k+1}^{n}(x_j - \bar{x}_n)^2\right\}}}, \quad k = 1, 2, \ldots, n-1,$$

where

$$\bar{x}_1 = \frac{1}{n-k}\sum_{j=1}^{n-k} x_j, \quad \text{and} \quad \bar{x}_n = \frac{1}{n-k}\sum_{j=k+1}^{n} x_j.$$

In practice, a less complicated formula is routinely used. This formula, which involves $\bar{x}$, the overall *mean, is

$$r_k = \frac{n\sum_{j=1}^{n-k}(x_j - \bar{x})(x_{k+j} - \bar{x})}{(n-k)\sum_{j=1}^{n}(x_j - \bar{x})^2}.$$

series structure *See* RELIABILITY THEORY.

sex-specific rate *See* INCIDENCE RATE; MORTALITY RATE.

Shannon index *See* DIVERSITY INDEX.

Shapiro, Samuel S. (1930–) An American statistician and engineer. A statistics graduate of City College (now City University), New York in 1952, Shapiro took an MS in industrial engineering at Columbia University in 1954. After a period working as a statistician in the Army Chemical Corps, Shapiro joined the General Electric Corporation, obtaining his MS (1960) and PhD (1963) in statistics at Rutgers University. Shapiro was a co-author of the 1965 paper that introduced the *Shapiro–Wilk test. In 1972 Shapiro joined the faculty at Florida International University, where is he now Professor of Statistics.

Shapiro–Wilk test A *test that the *population being sampled has a specified *distribution. It was introduced by *Shapiro and *Wilk in 1965. The test compares the ordered *sample values with the corresponding *order statistics from the specified distribution. The test is most commonly used to test for a *normal distribution, in which case the test statistic, W, is given by

$$W = \frac{\left(\sum_{j=1}^{n} w_j x_{(j)}\right)^2}{\sum_{j=1}^{n} \left(x_{(j)} - \bar{x}\right)^2},$$

where $x_{(j)}$ is the jth largest observation, $\bar{x}$ is the sample mean, and w_j is a function of the *means and *variances and *covariances of the order statistics. In the case where the hypothesized distribution is *exponential the statistic takes the simple form

$$W = \frac{n\left(\bar{x} - x_{(1)}\right)^2}{(n-1)\sum_{j=1}^{n} \left(x_{(j)} - \bar{x}\right)^2}.$$

shell sort An efficient *algorithm for arranging a set of n numbers in order of magnitude. It starts by applying a succession of *bubble or *shuttle sorts to carefully chosen subsets of the data. A shuttle sort is then applied to the resulting partially ordered data.

Shepard diagram A plot of two measurements of the distances between objects. One measurement is the true distance, and the other measurement is the apparent distance in some representation of the objects. For example, the apparent distance between objects in a photograph (two dimensions) and the real three-dimensional distance. The diagram is used in *multidimensional scaling to assess the extent of any distortion. No distortion would correspond to a set of collinear points.

Sheppard's corrections Adjustments to the values of the *moments of the *sample when these are calculated from grouped data. For example, with class intervals of width d, the correction to the calculated estimate of the *variance is to subtract $\frac{1}{12}d^2$.

Shewhart, Walter Andrew (1891–1967; b. New Canton, Illinois; d. Troy Hills, New Jersey) An American physicist and statistician. Shewhart studied physics at the Universities of Illinois and California. In 1918 he joined the Western Electric Company, which made hardware for Bell Telephone. A major problem was the reliability of the transmission systems. Study of this problem led to Shewhart's introduction of the *control chart, invented in 1924 and popularized in his 1931 book *Economic Control of Quality of Manufactured Products*. He was President of the *Institute of Mathematical Statistics in 1937 and again in 1944. He was President of the *American Statistical Association in 1945.

Shewhart chart *See* QUALITY CONTROL.

shuttle sort A simple, but not very efficient, *algorithm for arranging a set of n numbers in order of magnitude. The method starts with the left-hand pair of numbers, swapping them if necessary. The second and third numbers are now considered. If they are swapped then the first pair are reconsidered. Next the third and fourth numbers are considered. If swapped then previous pairs are again reconsidered, working from right to left. As with the *bubble sort, $\frac{1}{2}n(n-1)$ comparisons may be required.

16		8		13		4
	×					
8		16		13		4
			×			
8		13		16		4
	○					
8		13		16		4
					×	
8		13		4		16
			×			
8		4		13		16
	×					
4		8		13		16

Shuttle sort. A sorting method based on swaps of pairs of numbers. In the example, X indicates a swap and ○ that no swap is required. The sort works from left to right, reconsidering earlier pairs when a swap is made.

sigma (σ) The symbol usually used to denote the *standard deviation of a *population; σ^2 denoting the *variance.

sigma (Σ) Symbol used to denote a sum. Thus

$$\sum_{k=1}^{5} k = 1 + 2 + \cdots + 5,$$

$$\sum_{j=1}^{n} j^2 = 1^2 + 2^2 + \cdots + n^2,$$

$$\sum_{r=1}^{n} x_r = x_1 + x_2 + \cdots + x_n,$$

$$\sum_{k=1}^{3} a_{kj} x_k = a_{1j} x_1 + a_{2j} x_2 + a_{3j} x_3.$$

The sample *mean for n *observations $x_1, x_2, \ldots, x_n$ is

$$\frac{1}{n} \sum_{j=1}^{n} x_j$$

and the *unbiased estimate of the *population variance is

$$\frac{1}{n-1} \left\{ \sum_{j=1}^{n} x_j^2 - \frac{1}{n} \left(\sum_{j=1}^{n} x_j \right)^2 \right\}.$$

signal The value of a measurement that would be observed if the measurement were not contaminated by *random errors.

signed ranks; signed-rank test *See* NON-PARAMETRIC TESTS; WILCOXON SIGNED-RANK TESTS.

significance level; significant *See* HYPOTHESIS TEST.

sign test A simple *non-parametric test that is used in two situations:

1. A random *sample of n *observations $x_1, x_2, \ldots, x_n$ is taken on the *random variable X; the *null hypothesis is that the *population has *median m_0.
2. A random sample of n observations $(x_1, y_1), (x_2, y_2), \ldots, (x_n, y_n)$ is taken on the pair of random variables (X, Y); the null hypothesis is that the distribution of $X - Y$ has median 0.

In both cases the analysis begins by noting the signs of the differences $d_1, d_2, \ldots, d_n$, where in case (1) $d_j = x_j - m_0$ and in case (2) $d_j = x_j - y_j$. In either case the test statistic, r, is the number of differences that have a positive value.

If the null hypothesis is correct and there are no zero differences, r is an observation from a *binomial distribution with *parameters n and

0.5. If the *one-tail or two-tail probability (as appropriate) is unusually low, then the null hypothesis will be rejected.

If there are k differences equal to 0, it is conventional to ignore the corresponding observations and to use a binomial distribution with parameters $(n - k)$ and 0.5.

As an example, suppose that the null hypothesis is that the distribution of the weights of 20-year-old males has median 77 kg, the alternative being that this is not the case. A random sample of thirteen 20-year-old males have the following weights (in kg): 59, 84, 99, 83, 65, 70, 77, 69, 85, 66, 76, 73, 81. The question is whether these data support the null hypothesis. In the sample there is one value equal to the hypothesized median. This is ignored. Of the remaining twelve values, five exceed 77 kg. Using the binomial distribution with $n = 12$ and $p = 0.5$, the probability of five or fewer is 0.387. The two-tail probability is therefore 0.774: we conclude that there is no significant evidence to refute the null hypothesis.

simple random sample *See* SAMPLE.

simplex A generalized triangle or tetrahedron. Suppose $\mathbf{a}_1, \mathbf{a}_2, \ldots, \mathbf{a}_n$ are n *linearly independent *vectors, or points, in $\mathbb{R}^n$. The set of all points $\sum_{j=1}^n \lambda_j \mathbf{a}_j$, where each λ_j is non-negative and $\sum_{j=1}^n \lambda_j = 1$, is an n-simplex. *See* BARYCENTRIC COORDINATES.

simplex method An *algorithm introduced in 1947 by *Dantzig for the solution of a *linear programming problem. The algorithm works by moving from one vertex of the *feasible region to an adjacent one.

Simpson index *See* DIVERSITY INDEX.

Simpson's paradox An intriguing paradox illustrating how one may be misled when a relevant *variable is overlooked. The paradox is illustrated in the following example, which shows a cross-classification of three *dichotomous variables, A, B, and C (where, for example, A_1 and A_2 are the two categories of A).

	C_1					C_2							
	B_1	B_2	Total			B_1	B_2	Total			B_1	B_2	Total
A_1	95	800	895	+	A_1	400	5	405	=	A_1	495	805	1300
A_2	5	100	105		A_2	400	195	595		A_2	405	295	700
Total	100	900	1000		Total	800	200	1000		Total	900	1100	2000

In the *subpopulation corresponding to C_1, there is a strong positive association between A and B. The same is true for the subpopulation corresponding to C_2. However, when the information on these two very dissimilar subpopulations is pooled, the association for the entire population is strongly negative. *See* ECOLOGICAL FALLACY for a diagram that shows how this type of result can occur with continuous data.

Simpson's rule A method for finding an approximate value for an integral, which is usually more accurate than the *trapezium rule. Suppose we wish to find an approximate value for $\int_a^b f(x)dx$. The interval $a \leq x \leq b$ is divided into an even number, $2n$, of sub-intervals, each of length $(b-a)/(2n)$, at points $a = x_0, x_1, x_2, \ldots, x_{2n-1}, x_{2n} = b$, where $x_r = (a + rh)$ for $r = 0, 1, 2, \ldots, 2n-1, 2n$. Writing $y_r = f(x_r)$ the integral is approximated by

$$\frac{h}{6}\{y_0 + 4(y_1 + y_3 + \cdots + y_{2n-1}) + 2(y_2 + y_4 + \cdots + y_{2n-2}) + y_{2n}\}.$$

Simpson's rule gives the exact answer when f is a *polynomial of order 3 or less. In general, the error in using Simpson's Rule is approximately proportional to $1/n^4$, so that if the number of sub-intervals is doubled then the error is reduced by a factor of 16. Simpson's rule is based on fitting a quadratic graph in each adjacent pair of sub-intervals such that the graph passes through the points (x_{2r}, y_{2r}), (x_{2r+1}, y_{2r+1}) and (x_{2r+2}, y_{2r+2}).

simulated annealing A method for maximizing a function that permits an *algorithm to choose apparently suboptimal routes, but with decreasing *probability as the number of *iterations increases. The initially fluid choice of values becomes increasingly set as the algorithm progresses. The object is to reduce the chance of ending at a *local maximum.

simulation (Monte Carlo simulation) A procedure used when there is no analytic solution available for a problem involving *random variables. *Pseudo-random numbers are used to mimic the random variables involved. Two general methods are the *inverse transformation method and the *rejection method.

simultaneous equation model A collection of m *linear models in which each model involves k response variables ($k \leq m$) in addition to explanatory variables. Many *econometric models have this form. Often the explanatory variables are *latent variables, in which case the model is a **structural equation model**.

single linkage clustering A method of collecting *multivariate data into *clusters. *See* AGGLOMERATIVE CLUSTERING METHODS.

singular *See* MATRIX.

sink *See* NETWORK FLOW PROBLEMS.

size (of a sample) The number of *observations in the sample.

size (of a test) The *probability of a Type I error, i.e. that in a *hypothesis test the *null hypothesis is rejected when it is true; a synonym for significance level.

skew; skewness If the *distribution of a *variable is not symmetrical about the *median or the *mean it is said to be skew. The distribution has **positive skewness** if, in some sense, the tail of high values is longer than the tail of low values, and **negative skewness** if the reverse is true. Skewness is quantified by *Pearson's coefficient of skewness, the *quartile coefficient of skewness, or (preferably) the *moment coefficient of skewness. For another measure of the shape of a distribution, *see* KURTOSIS.

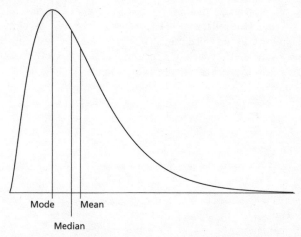

Mode   Mean
Median

Skew. An example of a positively skewed distribution. Here it is supposed that the random variable has a finite lower bound but an infinite upper bound. This typically results in the mean being bigger than the median and the median being bigger than the mode.

Sklar's theorem *See* COPULA.

slack variable *See* LINEAR PROGRAMMING.

slippage test *See* TUKEY'S QUICK TEST.

Slutzky, Evgeny Evgenievich (1880–1948; b. Novoe, Russia; d. Moscow, Russia) A Russian statistician. Slutzky entered the University of Kiev in 1899 to study mathematics. However, after twice being involved in student unrest, he was expelled and completed his studies in Munich. In 1905 he returned to Kiev, this time to study political economics. In 1913 he joined the staff of Kiev Institute of Commerce, moving to Moscow in 1926 to work in the government statistical offices. During his period there he worked on the theory of *time series and *stochastic processes. From 1938 he worked at the Institute of Mathematics at the USSR Academy of Sciences.

Slutzky–Yule effect An undesirable consequence of applying a *moving average to a *time series. Suppose a time series consists of randomly chosen *observations from the same *population. We would therefore hope that any averaging would bring out the fact that the *mean was constant. However, by chance some values will be larger than others. Let x_k be a particularly large value. When we apply a moving average, all the averages that involve x_k will be inflated. With most moving averages the inflation will be greatest for the average centred on the kth observation and will diminish on either side. Each extreme value will have a similar effect such that the series of averages will present oscillations that appear real but are due to chance.

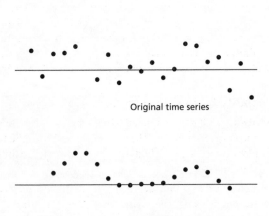

Original time series

Smoothed time series

Slutzky–Yule effect. Chance variations in single values translate into apparent cycles in moving averages.

Smirnov, Nikolai Vasil'evich (1900–66) A Russian mathematician. Smirnov obtained his first degree from Moscow University in 1926. His doctoral thesis (1938) served as a foundation for his subsequent work on the theory of *non-parametric tests. His subsequent career was spent at the Steklov Mathematical Institute of the USSR Academy of Sciences. Smirnov's forte lay in solving difficult computational problems using ingenious methods. Smirnov is remembered today for his 1939 work on the distribution of the test proposed by *Kolmogorov in 1933.

Snedecor, George Waddel (1881–1974; b. Memphis, Tennessee; d. Amherst, Massachusetts) An American biometrician. Snedecor received a BS in mathematics and physics from the University of Alabama in 1905 and an AM in physics from the University of Michigan in 1913. In that year he was one of the first faculty to join the Mathematics Department at Iowa State College and one of his first courses was on *Statistics. Snedecor spent his entire career at Iowa and the Department now occupies a building named in his honour. Iowa's first graduate student, in 1931, was Gertrude *Cox. Snedecor's *Statistical Methods Applied to Experiments in Agriculture and Biology*, co-authored with *Cochran, sold more than 125 000 copies.

snowflake curve An example of a curve having *fractal dimension. Starting with an equilateral triangle, the middle third, PQ, say, of a side is replaced by the two lines PR and RP so that P, Q, and R form the vertices of a smaller equilateral triangle, with R outside the original enclosed region. This process is repeatedly applied to each line segment. The resulting 'curve' is the snowflake curve and has infinite length but encloses a finite area. Its fractal dimension is defined to be ln 4/ln 3, since each 'edge' of the curve contains four copies each of 1/3 size. *See following diagram.*

social mobility table *See* MOBILITY TABLE.

social statistics *Statistics applied to the social sciences, particularly sociology and demography. The *Lexis diagram and the *life table are two examples of statistical methods developed specifically for this branch of the subject.

Società Italiana di Statistica The Italian Statistical Society which publishes the English language *Journal of the Italian Statistical Society*. The Society has about a thousand members.

Société Française de Statistique The French Statistical Society, formed from the amalgamation in 1997 of the Société de Statistique de

Snowflake curve. This is an example of a curve with fractal dimension $\neq 1$.

Paris (founded in 1860) with two related organizations. It publishes the *Journal de la SFdS*. The Society has about a thousand members.

Somers's d_{BA} *See* ORDINAL VARIABLE.

source *See* NETWORK FLOW PROBLEMS.

South African Statistical Association This association was founded in 1953 and has around two hundred members.

spatial autocorrelation *See* AUTOCORRELATION.

spatial process The manner in which values change from one spatial location to another.

SPC *See* STATISTICAL PROCESS CONTROL.

Spearman, Charles Edward (1863–1945; b. London, England; d. London, England) Spearman was a psychologist who was responsible for bringing the use of statistical methods into psychology and thereby changing the practice of that subject once and for all. Spearman was born in London and his initial choice of career was the army. He fought with distinction in the Burmese War and did not leave the army until 1897, when he went to Leipzig to study psychology. After jobs in various German universities he returned to London in 1907 as Professor of Psychology at University College, London, retiring in 1931. His first paper on correlation, introducing what is now called *Spearman's rho, appeared in 1904. In the same year a second paper laid the foundations for *factor analysis. Spearman was elected to the National Academy of Sciences in 1943.

Spearman–Brown formula *See* RELIABILITY.

Spearman's rho (ρ) A *rank correlation coefficient that may be used as an alternative to *Kendall's tau. Individuals are arranged in order according to two different criteria (or by two different people). The *null hypothesis is that the two orderings are *independent of one another. It is based on the differences in the ranks given in two orderings. Suppose that the jth individual is given rank x_j in one ordering and rank y_j in the second ordering. Define d_j by $d_j = x_j - y_j$. Then ρ (which lies in the interval -1 to 1 inclusive) is given by

$$\rho = 1 - \frac{6\Sigma d_j^2}{n(n^2 - 1)},$$

where n is the number of individuals. In fact, ρ is the *correlation coefficient for the pairs $(x_1, y_1), (x_2, y_2), \ldots, (x_n, y_n)$.

As an example, suppose that someone is asked to arrange in order of mass five similar boxes whose contents vary. The correct order is 1, 2, 3, 4, 5, but the order chosen is 2, 1, 3, 5, 4. The rank differences are -1, 1, 0, -1, 1, giving $\Sigma d_j^2 = 4$. The value of ρ is $1 - (6 \times 4)/(5 \times 24) = 0.8$. Comparing this value with the table of critical values (*see* APPENDIX XI) we find that, at the 5% significance level, there is no significant evidence to reject the null hypothesis that the boxes were arranged in a random order.

specificity *See* SENSITIVITY.

spectral density function; spectrum *See* PERIODOGRAM.

spherical data Directional data in three dimensions. The direction can be specified by the location of a point on a sphere of unit radius. Relative to axes at the centre of the sphere, the location can be expressed in either

*Cartesian coordinates, (x, y, z) or *spherical polar coordinates $(1, \theta, \phi)$. These are related as follows:

$$x = \sin\theta\cos\phi, \qquad y = \sin\theta\sin\phi, \qquad z = \cos\theta \qquad \begin{array}{l} 0° \leq \theta \leq 180°, \\ 0° \leq \phi < 360°. \end{array}$$

Spherical data are often represented in two dimensions using a *Schmidt net. A *distribution commonly used to model spherical data is the *Langevin distribution.

spherical polar coordinates Coordinates that specify the location of a point in three dimensions. The basis of these coordinates is a set of three mutually perpendicular lines, Ox, Oy, and Oz, that intersect at O. Consider a point P, in three-dimensional space, at a distance r from O, the angle $\widehat{zOP}$ being equal to θ. Let OM be the projection of OP on the xy-plane and let the angle $\widehat{xOM}$ be equal to ϕ. The spherical polar coordinates of P are (r, θ, ϕ), where $0° \leq \theta \leq 180°$, $0° \leq \phi < 360°$. These coordinates are related to the alternative Cartesian coordinates (x, y, z) by

$$x = r\sin\theta\cos\phi, \qquad y = r\sin\theta\sin\phi, \qquad z = r\cos\theta.$$

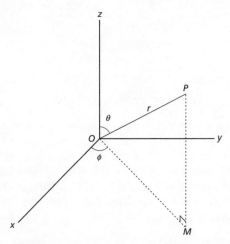

Spherical polar coordinates. Using Cartesian coordinates, the point P would be represented by (x, y, z), with $x = r\sin\theta\cos\phi$, $y = r\sin\theta\sin\phi$, $z = r\cos\theta$.

sphericity test A test of whether the null hypothesis of *independent observations with a constant *variance is valid. This hypothesis is of particular concern in the analysis of *longitudinal data, where the k

*repeated measures of individuals may display a noticeable *correlation. The most common test is **Mauchly's test**, for which the test statistic is W, given by

$$W = \det(\mathbf{S})\left(\frac{k+1}{\text{tr}(\mathbf{S})}\right)^{k+1},$$

where $\mathbf{S}$ is the $k \times k$ *sample covariance matrix, and $\det(\mathbf{S})$ and $\text{tr}(\mathbf{S})$ are, respectively, the determinant and the trace of $\mathbf{S}$ (*see* MATRIX).

Spjotvoll–Stoline test *See* TUKEY'S TEST.

spline A set of polynomials, one for each sub-interval, that give an approximation to the function f(x), defined on some interval $a \leq x \leq b$ where $a = x_0 < x_1 < \cdots < x_n = b$ is a subdivision of the interval $a \leq x \leq b$. The polynomials are all of the same degree, d, and are chosen so that the values of the polynomials and their first $(d-1)$ derivatives are continuous at the intermediate points of subdivision and the values of the polynomials agree with the value of f at each of the points.

split-half reliability *See* RELIABILITY.

split plot An *experimental design used when comparisons of different types of treatments can be made at different scales. For example, after ploughing, different fields (the **whole plots**) might be treated with different fertilizers. Subsequently, within each field, **sub-plots** might be planted with different varieties of vegetable. The analysis of the variability of the varieties uses within-field sub-plot variation, and the analysis for the fertilizers uses between-fields whole-plot variation.

S-PLUS A computer package with extensive graphical and statistical capabilities.

spreadsheet Computer software for recording and analysing numerical *data. Arithmetic and statistical operations can be carried out by simple instructions. Any associated graphs and diagrams that are required can also be automatically drawn.

SPREE *See* DEMING–STEPHAN ALGORITHM.

SPSS (Statistical Package for the Social Sciences) One of the major computer packages permitting many types of statistical analysis. Originally entitled 'Statistical Package for the Social Sciences'.

spurious correlation An alternative name for *nonsense correlation.

squared multiple correlation *See* ANOVA.

square matrix *See* MATRIX.

square table A *contingency table, having the same number of rows as columns, in which the definitions of the row and column categories are essentially the same, but differ, for example, by referring to different time points. If the *cell probability is denoted by p_{jk}, and parameters relating to rows, columns, diagonals, and individual cells by r_j, c_k, d_l, and m_{jk}, respectively, specialist models include:

$p_{jk} = p_{kj}$ for all j, k — the **symmetry model**;

$p_{jk} = \begin{cases} r_j c_k & \text{for } j \neq k \\ m_{jj} & \text{otherwise} \end{cases}$ — the **mover–stayer model**;

$p_{jk} = r_j c_k d_l$, where $l = j - k$ — the **diagonals model**;

$p_{jk} = \begin{cases} r_j c_k & \text{for } j \neq k \\ r_j c_j d_0 & \text{otherwise} \end{cases}$ — the **loyalty model**;

$p_{jk} = r_j c_k m_{jk}$ with $m_{jk} = m_{kj}$ — the **quasi-symmetry model**;

$p_{jk} = r_j c_k m_{jk}$ with m_{jk} equal to a product of terms corresponding

   to gaps between successive categories — the **distance model**.

SSAI *See* STATISTICAL SOCIETY OF AUSTRALIA, INC.

stabilizing the variance A phrase that is used to describe the use of a transformation to equalize the *variance of a set of *observations initially measured with unequal accuracy.

standard deviation The square root of the *variance. Karl *Pearson introduced the term in 1893, using the symbol σ in the following year.

standard error The square root of the *variance of a *statistic. For example, the standard error of the *mean of a *sample of n *observations taken from a *population with variance σ^2 is $\sigma/\sqrt{n}$. The same term is used for the corresponding sample estimate $s/\sqrt{n}$. *See* VARIANCE (DATA). The phrase 'standard error' was used by *Yule in 1897.

standardization The process of *standardizing.

standardized mortality ratio *See* MORTALITY RATE.

standardized residual *See* REGRESSION DIAGNOSTICS.

standardizing Converting a *random variable X with expectation μ

and *variance σ^2 to a random variable Y with expectation 0 and variance 1, using the transformation

$$Y = \frac{X - \mu}{\sigma}.$$

The same term is used for the process of converting a *data set $x_1, x_2, \ldots, x_n$, with *mean $\bar{x}$ and sample variance s^2, into the data set $y_1, y_2, \ldots, y_n$, with mean 0 and sample variance 1, by the transformation

$$y_j = \frac{x_j - \bar{x}}{s}, \qquad j = 1, 2, \ldots, n.$$

standard normal distribution A *normal distribution with *mean 0 and *variance 1.

standard normal variable A *random variable having a *standard normal distribution.

STARIMA models, STARMA models The space-time equivalents of *ARIMA models and *ARMA models.

Stata A powerful and versatile command-line statistical package well suited to *robust data analysis.

state vector *See* RELIABILITY THEORY.

STATGRAPHICS A computer package designed for interactive statistical data analysis.

stationarity; stationary A *time series displays stationarity if the *expected value at all time points is the same and if, additionally, the *correlation between the values at two time points, t and $t + \tau$, depends on the lag τ but not on t. The corresponding requirements hold for a *spatial process, with t replaced by the location of a point in space and τ replaced by a spatial lag. The time series or spatial process may be described as being stationary.

stationary point A point where a function has zero gradient. A function $f(x_1, x_2, \ldots, x_n)$ of one or more variables is said to have a **stationary value** at the point $(a_1, a_2, \ldots, a_n)$ if all the partial derivatives of f with respect to $x_1, x_2, \ldots, x_n$ vanish when $(x_1, x_2, \ldots, x_n) = (a_1, a_2, \ldots, a_n)$. The point $(a_1, a_2, \ldots, a_n)$ is a stationary point. The stationary point is a **maximum** (or **minimum**) if, for all neighbouring points, $f(x_1, x_2, \ldots, x_n)$ is less (or greater) than $f(a_1, a_2, \ldots, a_n)$. The stationary point is a **minimax** if there are points in the neighbourhood at which $f(x_1, x_2, \ldots, x_n) < f(a_1, a_2, \ldots, a_n)$, and points in the neighbourhood at which $f(x_1, x_2, \ldots, x_n) > f(a_1, a_2, \ldots, a_n)$. For example, $x^2 + 3y^2$ has a minimum (stationary) point at $(0, 0)$, $-2x^2 - 5y^2$ has a maximum point at $(0, 0)$, and $2x^2 - 5y^2$ has a minimax point at $(0, 0)$.

stationary process A *stochastic process is said to be stationary (or **strictly stationary**) if the joint distribution of the sequence of measurements $X_{1+l}, X_{2+l}, \ldots, X_{k+l}$ is, for all k, *independent of l. A less restrictive requirement is that the *expected value of the X-values should be constant and that, for all k, the *covariance between the values of X_k and X_{k+l} should depend on only the lag l — such a process is described as **weakly stationary** (or **second-order stationary**). *See also* MARKOV PROCESS.

stationary value *See* STATIONARY POINT.

statistic A function of the set of *random variables corresponding to a set of *observations. Often used to refer to the corresponding function of the *data. The word 'statistic' was introduced by Sir Ronald *Fisher in 1922.

statistical inference The process of drawing conclusions about the nature of some system on the basis of data subject to random variation. There are several distinguishable and apparently irreconcilable approaches to the process of inference; comfortingly, there are rarely any gross differences in the inferences that result. Approaches include *Bayesian inference and *fiducial inference; the approach first met by a student of *Statistics is usually that based on the *Neyman–Pearson lemma.

Statistical Methodology *See* JOURNAL OF THE ROYAL STATISTICAL SOCIETY.

statistical process control (SPC) The statistical methods used to monitor and improve the quality of the output of a manufacturing process. SPC includes the use of *quality control charts, *optimization, *reliability, and *experimental design.

Statistical Society of Australia, Inc. (SSAI) This society, founded in 1962, has nearly a thousand members. Together with the New Zealand Statistical Association it publishes the *Australia and New Zealand Journal of Statistics*.

Statistical Society of Canada This Society, founded in 1972, has about seven hundred members. It publishes the *Canadian Journal of Statistics*.

Statistical Theory and Methods Abstracts A publication of the *International Statistical Institute which provides comprehensive cross-referencing of new publications in *Statistics by topic area.

Statistica Neerlandica An English language journal, first published in 1946, with an emphasis on decision making. It is published by the *Netherlands Society for Statistics and Operations Research.

Statistica Sinica An English language journal published by the *International Chinese Statistical Association since 1991.

Statistics The science of collecting, displaying, and analysing *data.

statistics The plural of *statistic.

Statistics in Society See Journal of the Royal Statistical Society.

STATXACT A specialized statistical package for exact *non-parametric tests with *continuous variables or *categorical variables.

steepest ascent A method for finding the maximum (or minimum) of a function. The idea is to proceed in the direction for which the slope of the corresponding surface is greatest.

stem and leaf diagram A method of counting and ordering numerical *data without losing the detail of the individual data. The last significant digit for the whole data item is determined and each data item is represented by this digit (the leaf), with a stem consisting of the previous digits. For example, 58 is divided as 5|8, with a stem of 5 and a leaf of 8. Only the leaf is shown explicitly on the diagram.

```
5 | 8              Key: 5|8 means 58
4 |
3 | 9
2 | 2, 6
1 | 0, 0, 1, 2, 3, 4, 7
0 | 0, 0, 0, 1, 1, 1, 4, 6, 7
```

Stem and leaf diagram. The data are 22, 58, 12, 17, 4, 26, 10, 13, 1, 39, 0, 1, 10, 6, 0, 11, 14, 1, 0, 7. The diagram retains all the information in the data, while also giving an idea of the underlying distribution.

step diagram A diagram showing *data values for a numerical
*variable in which the *cumulative frequency of observations $\leq x$ is
plotted against x. The diagram consists of horizontal segments, with
jumps at the observed data values. The jump at x_j represents the
frequency of x_j in the sample. *See* CUMULATIVE FREQUENCY POLYGON; SAMPLE
DISTRIBUTION FUNCTION.

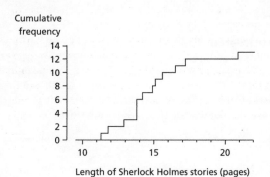

Length of Sherlock Holmes stories (pages)

Step diagram. The data illustrated here are the lengths of the stories in *The Return of
Sherlock Holmes*. Conan Doyle wrote the stories to fit regular slots in *Collier's Magazine*
and the *Strand Magazine* between September 1903 and December 1904. The lengths of the
stories are therefore very similar.

Stephan, Frederick Franklin (1903–71) An American social
statistician. A graduate of the University of Illinois, Stephan gained his
MS from the University of Chicago and joined the faculty of the
University of Pittsburgh in 1927. After a spell at Cornell University he
joined the faculty at Princeton University in 1947, initially as Professor of
Social Statistics in the Sociology department, subsequently transferring
to the Statistics department and retiring in 1971. Stephan was Editor of
the *Journal of the American Statistical Association* from 1935 to 1940. He
collaborated with *Deming on the 1940 paper that introduced the
*Deming–Stephan algorithm.

stepwise procedures Procedures for identifying an appropriate
model in the context of *multiple regression. The expectation of the
response variable, E(Y), is modelled as a linear combination of many
(p, say) explanatory X-variables. A natural question is whether all p of the
X-variables are required.

   Forward selection begins by determining which one of the X-variables
provides most information about Y. This variable is retained in all future

models. At the second stage the procedure considers the remaining
$(p-1)$ variables and determines which, in conjunction with the first
variable, provides most additional information about Y. This procedure
continues until there are no further variables that make worthwhile
extra contributions to the fit of the model. The successive contributions
are compared using an F-test: a contribution is worthwhile if the
observed F-value exceeds a critical value often referred to in the jargon of
computer packages as the **F to enter**.

Backward elimination mirrors forward selection by starting with the
model containing all p X-variables and removing ineffective variables
one by one. A variable is judged to be ineffective if its contribution results
in a value for the F-test that fails to exceed the **F to remove** value.

Forward selection and backward elimination are often referred to as
stepwise selection procedures because they move one variable at a time.
A general stepwise procedure would combine elements of the two; after
each removal stage there would be a check for possible additions.

Stirling's formula A formula providing an approximation, for large
values of the positive integer n, to $n!$ (*see* FACTORIAL):

$$n! \approx n^{n+\frac{1}{2}}e^{-n}\sqrt{2\pi}.$$

stochastic process A finite collection of related *random variables,
often ordered in time or space. The possible values of the random
variables are referred to as **states** and the possible configurations of
states constitute the **state space**. Examples of stochastic processes
include *Brownian motion, the *counting process, the *Markov chain,
the *Poisson process, and the *random walk. The phrase 'stochastic
process' was used by *Kolmogorov in 1932.

strata *See* STRATIFIED SAMPLING.

stratified sampling When a *population contains easily
recognizable subpopulations, or **strata**, of known sizes $(N_1, N_2, \ldots, N_s)$,
the method of stratified sampling will usually give better results than a
simple random *sample from the whole population. With stratified
sampling, simple random samples (of sizes $n_1, n_2, \ldots, n_s$) are taken from
each stratum. The method was introduced by *Neyman in 1934.

With **proportional allocation** the sizes of these samples satisfy

$$\frac{n_j}{N_j} = \frac{n_k}{N_k},$$

for all j and k.

If the *standard deviations of the values of the items in the various strata are known to be $S_1, S_2, \ldots, S_s$, then for a fixed sample size of n items, the **optimum allocation** (**Neyman allocation**) is obtained by choosing n_j so that

$$\frac{n_j}{n} \approx \frac{N_j S_j}{\sum_{m=1}^{s} N_m S_m}.$$

This allocation minimizes the *variance of the *estimator of the overall population *mean. If the standard deviations are not known then a *pilot study can be used to obtain estimates of their values.

stratum The singular of *strata. *See* STRATIFIED SAMPLING.

stress *See* MULTIDIMENSIONAL SCALING.

strictly stationary *See* STATIONARY PROCESS.

strong law of large numbers *See* LAWS OF LARGE NUMBERS.

structural equation model *See* SIMULTANEOUS EQUATION MODEL.

structural zero An entry in a *contingency table that is certain to be zero, whatever the sample size, because it corresponds to an impossible outcome. The table shown contains three structural zeros (top right) as well as one **random zero**—a cell whose entry is zero by chance.

Income of head of household

		<£20 000	£20 000–£30 000	> £30 000
Total household income	< £20 000	5	0	0
	£20 000–£30 000	2	8	0
	> £30 000	0	5	12

structure-preserving estimation *See* DEMING–STEPHAN ALGORITHM.

structure function *See* RELIABILITY THEORY.

Student The pen-name used by *Gosset.

Studentized range distribution The *distribution of the *statistic

$$\frac{\bar{x}_{(k)} - \bar{x}_{(1)}}{s/\sqrt{n}},$$

where random *samples of size n have been taken from k *independent and identically distributed *normal populations, with $\bar{x}_{(1)}$ and $\bar{x}_{(k)}$ being, respectively, the smallest and largest of the k sample means, and s^2 being the *pooled estimate of the common variance. This statistic is particularly used in *multiple comparison tests.

Studentized residual; Studentized deletion residual *See* REGRESSION DIAGNOSTICS.

Student's *t*-distribution *See* t-DISTRIBUTION.

study population *See* TARGET POPULATION.

sub-plot *See* SPLIT PLOT.

subpopulation A subset of the *population that has some characteristic in common.

subsample Part of a *sample.

SUDAAN A statistical package specifically designed for the analysis of correlated data from studies involving *longitudinal data, *repeated measures, and related complex surveys.

sufficient statistic *See* ESTIMATOR.

summary statistics *Measures of location, *measures of spread or sums, sums of squares and sums of cross-products for a set of *data. For a situation involving a dependent variable and explanatory variables, the entries in an *ANOVA table might also be referred to as summary statistics.

sum of squares *See* ANOVA.

supersaturated design *See* FACTORIAL EXPERIMENT.

surrogate variable A variable that can be measured (or is easy to measure) that is used in place of one that cannot be measured (or is difficult to measure). For example, whereas it may be difficult to assess the wealth of a household, it is relatively easy to assess the value of a house. *See also* PROXY VARIABLE.

survey *See* SAMPLE.

survival curve A plot of the *survivor function against time.

survival function *See* SURVIVOR FUNCTION.

survival time The time until the occurrence of a particular event such as death or the failure of a component.

survivor function (survival function) The probability that a component, or a device, survives until time t. It is equal to the **reliability function**, which is the probability that the component is still working at time t. The survivor function, $S(t)$, is given by $S(t) = 1 - F(t)$, where $F(t)$ is the lifetime *distribution function. *See also* HAZARD RATE.

symmetric confidence interval *See* CONFIDENCE INTERVAL.

symmetric distribution The *random variable X has a symmetric distribution if and only if there is a number c such that

$$P(X < c - k) = P(X > c + k) \qquad \text{for all } k.$$

In this case the distribution is symmetric about c. If X has a symmetric distribution and expectation μ, then $c = \mu$.

symmetric matrix *See* MATRIX.

symmetry model *See* SQUARE TABLE.

SYSTAT A computer package for statistical analysis which offers a wide range of graphical options.

systematic error An error that occurs when the result of measuring a variable whose actual value is x is $f(x)$, where f is a fixed function. A simple example would be to consistently truncate (so that, for example, 11.9 is reported as 11 rather than 12). *See* RANDOM ERROR.

systematic sampling A method of choosing a *sample from a sampling frame of size N. It is assumed that a list exists of the individuals in the frame, the two ends of the list being notionally joined. One unit is chosen at random from the list and then every kth unit thereafter is chosen for the sample (continuing counting from the start of the list when the end is reached) until the desired sample size is reached. The number k can be chosen so that N/k is approximately equal to the desired sample size.

Taguchi, Genichi (1924–; b. Niigata, Japan) A pioneer of modern
industrial quality control. During the Second World War Taguchi learnt
about *experimental design working in the *Institute of Statistical
Mathematics. From 1950 to 1962 he worked in the research and
development section of the Nippon Telephone and Telegraph Company.
During this period he met both Sir Ronald *Fisher and *Shewhart. In
1962 Taguchi was awarded a doctorate from Kyushu University. From
1964 to 1982 he was a professor at Aoyama Gakuin University in Tokyo.
Following his 1980 visit to Bell Laboratories, *Taguchi methods began to
be applied in the United States.

Taguchi methods Methods concerned with the optimization of
product and process before manufacture that concentrate on quality loss
rather than quality. Optimization involves *experimental design using
simple designs to estimate main effects. The success of Taguchi methods
is partly a consequence of the experimentation being tailored to the
application.

tail area; tail probability *See* HYPOTHESIS TEST.

tally chart A method of counting *frequencies, according to some
classification, in a set of *data. One line on a sheet of paper is assigned to
each category or number, in the case of a *discrete random variable, or
class, in the case of *grouped data. The data set is then worked through,
and each item is represented by a vertical stroke on the corresponding
line. For ease of counting, every fifth observation is represented by a

Duration (completed minutes)	Tally
1	II
2	I
3	III
4	ⅢⅡ II

Tally chart. The chart is a tally of the eruption times of Old Faithful on 1 August 1978.
Note the five-barred gate that makes counting the last row easy.

diagonal line crossing the previous four to make a **five-barred gate**. Tally charts are often used with grouped observations.

target population The population about which information is desired. The population that is actually surveyed is the **study population**.

tau (τ) A Greek letter used to signify *Kendall's measure of association. Also used as a symbol for time.

Taylor, Brook (1685–1731; b. Edmonton, England; d. London, England) An English mathematician. Taylor graduated from Cambridge University in 1709 and was elected a Fellow of the Royal Society in 1712. His work on *Taylor series was published in 1715. A lunar crater is named after him.

Taylor expansion *See* TAYLOR SERIES.

Taylor series (Taylor expansion) A series representation for a function f having continuous derivatives of all orders. The series is

$$f(a) + \frac{x}{1!}f'(a) + \frac{x^2}{2!}f''(a) + \frac{x^3}{3!}f'''(a) + \dots.$$

It is the *Maclaurin series of F, where $F(x) = f(a + x)$, so in the case $a = 0$ the Taylor and Maclaurin series are identical. As in the case of Maclaurin series, the Taylor series may or may not be convergent, and can vanish altogether.

Tchebycheff *See* CHEBYSHEV.

t-distribution (Student's t-distribution) The general *probability density function f for this distribution is

$$f(t) = \frac{1}{\sqrt{v}\,B(\frac{1}{2}, \frac{1}{2}v)}\left(1 + \frac{t^2}{v}\right)^{-\frac{1}{2}(v+1)}, \qquad -\infty < t < \infty,$$

where B is the *beta function and v is a positive *parameter (usually an integer) known as the number of *degrees of freedom. The distribution is symmetrical about its *mode at 0, which is therefore (for $v > 1$) also its mean. For $v > 2$ the distribution has variance $v/(v - 2)$.

When $v = 1$ the distribution is a *Cauchy distribution. As v increases, the distribution increasingly resembles the standard *normal distribution, which is its limit as $v \to \infty$. If X has a t-distribution with v degrees of freedom, then X^2 has an *F-distribution with one and v

degrees of freedom. A t-distribution with v degrees of freedom may be described as a t_v-distribution.

The form of the distribution was published in 1908 by *Gosset, writing under the pen-name 'Student', in the context of a random *sample of size n from a population having a *normal distribution. Gosset was finding the distribution of t, given by

$$t = \frac{\bar{x} - \mu}{s/\sqrt{n}},$$

where μ is the *population mean, and $\bar{x}$ and s are, respectively, the sample mean and *standard deviation with divisor $(n-1)$. In Gosset's case $v = (n-1)$.

The *percentage points (*see* APPENDIX VIII) of the t-distribution are used as *critical values in carrying out a t-test (*see* HYPOTHESIS TEST) based on the value of t when μ is replaced by μ_0, the value specified by the *null hypothesis.

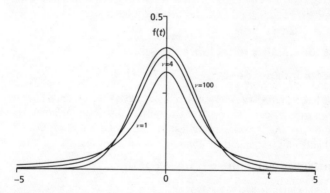

t-distribution. Distributions are illustrated for various values of the parameter v and all have mean 0. The case $v = \infty$ corresponds to the normal distribution, and the case $v = 1$ corresponds to the Cauchy distribution. The chance of a very extreme value is greater for a t-distribution than for the normal distribution, but decreases as v increases.

Technometrics A journal established in 1959 as a joint publication of the *American Statistical Association and the American Society for Quality Control. The purpose of the journal is to publish advances in the use of *Statistics in the physical, chemical, and engineering sciences.

ternary diagram *See* BARYCENTRIC COORDINATES.

tessellation A complete covering of a plane using a limited number of different shapes. Usually the shapes are polygons (as in the *Dirichlet tessellation). The plane can be tessellated with rectangles, or hexagons, or triangles (for example, using *Delaunay triangles). In a regular tessellation all the shapes are regular polygons (i.e. with all sides equal and all angles equal) of the same shape and size, and there are only three possible regular tessellations, using squares, equilateral triangles, or regular hexagons. Other semi-regular tessellations use two or more regular polygonal shapes, for example, squares and octagons. Many tessellations are periodic, i.e. the pattern repeats at regular intervals. A non-periodic tessellation, using two basic shapes, was invented by Sir Roger Penrose and is usually referred to as **Penrose tiling**.

test A means of determining whether a *hypothesis can be accepted. Typical hypotheses include 'the *distribution is *normal', 'the *mean is 5', or 'the *sample was unbiased'. The question is whether a hypothesis is acceptable, or should be rejected in favour of some alternative hypothesis. *Data are then examined and a judgement is made. *See* HYPOTHESIS TEST.

test for equality of variances *See* F-TEST.

test for independence *See* CHI-SQUARED TEST.

test for normality A test of the *hypothesis that a *sample has been drawn from a *population having a *normal distribution. Examples include the *Anderson–Darling test, the *Cramér–von Mises test, *D'Agostino's test, and the *Kolmogorov–Smirnov test.

test–retest reliability *See* RELIABILITY.

test statistic *See* HYPOTHESIS TEST.

tetrachoric correlation coefficient An estimate of the *correlation in a *sample drawn from a *population having a bivariate normal distribution. Consider the two-by-two table

$$\begin{array}{|cc|} \hline a & b \\ c & d \\ \hline \end{array}$$

where a, b, c, d are *cell frequencies. Now suppose that the classifying *random variables X and Y have a bivariate normal distribution with unknown *parameters. The values of X and Y have been reported with respect to some cut-off values of interest (e.g. '< 1'), and the precise

values are not available. The question of interest is the value of the
*correlation coefficient, ρ, in the *population. One approximate
estimate of ρ is provided by *Yule's Q. A better approximation, proposed
by Karl *Pearson, is the tetrachoric correlation coefficient

$$\sin\left\{\frac{\pi}{2}\left(\frac{\sqrt{ad}-\sqrt{bc}}{\sqrt{ad}+\sqrt{bc}}\right)\right\}.$$

theory of errors *See* NORMAL DISTRIBUTION.

The Statistician *See* JOURNAL OF THE ROYAL STATISTICAL SOCIETY.

Thiessen, Alfred H (1872–?) (b. Troy, New York) An American
meteorologist. He joined the US Weather Bureau at Pittsburgh in 1898.
He resigned in 1920 to join the US Army, retiring as a Major.

Thiessen polygon *See* DIRICHLET TESSELLATION.

Thurstone, Louis Leon (1887–1955; b. Chicago, Illinois; d. Chapel
Hill, Tennessee) An American psychometrician. Thurstone graduated
from Cornell University with a BS in electrical engineering in 1912 and
gained employment with Thomas Edison. Thurstone was interested in
the way people learn and this led to a PhD in psychology from the
University of Chicago in 1917. He then joined the faculty at Carnegie
Institute of Technology, moving, in 1924, to a chair at the University of
Chicago. In 1931 he introduced the concept of *factor analysis. He was
President of the American Psychological Association in 1933 and, in
1935, the first President of the Psychometric Society.

three-sigma rule An empirical rule stating that, for many reasonably
*symmetric *unimodal distributions, almost all of the *population lies
within three *standard deviations of the *mean. For the *normal
distribution about 99.7% of the population lies within three standard
deviations of the mean. *See also* TWO-SIGMA RULE.

tied ranks *See* RANKS.

tightened inspection *See* QUALITY CONTROL.

**time-homogeneous stationary chain; time-reversible
stationary chain** *See* MARKOV PROCESS.

time series A series of measurements over time, usually at regular
intervals, of a *random variable. A prime concern is the
*forecasting of future values using methods such as *exponential

smoothing, *Holt–Winters forecasting, or *Box–Jenkins methods. Models fitted include *autoregressive models and *moving average models. It is often necessary to *deseasonalize the data and to remove any underlying *trend before undertaking the analysis. *See also* PERIODOGRAM.

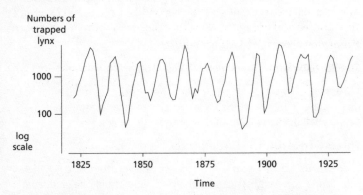

Time series. The famous series illustrated shows the variation in the numbers of lynx trapped in the neighbourhood of the Mackenzie River, in Canada, for the years 1821 to 1934. The numbers of lynx vary according to the amount of food available. A major component of this food is the snowshoe hare, which shows a similar cycle: a rise in the lynx population leads to a fall in the hare population, which leads to a fall in the lynx population, which leads to a rise in the hare population, and so on.

tobit analysis *See* CENSORED REGRESSION MODELS.

total probability law If the events $B_1, B_2, \ldots, B_n$ are *mutually exclusive and *exhaustive, then, for any event A,

$$P(A) = \sum_{j=1}^{n} P(B_j)P(A|B_j),$$

where $P(A)$ is the *probability of A and $P(A|B_j)$ is the *conditional probability of A given B_j.

total sum of squares *See* ANOVA.

trace *See* MATRIX.

traffic intensity A measure used in describing the behaviour of a *queue. The traffic intensity, ρ, is defined by

$$\rho = \frac{\text{Mean rate of arrival at the queue}}{\text{Mean rate of service}} = \frac{\text{Mean service time}}{\text{Mean inter-arrival time}}.$$

training set *See* DISCRIMINANT ANALYSIS.

transect In the context of *sampling natural *populations, a line drawn across the region of interest. Sampling may consist of examining the occurrence of organisms along the line, or visible from the line, or within a sequence of *quadrats centred on the line.

transformation A function of a *random variable. Examples are $\sqrt{X}$ or $\log X$, where X is the variable. Using the transformed variable may, for example, simplify a *model or *stabilize the variance.

transient state *See* MARKOV PROCESS.

transition matrix A square *matrix in which the rows and columns correspond to categories defined in equivalent ways. Usually the row categories refer to one time period and the column categories to a subsequent time period. The entries may be frequencies, probabilities, or conditional probabilities. *See also* MARKOV PROCESS.

transportation problem A *linear programming problem concerned with identifying the cheapest method for moving commodities (or people) from one set of locations to another. At a number of locations $(L_1, L_2, \ldots, L_l)$ specified amounts of a commodity are available. There are demands for specified amounts of the commodity at locations $(M_1, M_2, \ldots, M_m)$. The cost of transportation from L_j to M_k is known for all j and k. The problem is to satisfy the demands at minimum transportation cost.

transpose *See* MATRIX.

trapezium rule A method of finding an approximate value for an integral, based on finding the sum of the areas of trapezia. Suppose we wish to find an approximate value for $\int_a^b f(x)dx$. The interval $a \leq x \leq b$ is divided up into n sub-intervals, each of length $h = (b - a)/n$, and the integral is approximated by

$$\frac{1}{2}h(y_0 + 2y_1 + 2y_2 + \cdots + 2y_{n-1} + y_n),$$

where $y_r = f(a + rh)$. This is the sum of the areas of the individual trapezia, one of which is shown in the diagram. The error in using the trapezium rule is approximately proportional to $1/n^2$, so that if the number of sub-intervals is doubled, the error is reduced by a factor of 4. A more accurate method for approximating an integral is *Simpson's rule.

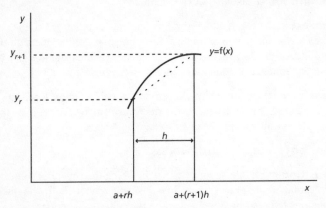

Trapezium rule. A method of approximate integration. A slight underestimate (as shown) will often be cancelled by a similar slight overestimate from another trapezium. Using narrower intervals will improve accuracy.

trapping state *See* MARKOV PROCESS.

travelling salesman problem A *network problem that can be formulated as a *combinatorial optimization problem. A salesman has to visit a number of interconnected destinations. The problem is to determine the route that minimizes the total distance travelled. *See following diagram.*

treatment A term used in the context of an *experimental design to refer to any prescribed combination of values of explanatory variables. In the original agricultural context a treatment was, for example, the application of a particular fertilizer. The term is particularly used in the context of *balanced incomplete blocks, *Latin squares, and *randomized blocks.

trend If the *mean of a *time series changes steadily over time then it is said to exhibit a trend.

Trial of the Pyx A *sampling procedure conducted by the English Royal Mint. The ceremony dates back to the thirteenth century. Typically, 100 golden guineas were randomly sampled and placed in a

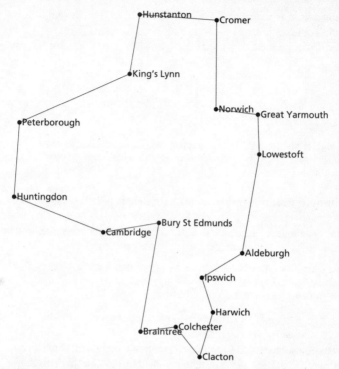

Travelling salesman problem. The route illustrated would be appropriate for a travelling salesman in East Anglia.

---

ceremonial box (the Pyx), which was then weighed. The Master of the Mint was responsible for the standard of the coinage. He was subject to severe penalties if the combined weight of the coins differed from its nominal weight by more than one part in 400.

triangular distribution If X and Y are *independent *random variables each having the same *uniform distribution then $(X + Y)$ has a triangular distribution. In the case of a *continuous random variable the graph of the *probability density function is an isosceles triangle. In the case of a *discrete random variable the graph of the *probability function has a triangular shape. The diagram shows this for the case of throwing two ordinary dice. The most probable score is 7 (which results in a 'Chance' card when starting from 'Go' on a Monopoly board).

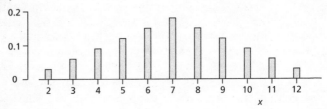

Triangular distribution. This is a graph of the distribution of the sum of the scores on two fair six-sided dice. The mode (corresponding to a probability of $\frac{6}{36}$) occurs when the sum is 7.

trimmed mean A *robust estimate of the *population mean. For a *sample of size n, with *data ordered so that $x_{(1)} \leq x_{(2)} \leq \cdots \leq x_{(n)}$, the trimmed mean, $\bar{x}_k$, is the mean of the data if the k smallest values and the k largest values are discarded (with $1 \leq k < \frac{1}{2}n$). So

$$\bar{x}_k = \frac{1}{n - 2k} \sum_{j=k+1}^{n-k} x_{(j)}.$$

The idea is to avoid the influence of extreme observations. The method was proposed by *Tukey in 1962.

Tukey also proposed the **Winsorized mean**, named after his former colleague Charles P. Winsor, who had died in 1951. In this case the k smallest values are replaced by $x_{(k)}$ and the k largest are replaced by $x_{(n-k+1)}$ to give the estimate w_k:

$$w_k = \frac{1}{n} \left(kx_{(k)} + \sum_{j=k+1}^{n-k} x_{(j)} + kx_{(n-k+1)} \right).$$

truncated distribution A *distribution in which values less than some critical value or more than some other critical value (or both) are not recorded. The most common situation is a distribution in which values of 0 are not reported. *Compare* CENSORED DATA.

t-test *See* HYPOTHESIS TEST.

Tukey, John Wilder (1915–2000; b. New Bedford, Massachusetts; d. New Brunswick, New Jersey) An influential American statistician. Tukey was a chemistry graduate at Brown University, gaining his MA in

1937. He followed this with a PhD in topology at Princeton University in 1939. During the Second World War he worked in the Fire Control Research Office alongside *Wilks and *Cochran. After the war he joined Wilks at Princeton University becoming a full professor at the age of 35. In 1946 he coined the word 'bit' as a shorthand for the 'binary digits' used by computers, and, in 1958, the word 'software' to describe the programs used by computers. In 1962 he introduced the *trimmed mean as one of a number of *robust *summary statistics. In 1965, with John Cooley, he introduced the *fast Fourier transform. His 1970 book *Exploratory Data Analysis* introduced the *stem and leaf diagram and the *boxplot. He was elected to membership of the National Academy of Sciences in 1961 and was awarded the National Medal of Science in 1973. He was President of the *Institute of Mathematical Statistics in 1960. He was awarded the *Wilks Medal of the *American Statistical Association in 1965 and was made an Honorary Fellow of the *Royal Statistical Society in 1986.

Tukey's quick test A simple two-sample *test. *Independent random *samples are taken from *populations with *cumulative distribution functions F_1 and F_2. It is assumed that, for some constant δ, $F_1(x) = F_2(x + \delta)$ for all x. The *null hypothesis is that $\delta = 0$, corresponding to the case where the samples have been drawn from a single population. The *alternative hypothesis is that $\delta \neq 0$, which corresponds to the case where one distribution has slipped relative to the other. This sort of test is called a **slippage test**.

Suppose the largest observation is in sample 1. Tukey's test statistic, T, is calculated as follows:

1. If the smallest observation is in sample 1, $T = 0$.
2. Otherwise T is the sum of the number of observations in sample 1 that are larger than the largest observation in sample 2 and the number of observations in sample 2 that are smaller than the smallest observation in sample 1.

Not only is the statistic easy to calculate, but its distribution is remarkably unaffected by the sample sizes. As a guide, T is significant at the 5% level if it is > 7; the corresponding values for the 1% and 0.1% levels are 9 and 13, respectively.

Tukey's test (honestly significant difference test; HSD test) A *multiple comparison test for use with *samples of equal size. The test states that two sample *means are significantly different if they differ by more than some specified *critical value corresponding to a given *upper-tail probability for the appropriate *Studentized range distribution. The **Spjotvoll–Stoline test** extends Tukey's test to the case of samples of unequal size.

Tukey window *See* PERIODOGRAM.

Turing, Alan Mathison (1912–54; b. London, England; d. Wilmslow, England) An English mathematician. Turing graduated from Cambridge University in 1934. The following year he was elected to fellowship of King's College, Cambridge as a consequence of his independent proof of the *central limit theorem. In 1936 he introduced the *Turing machine. In 1938 he obtained his PhD from Princeton University. During the Second World War he was involved in the breaking of the German codes produced by the Enigma machine. After the war he had posts at Cambridge and Manchester Universities. He was elected a Fellow of the Royal Society in 1951.

Turing machine A theoretical model of a computer, in which the machine functions in a sequence of discrete operations. The machine can be in only one of a finite list of internal states at any given moment. It consists of an infinite tape carrying symbols, which represent instructions, and a mechanism that can move the tape and read from, or write to, the tape. The mechanism can also change the internal state of the machine in accordance with instructions read from the tape.

two-by-two table *See* TWO-WAY TABLE.

two-sample test *See* HYPOTHESIS TEST.

two-sided test *See* HYPOTHESIS TEST.

two-sigma rule An empirical rule stating that, for many reasonably symmetric *unimodal distributions, approximately 95% of the *population lies within two *standard deviations of the *mean. *See also* THREE-SIGMA RULE.

two-stage least squares *See* MULTIPLE REGRESSION MODEL.

two-tail probability *See* ONE-TAIL PROBABILITY.

two-tailed test *See* HYPOTHESIS TEST.

two-way table A table with r rows and c columns in which the entry in cell (i, j) represents either the *frequency for that outcome (in the context of a *contingency table for *categorical variables) or a value

resulting from that row and column combination in the context of
*ANOVA. Such a table might be called an r-by-c table; if $r = c = 2$ then it
would be called a **two-by-two table**.

Type I error; Type II error *See* HYPOTHESIS TEST.

unbiased Fair. For example, a six-sided die is unbiased if it is equally likely to show any of its six sides.

unbiased estimate; unbiased estimator *See* ESTIMATOR.

uncorrected moment *See* MOMENT.

uncorrelated variables Variables displaying zero *correlation. *Independent variables are uncorrelated, though the converse is not always true.

undirected graphical model *See* GRAPHICAL MODEL.

uniform association model A model for a *contingency table that proposes that the *odds ratio for the categories in every component two-by-two table comprising adjacent cells should be the same. Thus, if $p_{j,k}$ denotes the *probability of an individual belonging to cell (j, k) then the model is

$$\frac{p_{j,k}\ p_{j+1,k+1}}{p_{j,k+1}\ p_{j+1,k}} = c,$$

where c is constant. The case $c = 1$ corresponds to *independence.

uniform distribution, continuous (rectangular distribution) The continuous distribution on (a, b) with *probability density function f given by

$$f(x) = \frac{1}{b-a}, \qquad a < x < b,$$

where a and b are constants. The distribution has *mean $\frac{1}{2}(a + b)$ and *variance $\frac{1}{12}(b - a)^2$.

The error when a decimal number is rounded to the nearest whole number has a uniform distribution on $(-0.5, 0.5)$, with mean error 0 and variance $\frac{1}{12}$.

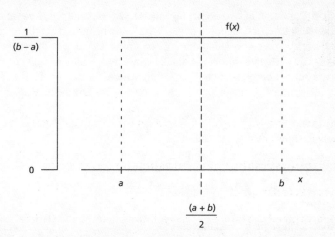

Continuous uniform distribution. The density function is constant for all values of the random variable in the given range, which implies that all intervals of the same width in this range are equally likely.

uniform distribution, discrete A distribution in which a *discrete random variable, X, say, can take a discrete set of possible values $x_1, x_2, \ldots, x_n$, with equal *probabilities, so that

$$P(X = x_j) = \frac{1}{n}, \qquad j = 1, 2, \ldots, n.$$

Often the set of possible values is the set of positive integers $1, 2, \ldots, n$, and in this case the *mean is $\frac{1}{2}(n+1)$ and the *variance is $\frac{1}{12}(n^2 - 1)$.

As an example, the score on a single throw of an *unbiased die has a uniform discrete distribution on 1, 2, 3, 4, 5, 6, with mean 7/2 and variance 35/12.

uniformly distributed Having a *uniform distribution.

uniformly most powerful test A test of a *null hypothesis which has power which is at least as great as that of any alternative test for all values of the *parameter under test. The concept was introduced by Sir Ronald *Fisher in 1934.

unimodal An adjective used to describe a *distribution or set of *data having a single *mode. If there is more than one mode then the distribution is *multimodal.

union The union of two events (*see* SAMPLE SPACE) A and B is the event 'either A or B occurs', denoted by $A \cup B$. It should be noted that the 'or' is inclusive and therefore the union includes the case when both events occur. The **exclusive union** of A and B is the event 'either A occurs or B occurs but not both', i.e. $(A \cap B') \cup (A' \cap B)$, where A' and B' are the complementary events to A and B, respectively. For any events A, B, C,

$$A \cup A = A, \quad A \cup S = S, \quad A \cup \phi = A,$$
$$A \cup B = B \cup A, \quad A \cup (B \cup C) = (A \cup B) \cup C,$$

where S is the sample space and ϕ is the empty set. The union of the n events $A_1, A_2, \ldots, A_n$ is the event 'at least one of $A_1, A_2, \ldots, A_n$ occurs'. It is denoted by $A_1 \cup A_2 \cup \cdots \cup A_n$. *See also* INTERSECTION; VENN DIAGRAM.

universal kriging *See* KRIGING.

universal set *See* SAMPLE SPACE.

unsaturated model *See* SATURATED MODEL.

upper percentage point *See* PERCENTAGE POINT.

upper quartile *See* QUARTILE.

upper-tail probability The *probability that the *random variable X takes values $\geq x$, where, usually, x is appreciably greater than the expectation of X.

urn model A model of a problem described in terms of balls being drawn from an urn. An example of a problem where a *hypergeometric distribution is appropriate is as follows. Three people are to be chosen, at random from a group of twenty. In the group of twenty are four named Smith. The event of interest is that all those chosen are named Smith. The corresponding urn model has an urn containing four black balls and sixteen white balls. Three balls are to be selected at random and without replacement. The event of interest is that all are black.

A trivial example of a problem where a *binomial distribution is appropriate concerns the chance of getting three heads when tossing a fair coin four times. The corresponding urn model involves an urn containing one black ball and one white ball. A ball is selected at random and replaced on four occasions. The event of interest is that a black ball is chosen on three occasions.

The phrases **without replacement** and **with replacement** can best be understood by imagining that the balls are selected one at a time. In the first case, after each selection the ball is placed on one side, so that the number in the urn has reduced by one. In the second case the ball is replaced in the urn, so that before each draw the urn contains the same mixture of balls.

utility A function that takes a numerical value for each possible state of a system (usually an economic system) and is intended as a measure of the benefit or usefulness of that state.

van der Waerden, Bartel Leendert (1903–96; b. Amsterdam, Netherlands; d. Zurich, Switzerland) An extraordinarily versatile Dutch mathematician. Van der Waerden studied at the Universities of Amsterdam and Göttingen, joining the staff at Göttingen in 1928. In 1931 he was appointed Professor of Mathematics at the University of Leipzig. After the Second World War he briefly held posts at Johns Hopkins University and Amsterdam University before moving in 1951 to Zurich University. His *non-parametric test for the equality of two *populations was published in 1952.

van der Waerden test A *non-parametric test that two *populations are the same, in the case where the *distributions are *continuous. Two *independent random samples $x_1, x_2, \ldots, x_m$ and $y_1, y_2, \ldots, y_n$ are drawn and the test statistic is W given by

$$W = \sum_{j=1}^{m} \Phi^{-1}\left(\frac{r_j}{m + n + 1}\right),$$

where r_j is the *rank of x_j when the $(m + n)$ *observations are arranged in increasing order of size, and Φ is the *cumulative distribution function of the *standard normal distribution. The restriction to continuous variables ensures that there are no ties. The test was proposed in 1952.

variable The characteristic measured or observed when an experiment is carried out or an observation is made. Variables may be non-numerical (*see* CATEGORICAL VARIABLE) or numerical. Since a non-numerical observation can always be coded numerically, a variable is usually taken to be numerical. *Statistics is concerned with *random variables and with variables whose measurement may involve *random errors.

variance (data) A measure of the variability of a set of *data. For data $x_1, x_2, \ldots, x_n$, with *mean $\bar{x}$ given by

$$\bar{x} = \frac{1}{n} \sum_{j=1}^{n} x_j,$$

the variance is defined to be

$$\frac{1}{n} \sum_{j=1}^{n} (x_j - \bar{x})^2 = \frac{1}{n} \left(\sum_{j=1}^{n} x_j^2 - n\bar{x}^2 \right) = \frac{1}{n} \left\{ \sum_{j=1}^{n} x_j^2 - \frac{1}{n} \left(\sum_{j=1}^{n} x_j \right)^2 \right\}.$$

The variance is never negative and can be zero only if all the data values are the same.

In the case where the *frequency of the *observation x_j is f_j, for $j = 1, 2, \ldots, m$, the variance can be calculated using

$$\frac{1}{n} \sum_{j=1}^{m} f_j (x_j - \bar{x})^2 = \frac{1}{n} \left(\sum_{j=1}^{m} f_j x_j^2 - n\bar{x}^2 \right) = \frac{1}{n} \left\{ \sum_{j=1}^{m} f_j x_j^2 - \frac{1}{n} \left(\sum_{j=1}^{m} f_j x_j \right)^2 \right\},$$

where n, the total *sample size, is given by

$$n = \sum_{j=1}^{m} f_j \quad \text{and} \quad \bar{x} = \frac{1}{n} \sum_{j=1}^{m} f_j x_j.$$

In these expressions for variance the divisor n is used. This is correct if the data set effectively constitutes the entire *population; for example, if the values $x_1, x_2, \ldots$ are the diameters of the planets of the solar system, or the lifetimes of all known patients with a rare disease. However, if the data constitute a random *sample from a population, and we are interested in the variance of the values in the population, as opposed to the variance of the values in the sample, then it is appropriate to use the divisor $(n - 1)$, since this leads to an *unbiased estimate of the population variance. This **sample variance** is given by

$$s^2 = \frac{1}{n-1} \sum_{j=1}^{n} (x_j - \bar{x})^2 \quad \text{or} \quad \frac{1}{n-1} \sum_{j=1}^{m} f_j (x_j - \bar{x})^2,$$

as appropriate.

variance (random variable) A measure of the variability in the values of a *random variable. It is defined as the expectation of the squared difference between the random variable and its expectation.

$$\text{Var}(X) = \text{E}\left[\{X - \text{E}(X)\}^2 \right] = \text{E}[X^2] - \{\text{E}(X)\}^2,$$

where $\text{E}(X)$ denotes the expectation of X. For a *discrete random variable X, taking values $x_1, x_2, \ldots, x_n$, the variance of X can be calculated as follows:

$$\text{Var}(X) = \sum_{j=1}^{n} P(X = x_j)\{x_j - E(X)\}^2,$$

$$= \sum_{j=1}^{n} x_j^2 P(X = x_j) - \{E(X)\}^2,$$

where

$$E(X) = \sum_{j=1}^{n} x_j P(X = x_j).$$

For a *continuous random variable X, with *probability density function f, the variance of X can be calculated as follows:

$$\text{Var}(X) = \int_{-\infty}^{\infty} \{x - E(X)\}^2 f(x)dx$$

$$= \int_{-\infty}^{\infty} x^2 f(x)dx - \{E(X)\}^2,$$

where

$$E(X) = \int_{-\infty}^{\infty} x f(x)dx.$$

The term 'variance' was coined by Sir Ronald *Fisher in 1918. Fisher used the symbol σ^2, since the variance was the square of the *standard deviation that Karl *Pearson had denoted by σ in 1894.

variance–covariance matrix (dispersion matrix) A square symmetric *matrix in which the elements on the main diagonal are *variances and the remaining elements are *covariances. Suppose $X_1, X_2, \ldots, X_p$ are *random variables and the variance of X_j is σ_j^2 and the covariance of X_j and X_k is c_{jk}. Then the variance–covariance matrix is Σ given by

$$\Sigma = \begin{pmatrix} \sigma_1^2 & c_{12} & \ldots & c_{1p} \\ c_{21} & \sigma_2^2 & \ldots & c_{2p} \\ \vdots & \vdots & \ddots & \vdots \\ c_{p1} & c_{p2} & \ldots & \sigma_p^2 \end{pmatrix}.$$

The same term is also used for the corresponding matrix based on sample values.

variance-ratio test See F-TEST.

variance reduction techniques Methods for reducing the size of a *simulation by efficient use of *pseudo-random numbers to reduce the *variance of the simulation estimates. Examples include the use of *antithetic variables and *stratified sampling.

variate A *random variable. A term used by Karl *Pearson in 1909.

varimax rotation A method used in *factor analysis. The aim is to rotate the *vector of factors so as to find a few key combinations that will simplify the analysis.

variogram *See* AUTOCORRELATION.

vector A *matrix with only one column (a column vector) or only one row (a row vector).

Venn, John (1834–1923; b. Hull, England; d. Cambridge, England) Following his graduation in mathematics from Cambridge University, Venn was elected to a fellowship of Gonville and Caius College, where a stained glass window now stands in his memory. In 1859 he was ordained as a priest and spent a year as a curate at Mortlake on the Thames. In 1862 he returned to Cambridge University as a lecturer specializing in logic. Venn had a general interest in all branches of *Statistics and a letter that he wrote in 1887 to the Editor of *Nature* stimulated an explosion of interest in the mathematical theory of statistics. His interests were not confined to mathematics. In 1909 he constructed a bowling machine that was used by the visiting Australian cricket team. In retirement he compiled the first three volumes of a history of Cambridge University. He was elected a Fellow of the Royal Society in 1883.

Venn diagram A simple diagram used to represent *unions and *intersections of sets. The diagram, described by *Venn in 1880 and

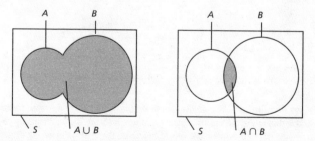

Venn diagram. These two diagrams illustrate the definitions of the union and intersection of the events A and B in the sample space S.

popularized by his 1881 book *Symbolic Logic*, was introduced by *Leibniz in the eighteenth century.

volatility An alternative name for *standard deviation which is used in the context of finance.

von Mises, Richard Martin Edler (1883–1953; b. L'vov, Ukraine; d. Boston, Massachusetts) An Austrian Jewish mathematician and engineer. Having studied machine engineering at the Technical University of Vienna (gaining his PhD in 1907), von Mises was appointed Professor of Applied Mathematics at the University of Strasbourg in 1909. During the First World War he was a pilot, designing and building his own plane. After the war he gave the first university course on the mechanics of powered flight. He was appointed Director of the Institute of Applied Mathematics in Berlin in 1920. However, in 1933, he left Hitler's Germany for Turkey. At the outbreak of the Second World War he moved to the United States, first to the Massachusetts Institute of Technology and then, in 1943, to Harvard University.

von Mises distribution (circular normal distribution) The principal *distribution used to model *cyclic data; derived by von Mises in 1918. The distribution has two parameters: μ ($-\pi < \mu \leq \pi$), the *circular mean, and κ (≥ 0), which is a measure of the *concentration of the distribution. If $\kappa = 0$ then the distribution degenerates to the *circular uniform distribution in which all directions are equally likely. As κ increases, the

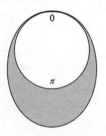

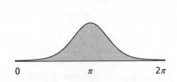

Von Mises distribution wrapped around the unit circle

Unwrapped von Mises distribution

Von Mises distribution. The left diagram shows the density function wrapped around a central circle to give the continuous analogue of a rose diagram. The right diagram gives a more conventional representation but fails to emphasize that the right-hand edge joins the left-hand edge of the diagram. In either diagram the extent of a portion of the shaded region is proportional to probability.

distribution becomes increasingly concentrated about μ. The *probability density function f is given (with directions in radians) by

$$f(\theta) = \frac{1}{2\pi I_0(\kappa)} \exp\{\kappa \cos(\theta - \mu)\}, \quad -\pi < \theta \leq \pi,$$

where $I_0(\kappa)$ is given by

$$I_0(\kappa) = \sum_{s=0}^{\infty} \left(\frac{\kappa^s}{2^s s!}\right)^2.$$

The density function can either be pictured around a circle or it can be 'unwrapped' on to a line—in which case it resembles a *normal distribution.

Voronoi, Georgy Fedoseevich (1868–1908; b. Zhuravka, Ukraine; d. Warsaw, Poland) A Russian algebraist. Voronoi was a graduate of the University of St Petersburg in 1889, and gained his PhD in 1897. By that time he was on the faculty of the University of Warsaw, where he became Professor of Pure Mathematics and where he remained until his premature death. He is remembered by statisticians for his 1908 examination of the properties of the regions surrounding each of a finite set of points in many-dimensional space. In two dimensions these are called Voronoi polygons.

Voronoi polygon *See* DIRICHLET TESSELLATION.

voter mobility table *See* MOBILITY TABLE.

voxel *See* PIXEL.

waiting time The time spent by an individual in a *queue waiting to be served.

Wald, Abraham (1902–50; b. Cluj, Romania; d. Travancore, India) A Hungarian Jewish geometer and statistician. Wald gained his PhD in mathematics in 1931 from the University of Vienna, where his research was in geometry. In 1938, on the Nazi seizure of Austria, he moved to Columbia University in the United States and turned his attention to statistical *decision theory and made important advances in the theory of *sequential sampling. He was President of the *Institute of Mathematical Statistics in 1948. He died in a plane crash in India.

Wald distribution See INVERSE NORMAL DISTRIBUTION.

Wald statistics *Statistics of the form

$$W = \{\mathbf{g}(\mathbf{T})\}'\mathbf{D}^{-1}\mathbf{g}(\mathbf{T}),$$

where $\mathbf{T}$ is an *estimator of a *vector parameter $\boldsymbol{\theta}$, $\mathbf{g}$ is some vector-valued function, and $\mathbf{D}$ is an estimator of the *variance–covariance matrix of the vector $\{\mathbf{g}(\mathbf{T}) - \mathbf{g}(\boldsymbol{\theta})\}$. The statistics are used to test the *null hypothesis that $\mathbf{g}(\boldsymbol{\theta}) = \mathbf{0}$, where $\mathbf{0}$ is a vector with all entries equal to 0. If $\mathbf{T}$ is the *maximum likelihood estimator then W has an approximate *chi-squared distribution with p *degrees of freedom (where p is the number of elements in $\boldsymbol{\theta}$).

Wald–Wolfowitz test See RUNS TEST.

Wallis, W. (Wilson) Allen (1912–98; b. Philadelphia, Minnesota; d. Rochester, New York) Wallis was a psychology graduate of the University of Minnesota in 1932. He then studied economics at Minnesota and at the University of Chicago. Subsequently he held posts in the Economics Departments at Columbia, Yale, and Stanford Universities. During the Second World War he headed a statistics think-tank. From 1946 to 1962 he was Professor of Statistics at the Business School of the University of Chicago. His paper with *Kruskal on the *Kruskal–Wallis test appeared in 1952. From 1951 to 1959 he was Editor of the *Journal of the American Statistical Association*. In 1962 he moved to the University of Rochester as President and then Chancellor (1975–82). On retirement from university life he was appointed Under Secretary of State for Economic Affairs (until 1989). He was awarded the *Wilks Medal of the *American Statistical Association in 1980.

Ward, Joe H. jun. (1926–) American mathematician and educational psychologist. Ward was educated at the University of Texas, gaining his BA in mathematics in 1947 and his PhD in educational psychology and mathematics in 1953. From 1951 to 1984 he was a personnel research psychologist with the United States Air Force, working with the University of Texas in the development of computer-based instructional systems.

Ward's method An *agglomerative clustering method proposed by *Ward in 1963. The clustering criterion is based on the error sum of squares, E, which is defined as the sum of the squared distances of individuals from the centre of gravity of the cluster to which they have been assigned. Initially, E is 0, since every individual is in a cluster of its own. At each stage the link created is the one that makes the least increase to E.

warning line *See* QUALITY CONTROL.

wash-out period *See* CROSSOVER TRIAL.

Watson, Geoffrey Stuart (1922–98; b. Bendigo, Australia; d. Princeton, New Jersey) An Australian statistician. A mathematics graduate of the University of Melbourne (1942), Watson worked for five years as a teacher before travelling to the United States, where he obtained his PhD from North Carolina State University in 1951. After a succession of posts in Australia, England, and Canada, in 1970 he took the chair of statistics at Princeton University, becoming Professor Emeritus in 1992. Watson is best known to statisticians for his introduction of the 1950 *Durbin–Watson statistic for testing for *serial correlation and the 1964 *Wheeler–Watson test for *cyclic data. He was a talented artist.

Watson's test A test of a specified *circular distribution, adapted from the *Cramér–von Mises test. Denote the n *observations by $\theta_1, \theta_2, \ldots, \theta_n$ and write

$$\bar{F} = \frac{1}{n} \sum_{j=1}^{n} F(\theta_j),$$

where $F(\theta)$ is the *probability of a value in the interval $(0, \theta)$ according to the *null hypothesis. The test statistic is U^2, given by

$$U^2 = \sum_{j=1}^{n} \{F(\theta_j)\}^2 - \frac{1}{n} \sum_{j=1}^{n} \{(2j-1)F(\theta_j)\} + n \left\{ \frac{1}{3} - \left(\bar{F} - \frac{1}{2} \right)^2 \right\}.$$

A large value of U^2 leads to rejection of the null hypothesis. A transformed version of U^2 is provided by U^*, given by

$$U^* = \left(U^2 - \frac{1}{10n} + \frac{1}{10n^2} \right) \left(1 + \frac{4}{5n} \right).$$

The distribution of U^* is approximately independent of n. The upper 10%, 5%, 2.5%, and 1% points of U^* are 0.152, 0.187, 0.222, and 0.268, respectively.

wavelet Functions that provide succinct and accurate representations of *time series and arrays of spatial *data. Wavelets have great potential as tools for data compression, when the data consist of, for example, images to be sent across computer networks. The wavelet representation of $g(t)$, a continuous function of time t, has the form

$$g(t) = a\varphi(t) + b\psi(t) + \sum_{n, m} c_{n, m} \psi(2^n t - m),$$

where n and m are integers, with $n \geq 1$ and $m \geq 0$. The function ψ is the **mother wavelet** and φ is the **scaling function** or **father wavelet**. This wavelet representation of $g(t)$ has similarities to a *Fourier series representation but is more flexible, since both φ and ψ are chosen to take the value 0 outside finite intervals. For a time series measured at N equi-spaced time points, the functions φ and ψ are related by the equations

$$\varphi(t) = \sum_{k=0}^{N} h_k \varphi(2t - k),$$

$$\psi(t) = \sum_{k=0}^{N} (-1)^k h_{N-k} \varphi(2t - k).$$

where $h_0, h_1, \ldots, h_N$ are constants, referred to as **filter coefficients**. The original N data values are represented by a sum of a weighted combination of the father and mother wavelets together with **daughter wavelets**. The daughter wavelet, ψ_{jk}, is related to the mother wavelet, ψ, by a simple formula that reflects a translation and dilation of the mother, so that the daughter looks like a compressed version of its mother. The relation is

$$\psi_{jk}(t) = 2^{\frac{1}{2}j}\psi\left(2^j t - k\right),$$

where $0 \leq t < 1$. The coefficient $2^{\frac{1}{2}j}$ simplifies subsequent analysis.

A summary of the wavelet decomposition for the case $N = 16$ is illustrated in the table.

Father wavelet, φ							
Mother wavelet, ψ							
ψ_{10}				ψ_{11}			
ψ_{20}		ψ_{21}		ψ_{22}		ψ_{23}	
ψ_{30}	ψ_{31}	ψ_{32}	ψ_{33}	ψ_{34}	ψ_{35}	ψ_{36}	ψ_{37}
1　2	3　4	5　6	7　8	9　10	11　12	13　14	15　16

Each of the ψ functions is a piecewise continuous function that takes the value 0 outside its range of influence. Thus, if observation 14 takes an unusually large value then this will be reflected in an unusually large coefficient of ψ_{36}; if observations 5 to 8 are unusually small then this will be reflected in an unusually small coefficient of ψ_{21}; and so on.

The first mother wavelet was given in the appendix of the 1909 PhD thesis by *Haar. This **Haar wavelet** is given by

$$\psi(t) = \begin{cases} -1 & 0 \leq t < \frac{1}{2} \\ 1 & \frac{1}{2} \leq t < 1 \\ 0 & \text{otherwise.} \end{cases}$$

Mother wavelets. The D_2 wavelet is the Haar wavelet. The wavelets illustrated here are the D_4, D_{12}, and D_{20} wavelets. All are members of the Daubechies family.

The family of wavelets that are currently most used were introduced by Ingrid *Daubechies in 1988. These wavelets have *fractal properties. Other families of wavelets include **symmlets** and **coiflets**.

weak law of large numbers *See* LAWS OF LARGE NUMBERS.

weakly stationary *See* STATIONARY PROCESS.

Wedderburn, Robert William Maclagan (1947–75; b. Edinburgh, Scotland) A Scottish statistician. Wedderburn was a graduate of Cambridge University. On graduation he joined the staff at Rothamsted Experimental Station where he collaborated with *Nelder on the influential 1972 paper that unified the treatment of *generalized linear models. By the time of his early death he had made valuable contributions to the *GENSTAT and *GLIM statistical packages.

Weibull, E. H. Wallodi (1887–1979; b. Schleswig-Holstein; d. Annecy, France) A Swedish engineer. Weibull studied at the Royal Institute of Technology in Stockholm, graduating in 1924 . He gained a doctorate from the University of Uppsala in 1932. He spent much of his career in the Royal Swedish Coast Guard, joining in 1904 as a midshipman and reaching the rank of major in 1940. Initially his job was concerned with studying the effects of underwater explosions and subsequently with the lifetimes of components. Weibull is best known for his 1939 paper introducing the *Weibull distribution. In 1972 the American Society of Mechanical Engineers awarded him their Gold Medal citing him as 'a pioneer in the study of fracture, fatigue and reliability'.

Weibull distribution A *random variable, X, that has a Weibull distribution can take any positive value and the *probability density function f is given by

$$f(x) = cx^{c-1}\exp(-x^c), \qquad x > 0,$$

where c is a positive constant. The distribution has *mean $\Gamma(\frac{1}{c} + 1)$ and *variance $\Gamma(\frac{2}{c} + 1) - \left\{\Gamma(\frac{1}{c} + 1)\right\}^2$, where Γ is the *gamma function. The Weibull distribution has been found to be very useful for describing the *distribution of the lifetimes of components and for analysing meteorological data—particularly in the context of extreme events. *See following diagram.*

weight *See* WEIGHTED AVERAGE.

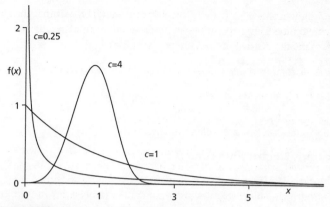

Weibull distribution. The shape depends on the parameter c. With $c < 1$ the distribution is often used to model extreme lifetimes.

---

weighted average (weighted mean) A weighted average is an average that can attach more importance (weight) to some *observations than to others. Suppose that the importance of observation x_j is represented by the non-negative numerical coefficient w_j (the **weight**) $(j = 1, 2, \ldots, n)$. The weighted average is

$$\frac{(w_1 x_1 + w_2 x_2 + \cdots + w_n x_n)}{(w_1 + w_2 + \cdots + w_n)}.$$

weighted least squares (WLS) A preferable alternative to *ordinary least squares (OLS) when estimating the *parameters of a *model using *independent *observations whose values are known to vary in accuracy in a specified way. *See also* GENERALIZED LEAST SQUARES.

weighted mean *See* WEIGHTED AVERAGE.

Welch's statistic A *statistic used in a *test for the equality of the *means of m *normal populations that cannot be assumed to have the same *variance. The test statistic, W, is given by

$$W = \frac{(m+1)u^2 \sum_{j=1}^{m} w_j (\bar{x}_j - \bar{x})^2}{\left\{ (m^2 - 1)u^2 + 2(m - 2) \sum_{j=1}^{m} (n_j - 1)(u - w_j)^2 \right\}},$$

where $\bar{x}_j$ is the mean of the n_j *observations in the jth sample, $w_j = n_j/s_j^2$, where s_j^2 is the variance (using the $(n - 1)$ divisor) of the jth sample,

$u = \sum_{j=1}^{m} w_j$ and $\bar{x} = \frac{1}{u} \sum_{j=1}^{m} w_j \bar{x}_j$. Under the *null hypothesis that the samples come from populations with the same mean, W has an approximate *F-distribution with $(m-1)$ and v *degrees of freedom, where v is the nearest integer to

$$\frac{1}{3} u^2 (m^2 - 1) \bigg/ \sum_{j=1}^{m} \frac{(u - w_j)^2}{n_j - 1}.$$

The special case $m = 2$ is called the *Behrens–Fisher problem.

Western Electric Rules *See* QUALITY CONTROL.

Wheeler–Watson test A *non-parametric test of the *null hypothesis that two *samples of *cyclic data have been drawn from the same *population. Suppose the samples contain n_1 and n_2 observations. These observations are first arranged in order of magnitude and then their actual values are replaced by coded values that (working in degrees) are multiples of $360/(n_1 + n_2)$, so that the kth largest value becomes

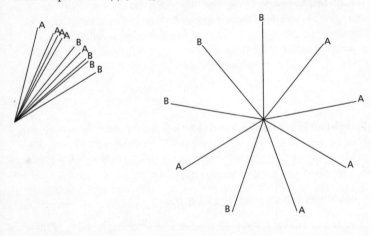

Original data   Coded data

Wheeler–Watson test. Two sets of birds are released from the same location. The five birds in set A fly off in the directions 15°, 25°, 28°, 31°, and 45°. The four birds in set B fly off in the directions 40°, 50°, 52°, and 60°. Arranging these data in order, we have AAAABABBB. Since $360/9 = 40$, the coded values for sample B are (in degrees) 200, 280, 320, and 360. The resulting value for R^2 is 4.88 and $T = 3.90$. We conclude that there is no significant evidence that the samples have been drawn from different populations.

$360k/(n_1 + n_2)$. Let the coded values for the first sample be $\theta_1, \theta_2, \ldots, \theta_{n_1}$, and calculate R^2 given by

$$R^2 = \left(\sum_{j=1}^{n_1} \sin \theta_j \right)^2 + \left(\sum_{j=1}^{n_1} \cos \theta_j \right)^2.$$

The value of R^2 for the second sample will be identical. The test statistic is T, given by

$$T = \frac{2(n_1 + n_2 - 1)R^2}{n_1 n_2}.$$

For large n_1 and n_2, T has an approximate *chi-squared distribution with two *degrees of freedom.

white noise *See* NOISE.

Whitney, Donald Ransom (1915–) An American statistician. Having obtained his MA at Princeton University, Whitney spent much of the Second World War teaching navigation to newly commissioned officers. After the war he obtained his doctorate at Ohio State University under the supervision of Henry *Mann. The subject of his research was the *non-parametric *Mann–Whitney test (published in 1947). Whitney remained at Ohio State University throughout his career, retiring as Professor Emeritus in 1983.

whole plot *See* SPLIT PLOT.

Wiener, Norbert (1894–1964; b. Columbia, Missouri; d. Stockholm, Sweden) An American mathematician. The son of Russian immigrants, Wiener was a child prodigy, obtaining his PhD (in mathematical logic) from Harvard University at the age of eighteen. He developed the mathematics of **Wiener processes**, which are fundamental to an understanding of *Brownian motion. His interests were wide-ranging: he became well known to the general scientific public for his philosophical discussion of cybernetics (a term he coined in 1945). Amongst his sayings is 'A professor is one who can speak on any subject—for precisely fifty minutes', though some might disagree with this observation! A lunar crater is named after him.

Wiener process *See* WIENER.

Wilcoxon, Frank (1892–1965; b. Glengarriffe Castle, Ireland; d. Tallahassee, Florida) An American chemist and self-taught statistician.

Wilcoxon was a teenage rebel who was in turn a merchant seaman, a petrol-station attendant, and a tree surgeon. Subsequently conforming, he obtained his doctorate in physical chemistry from Cornell University in 1924. For the next 25 years he worked as a chemist in various of the larger chemical firms in the USA. Only for the last seven years of his working life was he officially employed as a statistician. However, his interest in the subject dated back to 1925, when he wished to devise statistical tests of the effectiveness of various types of insecticide and fungicide. He led the research group that worked on the development of various of the pyrethrin-based insecticides, including Malathion. His statistical work concentrated on devising methods of testing that were simple and easy to understand (the hallmarks of *non-parametric tests) and he is remembered now in the context of the *Wilcoxon signed-rank tests introduced in a 1945 paper.

Wilcoxon rank-sum test *See* Mann–Whitney test.

Wilcoxon signed-rank tests *Non-parametric tests that extend the *sign tests. The single-sample version (*observations $x_1, x_2, \ldots$), suitable for a *symmetric distribution, tests the *null hypothesis that the *population median has a specified value (m_0). The **matched-pair** (or **paired-sample**) version (observation pairs $(x_1, y_1), (x_2, y_2), \ldots$) is concerned with the differences $(x_1 - y_1), (x_2 - y_2), \ldots$. With the assumption that these differences are *independent observations from a symmetric distribution, the null hypothesis is that this distribution has median zero.

To determine the value of the test statistic, z, the first step is to calculate the differences $d_1, d_2, \ldots$, where $d_j = x_j - m_0$ (single sample) or $d_j = x_j - y_j$ (matched pairs). After zero differences have been discarded, the remaining n are arranged in ascending order of $|d_j|$. The magnitudes are replaced by the corresponding *ranks, with tied ranks where necessary. The signs of $d_1, d_2, \ldots$, are now attributed to the ranks, resulting in **signed ranks**. Let P be the sum of the positive signed ranks and let T be the smaller of P and $\frac{1}{2}n(n+1) - P$. The test statistic, given by

$$z = \frac{\frac{1}{4}n(n+1) - T - \frac{1}{2}}{\sqrt{\frac{1}{24}n(n+1)(2n+1)}},$$

is an observation from the upper half of an approximate *standard normal distribution. The $\frac{1}{2}$ is a *continuity correction.

As an example, suppose that a symmetric distribution is believed to have median 100. A random sample of eight observations is reported as consisting of the values 92.3, 57.6, 88.8, 110.5, 100.0, 181.0, 96.0, 105.7. The supposed median is subtracted from each observation to give -7.7, -42.4, -11.2, 10.5, 0, 81.0, -4.0, 5.7. The value 0 is discarded and the remainder are arranged in order of ascending absolute magnitude: -4.0, 5.7, -7.7, 10.5, -11.2, -42.4, 81.0. Retaining the signs while replacing the values by ranks gives -1, 2, -3, 4, -5, -6, 7, so that $p = 13$, $\frac{1}{2}n(n + 1) - P = 15$ (since $n = 7$) and $T = 13$. The test statistic is

$$z = \frac{14 - 13 - 0.5}{\sqrt{35}} = 0.085.$$

Comparing with the tables of upper-tail percentage points of the standard normal distribution (Appendix VI), we see that the null hypothesis that the median is 100 should not be rejected.

Wilk, Martin Bradbury (1922–) In 1953 Wilk received his master's degree from Iowa State University, gaining his doctorate in 1955. Wilk was a co-author of the 1965 paper that introduced the *Shapiro–Wilk test. After teaching at Rutgers University, he joined AT&T. In 1981 he was appointed Chief Statistician of the Canadian government statistics office, Statistics Canada. Subsequently he was an adjunct professor at Carleton University, Ottawa. He was made an Honorary Fellow of the *Royal Statistical Society in 1986 and an Honorary Member of the *Statistical Society of Canada in 1988.

Wilks, Samuel Stanley (1906–64; Little Elm, Texas; d. Princeton, New Jersey) An American mathematical statistician. Wilks's first degree was in architecture (in 1926 at the North Texas Teacher's College). He then obtained a BSc in mathematics at the University of Texas (1928) and a PhD in statistics at Iowa State University in 1931. In 1933, after a year in England working with Sir Ronald *Fisher and John *Wishart, Wilks joined the staff at Princeton University, where he remained until his early death. In 1935 he was one of the founder members of the *Institute of Mathematical Statistics and served as its President in 1940. From 1938 to 1949 he was Editor of the *Annals of Mathematical Statistics*. He was President of the *American Statistical Association in 1950.

Wilks's lambda (Λ) *See* MULTIVARIATE ANALYSIS OF VARIANCE.

Wilks Medal A prestigious award made annually by the *American Statistical Association in memory of their former President, Samuel *Wilks.

window width *See* KERNEL METHODS.

Winsorized mean *See* TRIMMED MEAN.

Wishart, John (1898–1956; b. Montrose, Scotland; d. Acapulco, Mexico) A Scottish statistician. Either side of the First World War (when he served in the Black Watch regiment), Wishart studied mathematics and physics at Edinburgh University. In 1924 he joined Karl *Pearson at University College, London, but then in 1927 he joined Sir Ronald *Fisher at Rothamsted. His work on the *Wishart distribution appeared in 1928. In 1931 he moved to Cambridge University, becoming Head of the Statistical Laboratory in 1953.

Wishart distribution The joint distribution of the *variances and *covariances in *samples from a *multivariate normal distribution.

without replacement; with replacement *See* URN MODEL.

WLS *See* WEIGHTED LEAST SQUARES.

Wolfowitz, Jacob (1910–1981; b. Warsaw, Poland; d. Tampa, Florida) An American statistician. Wolfowitz joined his father in New York when he was ten. He graduated in mathematics from City College (now City University), New York in 1932 spending the next ten years as a mathematics teacher. During this period he worked on *non-parametric tests (he coined the word 'non-parametric') and began his collaboration with *Wald; the *Wald–Wolfowitz test was published in 1940. He also made pioneering contributions to *dynamic programming. He received his PhD in 1942 from New York University. After a period of war-related research, Wolfowitz held posts at several institutions, including Cornell University (1951), the University of Illinois at Urbana (1970), and the University of South Florida in Tampa (1978). Elected to membership of the National Academy of Sciences and the American Academy of Arts and Sciences, he was President of the *Institute of Mathematical Statistics in 1959.

wrapped Cauchy distribution A distribution used to model *circular data. It can be regarded as wrapping a *Cauchy distribution around the circle. The *probability density function f is given by

$$f(\theta) = \frac{1 - \rho^2}{2\pi(1 + \rho^2 - 2\rho\cos\theta)}, \quad -\pi < \theta \leq \pi,$$

where ρ $(0 < \rho < 1)$ is a constant. The distribution closely resembles the *von Mises distribution.

X *See* MULTIPLE REGRESSION MODEL.

x̄-chart *See* QUALITY CONTROL.

Yates, Frank (1902–94; Manchester, England; d. Harpenden, England) An English mathematician and statistician. In 1924 Yates left Cambridge University with a first class degree in mathematics. After two years as a mathematics teacher, he joined the Gold Coast survey as a mathematics advisor. However, the climate affected his health and he returned to England, joining the statistics staff at Rothamsted (the agricultural research institute in Hertfordshire) in 1931. He was a keen user of computers, writing that 'to be a good theoretical statistician one must also compute'. He was elected a Fellow of the Royal Society in 1948. Although he retired in 1967, he maintained his links with Rothamsted and his last paper was published in 1990, when he was 88. He was President of the *Royal Statistical Society in 1967 and was awarded its *Guy Medal in Gold in 1960.

Yates-corrected chi-squared test A test of the *null hypothesis of *independence between the classifying *variables for a two-by-two table. If the *cell frequencies are denoted by a, b, c, and d, the row totals by m and n, the column totals by r and s, and the grand total by N, the table is

$$
\begin{array}{cc|c}
a & b & m \\
c & d & n \\
\hline
r & s & N
\end{array}
$$

The (uncorrected) *chi-squared test statistic is

$$ X^2 = \frac{N(ad - bc)^2}{mnrs}, $$

with an approximate *chi-squared distribution with one *degree of freedom. An equivalent test compares X, which can only take a limited number of values, with the *continuous standard *normal distribution. This is a case for incorporating a *continuity correction, and if we square the continuity-corrected test we get the Yates-corrected chi-squared test statistic:

$$X_c^2 = \frac{N\left(|ad - bc| - \frac{1}{2}N\right)^2}{mnrs},$$

in which the $\frac{1}{2}N$ is **Yates's correction**.

Yates's correction *See* YATES-CORRECTED CHI-SQUARED TEST.

Youden, William John (1900–71; b. Townsville, Australia; d. Washington, DC) An American chemical engineer and statistician. Youden's parents took him to the United States when he was seven. Youden's first degree, at Rochester University, was in chemical engineering and his MSc and PhD (in 1924) were in chemistry, at Columbia University. In 1928 Youden, working in an institute for plant research, came across Sir Ronald *Fisher's *Statistical Methods*, and this book engendered his subsequent interest in experimental design. During the Second World War Youden worked as a systems analyst. After the war, he joined the National Bureau of Standards, continuing to work on experimental design. Youden was awarded the *Wilks Medal of the *American Statistical Association in 1969.

Youden square *See* LATIN SQUARE.

Yule, George Udny (1871–1951; b. Morham, Scotland; d. Cambridge, England) A Scottish statistician. In 1890 Yule graduated from University College, London, where he studied engineering. After two years as an engineer, and a year researching in physics, he was appointed by Karl *Pearson to a demonstratorship at University College. Yule's first paper on statistics appeared in 1895. In 1911 he published *Introduction to the Theory of Statistics*, which was the standard reference book on mathematical statistics for the next forty years (the fourteenth edition, written jointly with Maurice *Kendall, was published in 1950). In 1912 he moved to Cambridge University, where he spent most of the rest of his life, retiring in 1931 (when he learnt to fly and obtained his pilot's licence). During his time at Cambridge he worked on the theory of *time series, introducing the terms *correlogram and *autoregressive models. He was elected a Fellow of the Royal Society in 1921. He was President of the *Royal Statistical Society in 1924, having been awarded its *Guy Medal in Gold in 1911.

Yule's Q A *measure of association for a two-by-two table, suggested by *Yule in 1900. For the table

$$
\begin{array}{|cc|}
\hline
a & b \\
c & d \\
\hline
\end{array}
$$

Q is given by $Q = (ad - bc)/(ad + bc)$. If the *null hypothesis of independence between the classifying variables is correct and if the cell frequencies are not too small, Q has an approximate *normal distribution with *mean 0 and *variance

$$
\frac{1}{4}(1 - Q^2)^2 \left(\frac{1}{a} + \frac{1}{b} + \frac{1}{c} + \frac{1}{d} \right).
$$

See TETRACHORIC CORRELATION COEFFICIENT.

Yule–Walker equations *See* AUTOREGRESSIVE MODELS.

Z

zero-sum game *See* GAME THEORY.

z-test A test of the *null hypothesis that a *sample of n *independent observations, taken from a *normal distribution with *variance σ^2, has *mean μ. Writing the sample mean as $\bar{x}$, the test statistic is z, given by

$$z = \frac{\bar{x} - \mu}{\sigma/\sqrt{n}}.$$

If the *alternative hypothesis is that the mean is not μ, then, if $|z| > 1.96$, there is evidence to reject the null hypothesis at the 5% significance level in favour of the alternative hypothesis. *See* HYPOTHESIS TEST.

z-value A value, usually of a test statistic, that is hypothesized to have come from a *normal distribution with *mean 0 and *variance 1. The z-statistic was introduced by Sir Ronald *Fisher in 1924.

Appendices

Appendix I: Statistical Notation

Symbol	Reference
$\sim$	has the same distribution as
|	conditional, e.g. A occurs given that B occurs
$B(n, p)$	binomial distribution with parameters n and p
$Cov(X, Y)$	covariance of the random variables X and Y
$E(X)$	expected value of the random variable X
F	cumulative distribution function
f	probability density function
G	probability-generating function
G^2	likelihood-ratio goodness-of-fit statistic
H_0, H_1	null and alternative hypotheses
M	moment-generating function
$N(\mu, \sigma^2)$	normal distribution (mean μ, variance σ^2)
n	sample size
P	probability (of an event)
p	probability value
r	product-moment correlation coefficient; residual
s^2	sample variance given by $\sum_j (x_j - \bar{x})^2 / (n - 1)$
$Var(X)$	variance of the random variable X
X, Y	random variables
X^2	chi-squared test statistic
x	an observation
$\bar{x}$	sample mean
$x_{(j)}$	an ordered observation, e.g. $x_{(1)} \leq x_{(2)} \leq \ldots$
$\bar{x}_{-j}, \hat{x}_{-j}$	values calculated omitting observation j
$\hat{y}$	an estimate of y
Z	random variable (often standard normal random variable)

Appendix II: Mathematical Notation

Symbol	Reference
∞	infinity
!	factorial
$'$	For a function f, $f'(x)$ is the derivative of $f(x)$ with respect to x
$'$	For a set, *see* COMPLEMENTARY EVENT
$'$	For a matrix, *see* MATRIX
(a, b)	an open interval, i.e. the set of real numbers between a and b: $a < x < b$
$\mathbb{R}$	the set of all real numbers
$\mathbb{R}^n$	the set of all row vectors, or column vectors, with n real elements
$\cup$	union
$\cap$	intersection
$\approx$	approximately equal to
$<$	less than
$\leq$	less than or equal to
$>$	greater than
$\geq$	greater than or equal to
$\rightarrow$	tends to, e.g. $n \rightarrow \infty$
$|x|$	absolute value of x
a^b	a raised to the power b
$a^{\frac{1}{2}}$	$\sqrt{a}$, the (positive) square root of a
B	beta function, *see* BETA FUNCTION
$\partial y / \partial x$	partial derivative of y with respect to x. See HESSIAN MATRIX
$\max_j\{x_1, x_2, \ldots, x_n\}$	the greatest of $\{x_1, x_2, \ldots, x_n\}$
$\min_j\{x_1, x_2, \ldots, x_n\}$	the least of $\{x_1, x_2, \ldots, x_n\}$
$\mathrm{sign}(x)$	the sign of x, i.e. $$\mathrm{sign}(x) = \begin{cases} +1 & \text{if } x > 0 \\ 0 & \text{if } x = 0 \\ -1 & \text{if } x < 0 \end{cases}$$
X	a matrix
x	a vector
$\{r_j\}$	the set $r_1, r_2, \ldots$
$\binom{n}{r}$	binomial coefficient

Appendix III: Greek Letters

	Symbol	Reference
α	alpha	*See* ALPHA; LINEAR REGRESSION
β	beta	*See* BETA; LINEAR REGRESSION
$\boldsymbol{\beta}$		*See* MULTIPLE REGRESSION MODEL
γ	gamma	*See* ORDINAL VARIABLE
Γ	capital gamma	*See* GAMMA FUNCTION
δ	delta	
ε	epsilon	
θ	theta	an angle; a parameter
κ	kappa	*See* COHEN'S KAPPA; KURTOSIS
λ	lambda	*See* GOODMAN AND KRUSKAL'S LAMBDA
Λ	capital lambda	*See* MULTIVARIATE ANALYSIS OF VARIANCE
μ	mu	*See* POPULATION MEAN; MOMENT
ν	nu	*See* DEGREES OF FREEDOM; CHI-SQUARED DISTRIBUTION; F-DISTRIBUTION; t-DISTRIBUTION
π	pi	*See* PI
$\prod$	capital pi	a product *e.g.* $\prod_{n=1}^{N} n = 1 \times 2 \times \cdots \times N = N!$
ρ	rho	*See* AUTOCORRELATION; CORRELATION COEFFICIENT; SPEARMAN'S RHO
σ	sigma	standard deviation
Σ	capital sigma	summation (*see* SIGMA)
Σ		a variance–covariance matrix
τ	tau	*See* KENDALL'S TAU
ϕ	phi	probability density function of a standard normal distribution; also the empty set
Φ	capital phi	the cumulative distribution function of a standard normal distribution
χ	chi	For χ^2, *see* CHI-SQUARED DISTRIBUTION
ψ	psi	*See* WAVELET

Appendix IV: Cumulative Probabilities for the Binomial Distribution

For the given values of the sample size, n, and the probability of success, p, the table gives the values of $P(X \leq r)$.

n	r	p									
		0.05	0.10	0.15	0.20	0.25	0.30	0.35	0.40	0.45	0.50
2	0	.9025	.8100	.7225	.6400	.5625	.4900	.4225	.3600	.3025	.2500
	1	.9975	.9900	.9775	.9600	.9375	.9100	.8775	.8400	.7975	.7500
3	0	.8574	.7290	.6141	.5120	.4219	.3430	.2746	.2160	.1664	.1250
	1	.9928	.9720	.9393	.8960	.8438	.7840	.7183	.6480	.5748	.5000
	2	.9999	.9990	.9966	.9920	.9844	.9730	.9571	.9360	.9089	.8750
4	0	.8145	.6561	.5220	.4096	.3164	.2401	.1785	.1296	.0915	.0625
	1	.9860	.9477	.8905	.8192	.7383	.6517	.5630	.4752	.3910	.3125
	2	.9995	.9963	.9880	.9728	.9492	.9163	.8735	.8208	.7585	.6875
	3		.9999	.9995	.9984	.9961	.9919	.9850	.9744	.9590	.9375
5	0	.7738	.5905	.4437	.3277	.2373	.1681	.1160	.0778	.0503	.0313
	1	.9774	.9185	.8352	.7373	.6328	.5282	.4284	.3370	.2562	.1875
	2	.9988	.9914	.9734	.9421	.8965	.8369	.7648	.6826	.5931	.5000
	3		.9995	.9978	.9933	.9844	.9692	.9460	.9130	.8688	.8125
	4			.9999	.9997	.9990	.9976	.9947	.9898	.9815	.9688
6	0	.7351	.5314	.3771	.2621	.1780	.1176	.0754	.0467	.0277	.0156
	1	.9672	.8857	.7765	.6554	.5339	.4202	.3191	.2333	.1636	.1094
	2	.9978	.9842	.9527	.9011	.8306	.7443	.6471	.5443	.4415	.3438
	3	.9999	.9987	.9941	.9830	.9624	.9295	.8826	.8208	.7447	.6563
	4		.9999	.9996	.9984	.9954	.9891	.9777	.9590	.9308	.8906
	5				.9999	.9998	.9993	.9982	.9959	.9917	.9844
7	0	.6983	.4783	.3206	.2097	.1335	.0824	.0490	.0280	.0152	.0078
	1	.9556	.8503	.7166	.5767	.4449	.3294	.2338	.1586	.1024	.0625
	2	.9962	.9743	.9262	.8520	.7564	.6471	.5323	.4199	.3164	.2266
	3	.9998	.9973	.9879	.9667	.9294	.8740	.8002	.7102	.6083	.5000
	4		.9998	.9988	.9953	.9871	.9712	.9444	.9037	.8471	.7734
	5			.9999	.9996	.9987	.9962	.9910	.9812	.9643	.9375
	6					.9999	.9998	.9994	.9984	.9963	.9922

n	r	p									
		0.05	0.10	0.15	0.20	0.25	0.30	0.35	0.40	0.45	0.50
8	0	.6634	.4305	.2725	.1678	.1001	.0576	.0319	.0168	.0084	.0039
	1	.9428	.8131	.6572	.5033	.3671	.2553	.1691	.1064	.0632	.0352
	2	.9942	.9619	.8948	.7969	.6785	.5518	.4278	.3154	.2201	.1445
	3	.9996	.9950	.9786	.9437	.8862	.8059	.7064	.5941	.4770	.3633
	4		.9996	.9971	.9896	.9727	.9420	.8939	.8263	.7396	.6367
	5			.9998	.9988	.9958	.9887	.9747	.9502	.9115	.8555
	6				.9999	.9996	.9987	.9964	.9915	.9819	.9648
	7						.9999	.9998	.9993	.9983	.9961
10	0	.5987	.3487	.1969	.1074	.0563	.0282	.0135	.0060	.0025	.0010
	1	.9139	.7361	.5443	.3758	.2440	.1493	.0860	.0464	.0233	.0107
	2	.9885	.9298	.8202	.6778	.5256	.3828	.2616	.1673	.0996	.0547
	3	.9990	.9872	.9500	.8791	.7759	.6496	.5138	.3823	.2660	.1719
	4	.9999	.9984	.9901	.9672	.9219	.8497	.7515	.6331	.5044	.3770
	5		.9999	.9986	.9936	.9803	.9527	.9051	.8338	.7384	.6230
	6			.9999	.9991	.9965	.9894	.9740	.9452	.8980	.8281
	7				.9999	.9996	.9984	.9952	.9877	.9726	.9453
	8						.9999	.9995	.9983	.9955	.9893
	9								.9999	.9997	.9990

NB. Missing values correspond to probabilities equal to 1.0000, correct to four decimal places.

Appendix V: Cumulative Probabilities for the Poisson Distribution

For the given values of the parameter λ (equal to the mean of the distribution) the table gives the values of $P(X \leq r)$.

r	λ																			
	0.1	0.2	0.3	0.4	0.5	0.6	0.7	0.8	0.9	1.0	1.2	1.4	1.6	1.8	2.0	2.5	3.0	3.5	4.0	5.0
0	.9048	.8187	.7408	.6703	.6065	.5488	.4966	.4493	.4066	.3679	.3012	.2466	.2019	.1653	.1353	.0821	.0498	.0302	.0183	.0067
1	.9953	.9825	.9631	.9384	.9098	.8781	.8442	.8088	.7725	.7358	.6626	.5918	.5249	.4628	.4060	.2873	.1991	.1359	.0916	.0404
2	.9998	.9989	.9964	.9921	.9856	.9769	.9659	.9526	.9371	.9197	.8795	.8335	.7834	.7306	.6767	.5438	.4232	.3208	.2381	.1247
3		.9999	.9997	.9992	.9982	.9966	.9942	.9909	.9865	.9810	.9662	.9463	.9212	.8913	.8571	.7576	.6472	.5366	.4335	.2650
4				.9999	.9998	.9996	.9992	.9986	.9977	.9963	.9923	.9857	.9763	.9636	.9473	.8912	.8153	.7254	.6288	.4405
5							.9999	.9998	.9997	.9994	.9985	.9968	.9940	.9896	.9834	.9580	.9161	.8576	.7851	.6160
6										.9999	.9997	.9994	.9987	.9974	.9955	.9858	.9665	.9347	.8893	.7622
7												.9999	.9997	.9994	.9989	.9958	.9881	.9733	.9489	.8666
8														.9999	.9998	.9989	.9962	.9901	.9786	.9319
9																.9997	.9989	.9967	.9919	.9682
10																.9999	.9997	.9990	.9972	.9863
11																	.9999	.9997	.9991	.9945
12																		.9999	.9997	.9980
13																			.9999	.9993
14																				.9998
15																				.9999

NB. Missing values correspond to probabilities equal to 1.0000, correct to four decimal places.

Appendix VI: Upper-Tail Percentage Points for the Standard Normal Distribution

The table gives the values of z for which $P(Z > z) = q\%$, where Z has a normal distribution with mean 0 and variance 1.

$q(\%)$	z	$q(\%)$	z	$q(\%)$	z	$q(\%)$	z	$q(\%)$	z
50	0.000	15	1.036	2.5	1.960	1.0	2.326	0.04	3.353
45	0.126	14	1.080	2.4	1.977	0.9	2.366	0.03	3.432
40	0.253	13	1.126	2.3	1.995	0.8	2.409	0.02	3.540
35	0.385	12	1.175	2.2	2.014	0.7	2.457	0.01	3.719
30	0.524	11	1.227	2.1	2.034	0.6	2.512	$0.0^2 5$	3.891
25	0.674	10	1.282	2.0	2.054	0.5	2.576	$0.0^2 1$	4.265
24	0.706	9	1.341	1.9	2.075	0.4	2.652	$0.0^3 5$	4.417
23	0.739	8	1.405	1.8	2.097	0.3	2.748	$0.0^3 1$	4.753
22	0.772	7	1.476	1.7	2.120	0.2	2.878	$0.0^4 5$	4.892
21	0.806	6	1.555	1.6	2.144	0.1	3.090	$0.0^4 1$	5.199
20	0.842	5	1.645	1.5	2.170	0.09	3.121	$0.0^5 5$	5.327
19	0.878	4.5	1.695	1.4	2.197	0.08	3.156	$0.0^5 1$	5.612
18	0.915	4	1.751	1.3	2.226	0.07	3.195	$0.0^6 5$	5.731
17	0.954	3.5	1.812	1.2	2.257	0.06	3.239	$0.0^6 1$	5.998
16	0.994	3	1.881	1.1	2.290	0.05	3.291	$0.0^7 5$	6.109

In the table, the notation $0.0^5 1$ means $0.000\,001$

Appendix VII: The Standard Normal Distribution Function

The table gives the values of $\Phi(z) = P(Z < z)$, where Z has a normal distribution with mean 0 and variance 1.

$\Phi(z)$

z	0	1	2	3	4	5	6	7	8	9
0.0	.5000	.5040	.5080	.5120	.5160	.5199	.5239	.5279	.5319	.5359
0.1	.5398	.5438	.5478	.5517	.5557	.5596	.5636	.5675	.5714	.5753
0.2	.5793	.5832	.5871	.5910	.5948	.5987	.6026	.6064	.6103	.6141
0.3	.6179	.6217	.6255	.6293	.6331	.6368	.6406	.6443	.6480	.6517
0.4	.6554	.6591	.6628	.6664	.6700	.6736	.6772	.6808	.6844	.6879
0.5	.6915	.6950	.6985	.7019	.7054	.7088	.7123	.7157	.7190	.7224
0.6	.7257	.7291	.7324	.7357	.7389	.7422	.7454	.7486	.7517	.7549
0.7	.7580	.7611	.7642	.7673	.7704	.7734	.7764	.7794	.7823	.7852
0.8	.7881	.7910	.7939	.7967	.7995	.8023	.8051	.8078	.8106	.8133
0.9	.8159	.8186	.8212	.8238	.8264	.8289	.8315	.8340	.8365	.8389
1.0	.8413	.8438	.8461	.8485	.8508	.8531	.8554	.8577	.8599	.8621
1.1	.8643	.8665	.8686	.8708	.8729	.8749	.8770	.8790	.8810	.8830
1.2	.8849	.8869	.8888	.8907	.8925	.8944	.8962	.8980	.8997	.9015
1.3	.9032	.9049	.9066	.9082	.9099	.9115	.9131	.9147	.9162	.9177

ADD

	1	2	3	4	5	6	7	8	9
	4	8	12	16	20	24	28	32	36
	4	8	12	16	20	24	28	32	36
	4	8	12	15	19	23	27	31	35
	4	7	11	15	19	22	26	30	34
	4	7	11	14	18	22	25	29	32
	3	7	10	14	17	20	24	27	31
	3	7	10	13	16	19	23	26	29
	3	6	9	12	15	18	21	24	27
	3	5	8	11	14	16	19	22	25
	3	5	8	10	13	15	18	20	23
	2	5	7	9	12	14	16	19	21
	2	4	6	8	10	12	14	16	18
	2	4	6	7	9	11	13	15	17
	2	3	5	6	8	10	11	13	14

Example: $\Phi(0.657) = 0.7422 + 0.0023 = 0.7445$

$\Phi(z)$

z	0	1	2	3	4	5	6	7	8	9	ADD 1	2	3	4	5	6	7	8	9
1.4	.9192	.9207	.9222	.9236	.9251	.9265	.9279	.9292	.9306	.9319	1	3	4	6	7	8	10	11	13
1.5	.9332	.9345	.9357	.9370	.9382	.9394	.9406	.9418	.9429	.9441	1	2	4	5	6	7	8	10	11
1.6	.9452	.9463	.9474	.9484	.9495	.9505	.9515	.9525	.9535	.9545	1	2	3	4	5	6	7	8	9
1.7	.9554	.9564	.9573	.9582	.9591	.9599	.9608	.9616	.9625	.9633	1	2	3	4	4	5	6	7	8
1.8	.9641	.9649	.9656	.9664	.9671	.9678	.9686	.9693	.9699	.9706	1	1	2	3	4	4	5	6	6
1.9	.9713	.9719	.9726	.9732	.9738	.9744	.9750	.9756	.9761	.9767	1	1	2	2	3	4	4	5	5
2.0	.9772	.9778	.9783	.9788	.9793	.9798	.9803	.9808	.9812	.9817	0	1	1	2	2	3	3	4	4
2.1	.9821	.9826	.9830	.9834	.9838	.9842	.9846	.9850	.9854	.9857	0	1	1	2	2	2	3	3	4
2.2	.9861	.9864	.9868	.9871	.9875	.9878	.9881	.9884	.9887	.9890	0	1	1	1	2	2	2	3	3
2.3	.9893	.9896	.9898	.9901	.9904	.9906	.9909	.9911	.9913	.9916	0	1	1	1	1	1	2	2	2
2.4	.9918	.9920	.9922	.9924	.9927	.9929	.9931	.9932	.9934	.9936	0	0	1	1	1	1	1	2	2
2.5	.9938	.9940	.9941	.9943	.9945	.9946	.9948	.9949	.9951	.9952	0	0	0	1	1	1	1	1	1
2.6	.9953	.9955	.9956	.9957	.9958	.9960	.9961	.9962	.9963	.9964	0	0	0	0	1	1	1	1	1
2.7	.9965	.9966	.9967	.9968	.9969	.9970	.9971	.9972	.9973	.9974	0	0	0	0	0	1	1	1	1
2.8	.9974	.9975	.9976	.9977	.9977	.9978	.9979	.9979	.9980	.9981	0	0	0	0	0	0	0	1	1
2.9	.9981	.9982	.9982	.9983	.9984	.9984	.9985	.9985	.9986	.9986	0	0	0	0	0	0	0	0	0

Example: $\Phi(1.846) = 0.9671 + 0.0004 = 0.9675$

Appendix VIII: Percentage Points for the *t*-Distribution

If T has a t-distribution with v degrees of freedom then a tabulated value, t, is such that $P(T < t) = p\%$

v	$p(\%)$								
	75	90	95	97.5	99	99.5	99.75	99.9	99.95
1	1.000	3.078	6.314	12.71	31.82	63.66	127.3	318.3	636.6
2	0.816	1.886	2.920	4.303	6.965	9.925	14.09	22.33	31.60
3	0.765	1.638	2.353	3.182	4.541	5.841	7.453	10.21	12.92
4	0.741	1.533	2.132	2.776	3.747	4.604	5.598	7.173	8.610
5	0.727	1.476	2.015	2.571	3.365	4.032	4.773	5.893	6.869
6	0.718	1.440	1.943	2.447	3.143	3.707	4.317	5.208	5.959
7	0.711	1.415	1.895	2.365	2.998	3.499	4.029	4.785	5.408
8	0.706	1.397	1.860	2.306	2.896	3.355	3.833	4.501	5.041
9	0.703	1.383	1.833	2.262	2.821	3.250	3.690	4.297	4.781
10	0.700	1.372	1.812	2.228	2.764	3.169	3.581	4.144	4.587
11	0.697	1.363	1.796	2.201	2.718	3.106	3.497	4.025	4.437
12	0.695	1.356	1.782	2.179	2.681	3.055	3.428	3.930	4.318
13	0.694	1.350	1.771	2.160	2.650	3.012	3.372	3.852	4.221
14	0.692	1.345	1.761	2.145	2.624	2.977	3.326	3.787	4.140
15	0.691	1.341	1.753	2.131	2.602	2.947	3.286	3.733	4.073
16	0.690	1.337	1.746	2.120	2.583	2.921	3.252	3.686	4.015
17	0.689	1.333	1.740	2.110	2.567	2.898	3.222	3.646	3.965
18	0.688	1.330	1.734	2.101	2.552	2.878	3.197	3.610	3.922
19	0.688	1.328	1.729	2.093	2.539	2.861	3.174	3.579	3.883
20	0.687	1.325	1.725	2.086	2.528	2.845	3.153	3.552	3.850
21	0.686	1.323	1.721	2.080	2.518	2.831	3.135	3.527	3.819
22	0.686	1.321	1.717	2.074	2.508	2.819	3.119	3.505	3.792
23	0.685	1.319	1.714	2.069	2.500	2.807	3.104	3.485	3.768
24	0.685	1.318	1.711	2.064	2.492	2.797	3.091	3.467	3.745
25	0.684	1.316	1.708	2.060	2.485	2.787	3.078	3.450	3.725
26	0.684	1.315	1.706	2.056	2.479	2.779	3.067	3.435	3.707
27	0.684	1.314	1.703	2.052	2.473	2.771	3.057	3.421	3.690
28	0.683	1.313	1.701	2.048	2.467	2.763	3.047	3.408	3.674
29	0.683	1.311	1.699	2.045	2.462	2.756	3.038	3.396	3.659
30	0.683	1.310	1.697	2.042	2.457	2.750	3.030	3.385	3.646
40	0.681	1.303	1.684	2.021	2.423	2.704	2.971	3.307	3.551
60	0.679	1.296	1.671	2.000	2.390	2.660	2.915	3.232	3.460
120	0.677	1.289	1.658	1.980	2.358	2.617	2.860	3.160	3.373
∞	0.674	1.282	1.645	1.960	2.326	2.576	2.807	3.090	3.291

Appendix IX: Percentage Points for the *F*-Distribution

Upper 5% points

ν_2	ν_1 1	2	3	4	5	6	7	8	12	24	40	∞
1	161.4	199.5	215.7	224.6	230.2	234.0	236.8	238.9	243.9	249.1	251.1	254.3
2	18.51	19.00	19.16	19.25	19.30	19.33	19.35	19.37	19.41	19.45	19.47	19.50
3	10.13	9.55	9.28	9.12	9.01	8.94	8.89	8.85	8.74	8.64	8.59	8.53
4	7.71	6.94	6.59	6.39	6.26	6.16	6.09	6.04	5.91	5.77	5.72	5.63
5	6.61	5.79	5.41	5.19	5.05	4.95	4.88	4.82	4.68	4.53	4.46	4.36
6	5.99	5.14	4.76	4.53	4.39	4.28	4.21	4.15	4.00	3.84	3.77	3.67
7	5.59	4.74	4.35	4.12	3.97	3.87	3.79	3.73	3.57	3.41	3.34	3.23
8	5.32	4.46	4.07	3.84	3.69	3.58	3.50	3.44	3.28	3.12	3.04	2.93
9	5.12	4.26	3.86	3.63	3.48	3.37	3.29	3.23	3.07	2.90	2.83	2.71
10	4.96	4.10	3.71	3.48	3.33	3.22	3.14	3.07	2.91	2.74	2.66	2.54
12	4.75	3.89	3.49	3.26	3.11	3.00	2.91	2.85	2.69	2.51	2.43	2.30
15	4.54	3.68	3.29	3.06	2.90	2.79	2.71	2.64	2.48	2.29	2.20	2.07
18	4.41	3.55	3.16	2.93	2.77	2.66	2.58	2.51	2.34	2.15	2.06	1.92
20	4.35	3.49	3.10	2.87	2.71	2.60	2.51	2.45	2.28	2.08	1.99	1.84
25	4.24	3.39	2.99	2.76	2.60	2.49	2.40	2.34	2.16	1.96	1.87	1.71
30	4.17	3.32	2.92	2.69	2.53	2.42	2.33	2.27	2.09	1.89	1.79	1.62
40	4.08	3.23	2.84	2.61	2.45	2.34	2.25	2.18	2.00	1.79	1.69	1.51
60	4.00	3.15	2.76	2.53	2.37	2.25	2.17	2.10	1.92	1.70	1.59	1.39
∞	3.84	3.00	2.60	2.37	2.21	2.10	2.01	1.94	1.75	1.52	1.39	1.00

NB. The lower 5% point of an F_{ν_1,ν_2} distribution is the reciprocal of the upper 5% point of an F_{ν_2,ν_1} distribution.

Appendix IX (*cont.*): Percentage Points for the *F*-Distribution
Upper 2.5% points

ν_2	ν_1											
	1	2	3	4	5	6	7	8	12	24	40	∞
1	647.8	799.5	864.2	899.6	921.8	937.1	948.2	956.7	967.7	997.2	1006	1018
2	38.51	39.00	39.17	39.25	39.30	39.33	39.36	39.37	39.41	39.46	39.47	39.50
3	17.44	16.04	15.44	15.10	14.88	14.73	14.62	14.54	14.34	14.12	14.04	13.90
4	12.22	10.65	9.98	9.60	9.36	9.20	9.07	8.98	8.75	8.51	8.41	8.26
5	10.01	8.43	7.76	7.39	7.15	6.98	6.85	6.76	6.52	6.28	6.18	6.02
6	8.81	7.26	6.60	6.23	5.99	5.82	5.70	5.60	5.37	5.12	5.01	4.85
7	8.07	6.54	5.89	5.52	5.29	5.12	4.99	4.90	4.67	4.42	4.31	4.14
8	7.57	6.06	5.42	5.05	4.82	4.65	4.53	4.43	4.20	3.95	3.84	3.67
9	7.21	5.71	5.08	4.72	4.48	4.32	4.20	4.10	3.87	3.61	3.51	3.33
10	6.94	5.46	4.83	4.47	4.24	4.07	3.95	3.85	3.62	3.37	3.26	3.08
12	6.55	5.10	4.47	4.12	3.89	3.73	3.61	3.51	3.28	3.02	2.91	2.72
15	6.20	4.77	4.15	3.80	3.58	3.41	3.29	3.20	2.96	2.70	2.59	2.40
18	5.98	4.56	3.95	3.61	3.38	3.22	3.10	3.01	2.77	2.50	2.38	2.19
20	5.87	4.46	3.86	3.51	3.29	3.13	3.01	2.91	2.68	2.41	2.29	2.09
25	5.69	4.29	3.69	3.35	3.13	2.97	2.85	2.75	2.51	2.24	2.12	1.91
30	5.57	4.18	3.59	3.25	3.03	2.87	2.75	2.65	2.41	2.14	2.01	1.79
40	5.42	4.05	3.46	3.13	2.90	2.74	2.62	2.53	2.29	2.01	1.88	1.64
60	5.29	3.93	3.34	3.01	2.79	2.63	2.51	2.41	2.17	1.88	1.74	1.48
∞	5.02	3.69	3.12	2.79	2.57	2.41	2.29	2.19	1.94	1.64	1.48	1.00

NB. The lower 2.5% point of an F_{ν_1, ν_2} distribution is the reciprocal of the upper 2.5% point of an F_{ν_2, ν_1} distribution.

Upper 1% points

ν_2	ν_1											
	1	2	3	4	5	6	7	8	12	24	40	∞
1	4052	4999	5403	5625	5764	5859	5928	5981	6106	6235	6287	6366
2	98.50	99.00	99.17	99.25	99.30	99.33	99.36	99.37	99.42	99.46	99.47	99.50
3	34.12	30.82	29.46	28.71	28.24	27.91	27.67	27.49	27.05	26.60	26.41	26.13
4	21.20	18.00	16.69	15.98	15.52	15.21	14.98	14.80	14.37	13.93	13.75	13.46
5	16.26	13.27	12.06	11.39	10.97	10.67	10.46	10.29	9.89	9.47	9.29	9.02
6	13.75	10.92	9.78	9.15	8.75	8.47	8.26	8.10	7.72	7.31	7.14	6.88
7	12.25	9.55	8.45	7.85	7.46	7.19	6.99	6.84	6.47	6.07	5.91	5.65
8	11.26	8.65	7.59	7.01	6.63	6.37	6.18	6.03	5.67	5.28	5.12	4.86
9	10.56	8.02	6.99	6.42	6.06	5.80	5.61	5.47	5.11	4.73	4.57	4.31
10	10.04	7.56	6.55	5.99	5.64	5.39	5.20	5.06	4.71	4.33	4.17	3.91
12	9.33	6.93	5.95	5.41	5.06	4.82	4.64	4.50	4.16	3.78	3.62	3.36
15	8.68	6.36	5.42	4.89	4.56	4.32	4.14	4.00	3.67	3.29	3.13	2.87
18	8.29	6.01	5.09	4.58	4.25	4.01	3.84	3.71	3.37	3.00	2.84	2.57
20	8.10	5.85	4.94	4.43	4.10	3.87	3.70	3.56	3.23	2.86	2.69	2.42
25	7.77	5.57	4.68	4.18	3.85	3.63	3.46	3.32	2.99	2.62	2.45	2.17
30	7.56	5.39	4.51	4.02	3.70	3.47	3.30	3.17	2.84	2.47	2.30	2.01
40	7.31	5.18	4.31	3.83	3.51	3.29	3.12	2.99	2.66	2.29	2.11	1.80
60	7.08	4.98	4.13	3.65	3.34	3.12	2.95	2.82	2.50	2.12	1.94	1.60
∞	6.63	4.61	3.78	3.32	3.02	2.80	2.64	2.51	2.18	1.79	1.59	1.00

NB. The lower 1% point of an F_{ν_1, ν_2} distribution is the reciprocal of the upper 1% point of an F_{ν_2, ν_1} distribution.

Appendix X: Percentage Points for the Chi-Squared Distribution

If X has a χ^2 distribution with ν degrees of freedom, then a tabulated value, x, is such that $P(X < x) = p\%$.

		$p(\%)$								
	Lower tail			Upper tail						
ν	0.5	2.5	5	90	95	97.5	99	99.5	99.9	
1	$0.0^4 3927$	$0.0^3 9821$	$0.0^2 3932$	2.706	3.841	5.024	6.635	7.879	10.83	
2	0.01003	0.05064	0.1026	4.605	5.991	7.378	9.210	10.60	13.82	
3	0.07172	0.2158	0.3518	6.251	7.815	9.348	11.34	12.84	16.27	
4	0.2070	0.4844	0.7107	7.779	9.488	11.14	13.28	14.86	18.47	
5	0.4117	0.8312	1.145	9.236	11.07	12.83	15.09	16.75	20.52	
6	0.6757	1.237	1.635	10.64	12.59	14.45	16.81	18.55	22.46	
7	0.9893	1.690	2.167	12.02	14.07	16.01	18.48	20.28	24.32	
8	1.344	2.180	2.733	13.36	15.51	17.53	20.09	21.95	26.12	
9	1.735	2.700	3.325	14.68	16.92	19.02	21.67	23.59	27.88	
10	2.156	3.247	3.940	15.99	18.31	20.48	23.21	25.19	29.59	
11	2.603	3.816	4.575	17.28	19.68	21.92	24.72	26.76	31.26	
12	3.074	4.404	5.226	18.55	21.03	23.34	26.22	28.30	32.91	
13	3.565	5.009	5.892	19.81	22.36	24.74	27.69	29.82	34.53	
14	4.075	5.629	6.571	21.06	23.68	26.12	29.14	31.32	36.12	

				$p(\%)$							
ν	Lower tail					Upper tail					
	0.5	2.5	5	90	95	97.5	99	99.5	99.9
15	4.601	6.262	7.261	22.31	25.00	27.49	30.58	32.80	37.70
16	5.142	6.908	7.962	23.54	26.30	28.85	32.00	34.27	39.25
17	5.697	7.564	8.672	24.77	27.59	30.19	33.41	35.72	40.79
18	6.265	8.231	9.390	25.99	28.87	31.53	34.81	37.16	42.31
19	6.844	8.907	10.12	27.20	30.14	32.85	36.19	38.58	43.82
20	7.434	9.591	10.85	28.41	31.41	34.17	37.57	40.00	45.31
21	8.034	10.28	11.59	29.62	32.67	35.48	38.93	41.40	46.80
22	8.643	10.98	12.34	30.81	33.92	36.78	40.29	42.80	48.27
23	9.260	11.69	13.09	32.01	35.17	38.08	41.64	44.18	49.73
24	9.886	12.40	13.85	33.20	36.42	39.36	42.98	45.56	51.18
25	10.52	13.12	14.61	34.38	37.65	40.65	44.31	46.93	52.62
30	13.79	16.79	18.49	40.26	43.77	46.98	50.89	53.67	59.70
40	20.71	24.43	26.51	51.81	55.76	59.34	63.69	66.77	73.40
50	27.99	32.36	34.76	63.17	67.50	71.42	76.15	79.49	86.66
60	35.53	40.48	43.19	74.40	79.08	83.30	88.38	91.95	99.61
70	43.28	48.76	51.74	85.53	90.53	95.02	100.4	104.2	112.3
80	51.17	57.15	60.39	96.58	101.9	106.6	112.3	116.3	124.8
90	59.20	65.65	69.13	107.6	113.1	118.1	124.1	128.3	137.2
100	67.33	74.22	77.93	118.5	124.3	129.6	135.8	140.2	149.4

NB. For values of $\nu > 100$ use the result that $\sqrt{2X}$ has an approximate normal distribution with mean $\sqrt{2\nu - 1}$ and variance 1.

Appendix XI: Critical Values for Spearman's Rank Correlation Coefficient, ρ

It is assumed that at least one ranking consists of a random permutation of the numbers 1 to n.

For $n > 40$, assuming H_0, ρ is approximately an observation from a normal distribution with mean 0 and variance $1/(n-1)$.

Critical values for one-tailed tests using ρ

The entries in the table are the smallest values of ρ (to three decimal places) that correspond to one-tail probabilities $\leq 5\%$ (or 1%). The observed value is significant *if it is equal to, or greater than*, the value in the table. The exact significance level never exceeds the nominal value (5% or 1%). The table can also be used to provide 10% and 2% critical values for two-tailed tests for ρ. The asterisk indicates that significance at this level cannot be achieved in this case.

n	5%	1%	n	5%	1%
4	1.000	$\star$	18	.401	.550
5	.900	1.000	19	.391	.535
6	.829	.943	20	.380	.522
7	.714	.893	21	.370	.509
8	.643	.833	22	.361	.497
9	.600	.783	23	.353	.486
10	.564	.745	24	.344	.476
11	.536	.709	25	.337	.466
12	.503	.678	26	.331	.457
13	.484	.648	27	.324	.449
14	.464	.626	28	.318	.441
15	.446	.604	29	.312	.433
16	.429	.582	30	.306	.425
17	.414	.566	40	.264	.368

Critical values for two-tailed tests using ρ

The entries in the table are the smallest values of ρ (to three decimal places) that correspond to two-tail probabilities $\leq 5\%$ (or 1%). The observed value is significant *if it is equal to, or greater, than* the value in the table. The exact significance level never exceeds the nominal value (5% or 1%). The table can also be used to provide 2.5% and 0.5% critical values for one-tailed tests for ρ. The asterisks indicate that significance at this level cannot be achieved in these cases.

n	5%	1%	n	5%	1%
4	*	*	18	.472	.600
5	1.000	*	19	.460	.584
6	.886	1.000	20	.447	.570
7	.786	.929	21	.436	.556
8	.738	.881	22	.425	.544
9	.700	.883	23	.416	.532
10	.648	.794	24	.407	.521
11	.618	.755	25	.398	.511
12	.587	.727	26	.390	.501
13	.560	.703	27	.383	.492
14	.538	.679	28	.375	.483
15	.521	.654	29	.368	.475
16	.503	.635	30	.362	.467
17	.488	.618	40	.313	.405

Appendix XII: Critical values for Kendall's τ

It is assumed that at least one ranking consists of a random permutation of the numbers 1 to n.

For $n > 40$, assuming H_0, τ is approximately an observation from a normal distribution with mean 0 and variance $\{2(2n+5)\}/\{9n(n-1)\}$.

Critical values for one-tailed tests using τ

The entries in the table are the smallest values of τ (to three decimal places) that correspond to one-tail probabilities $\leq 5\%$ (or 1%). The observed value is significant *if it is equal to, or greater than*, the value in the table. The exact significance level never exceeds the nominal value (5% or 1%). The table can also be used to provide 10% and 2% critical values for two-tailed tests for τ. The asterisk indicates that significance at this level cannot be achieved in this case.

n	5%	1%	n	5%	1%
4	1.000	*	18	.294	.412
5	.800	1.000	19	.287	.392
6	.733	.867	20	.274	.379
7	.619	.810	21	.267	.371
8	.571	.714	22	.264	.359
9	.500	.667	23	.257	.352
10	.467	.600	24	.246	.341
11	.418	.564	25	.240	.333
12	.394	.545	26	.237	.329
13	.359	.513	27	.231	.322
14	.363	.473	28	.228	.312
15	.333	.467	29	.222	.310
16	.317	.433	30	.218	.301
17	.309	.426	40	.185	.256

Critical values for two-tailed tests using τ

The entries in the table are the smallest values of τ (to three decimal places) that correspond to two-tail probabilities $\leq$ 5% (or 1%). The observed value is significant *if it is equal to, or greater than*, the value in the table. The exact significance level never exceeds the nominal value (5% or 1%). The table can also be used to provide 2.5% and 0.5% critical values for one-tailed tests for τ. The asterisks indicate that significance at this level cannot be achieved in these cases.

n	5%	1%	n	5%	1%
4	*	*	18	.346	.451
5	1.000	*	19	.333	.439
6	.867	1.000	20	.326	.421
7	.714	.905	21	.314	.410
8	.643	.786	22	.307	.394
9	.556	.722	23	.296	.391
10	.511	.644	24	.290	.377
11	.491	.600	25	.287	.367
12	.455	.576	26	.280	.360
13	.436	.564	27	.271	.356
14	.407	.516	28	.265	.344
15	.390	.505	29	.261	.340
16	.383	.483	30	.255	.333
17	.368	.471	40	.218	.285

Appendix XIII: Critical Values for the Product-Moment Correlation Coefficient, r

It is assumed that X and Y are uncorrelated and have normal distributions.

Critical values for two-tailed tests

The values in the table are the two-tailed 5% and 1% points of the distribution of r and hence are appropriate for upper-tail 2.5% and 0.5% tests.

n	5%	1%	n	5%	1%	n	5%	1%	n	5%	1%
4	.950	.990	13	.553	.684	22	.423	.537	40	.312	.403
5	.878	.959	14	.532	.661	23	.413	.526	50	.279	.361
6	.811	.917	15	.514	.641	24	.404	.515	60	.254	.330
7	.754	.874	16	.497	.623	25	.396	.505	70	.235	.306
8	.707	.834	17	.482	.606	26	.388	.496	80	.220	.286
9	.666	.798	18	.468	.590	27	.381	.487	90	.207	.270
10	.632	.765	19	.456	.575	28	.374	.478	100	.197	.256
11	.602	.735	20	.444	.561	29	.367	.470	110	.187	.245
12	.576	.708	21	.433	.549	30	.361	.463	120	.179	.234

Critical values for one-tailed tests

The values in the table are the upper-tail 5% and 1% points of the distribution of r and hence are appropriate for two-tailed 10% and 2% tests.

n	5%	1%	n	5%	1%	n	5%	1%	n	5%	1%
4	.900	.980	13	.476	.634	22	.360	.492	40	.264	.367
5	.805	.934	14	.458	.612	23	.352	.482	50	.235	.328
6	.729	.882	15	.441	.592	24	.344	.472	60	.214	.300
7	.669	.833	16	.426	.574	25	.337	.462	70	.198	.278
8	.621	.789	17	.412	.558	26	.330	.453	80	.185	.260
9	.582	.750	18	.400	.543	27	.323	.445	90	.174	.245
10	.549	.715	19	.389	.529	28	.317	.437	100	.165	.232
11	.521	.685	20	.378	.516	29	.312	.430	110	.158	.222
12	.497	.658	21	.369	.503	30	.306	.423	120	.151	.212

For values outside the range of the tables, use the fact that, assuming H_0, $r\sqrt{(n-2)/(1-r^2)}$ is an observation from a t_{n-2} distribution. Alternatively, use the result that $r\sqrt{n-1}$ is approximately an observation from a $N(0, 1)$ distribution.

Appendix XIV: Pseudo-Random Numbers

07552	37078	70487	39809	35705	42662
28859	92692	51960	51172	02339	94211
64473	62150	49273	29664	05698	05946
55434	20290	33414	26519	65317	47580
20131	05658	01643	17950	74442	30519
04287	26200	37224	23042	85793	50649
19631	42910	35954	88679	34461	45854
52646	83321	52538	41676	71829	00734
11107	55247	73970	67044	29864	72349
16311	04954	92332	51595	96460	77412
37057	83986	98419	76401	15412	68418
33724	28633	85953	82213	07827	48740
43737	15929	19659	52804	72335	25208
16929	84478	31341	60265	19404	27881
10131	98571	20877	34585	22353	54505
29998	48921	60361	12353	28334	84764
96525	74926	82302	97562	57805	40464
49955	60120	14557	04036	55397	54710
27936	70742	69960	69090	25800	53457
43045	75684	77671	70298	21292	27677
38782	35325	61068	64149	73456	06831
47347	47512	09263	83713	04450	31376
98561	93657	76725	55243	95540	31611
30674	43720	80477	82488	44328	55607
20293	63332	24626	56001	23528	85302

Appendix XV: Selected Landmarks in the Development of Statistics

Date	Event
1657	The first treatise on *probability, *Van Rekeningh in Spelen van Geluck (Calculation in Games of Chance)*, written by *Huygens.
1662	*Graunt publishes *Natural and Political Observations Mentioned in a Following Index and Made upon the Bills of Mortality*, introducing the *life table.
1711	*De Moivre publishes a (largely overlooked) derivation of the *Poisson distribution (*Poisson's better-known derivation was published in 1837).
1713	Jacob *Bernoulli publishes *Ars Conjectandi (The Art of Conjecture)*, containing a derivation of the *binomial distribution.
1733	*De Moivre published *The Doctrine of Chances* in 1718. The second (1738) edition contains a supplement dated 12 November 1733 which gives the formula for the *probability density function of the *normal distribution.
1763	*Bayes introduces the idea of a *prior distribution.
1805	*Legendre publishes the first account of the method of *least squares.
1812	*Laplace uses *generating functions in his *Théorie Analytique des Probabilités*.
1834	*Royal Statistical Society founded.
1835	*Quetelet applies the *normal distribution to describe the 'normal man'.
1839	*American Statistical Association founded in Boston, Massachusetts.
1847	*De Morgan publishes his laws of probability.
1863	*Abbe publishes a derivation of the *chi-squared distribution.
1869	*Galton uses the term *correlation in its statistical sense in his book *Hereditary Genius*.
1877	*Galton uses the term *regression in a lecture, entitled 'Typical Laws of Heredity in Man', on 9 February to the Royal Institution.
1880	*Venn introduces *Venn diagrams.
1885	Establishment of the *International Statistical Institute.
1896	Karl *Pearson introduces the product-moment correlation coefficient.
1900	Karl *Pearson introduces the *chi-squared test.
1901	First issue of *Biometrika*, edited by Karl *Pearson.
1902	Publication of *Elementary Principles of Statistical Mechanics*, by *Gibbs.
1908	Publication of *The Probable Error of a Mean*, by *Gossett. This introduces the *t-test.

1908	*Hardy and Weinberg publish *Population Genetics*.
1911	*Yule publishes *An Introduction to the Theory of Statistics*. The fourteenth edition was published in 1950.
1912	Sir Ronald *Fisher introduces the method of *maximum likelihood for parameter estimation.
1922	Sir Ronald *Fisher introduces the *F*-test for the comparison of variance estimates.
1924	*Shewhart introduces the *control chart.
1925	Sir Ronald *Fisher publishes the first edition of *Statistical Methods for Research Workers*, setting out *inter alia* *ANOVA tables. The thirteenth edition was published in 1970.
1928	*Neyman and Egon *Pearson introduce the idea of a *confidence interval.
1931	*Von Mises introduces the idea of sample space.
1933	*Kolmogorov publishes his axiomatic treatment of probability, *Foundations of the Theory of Probability*.
1933	*Kolmogorov introduces the *Kolmogorov–Smirnov test.
1933	*Neyman and Egon *Pearson introduce the procedure for *hypothesis testing.
1935	Foundation of the *Institute of Mathematical Statistics.
1938	*Kolmogorov publishes *Analytic Methods in Probability Theory*, which sets out the foundations of *Markov processes.
1947	*Dantzig introduces the *simplex method for constrained *optimization.
1947	Foundation of the *International Biometric Society and the journal *Biometrics*.
1948	*Wiener publishes *Cybernetics: or Control and Communication in the Animal and the Machine*.
1950	*Feller publishes the first volume of *An Introduction to Probability Theory and its Applications*, the definitive text on *stochastic processes.
1963	*Barnard suggests the *Monte Carlo approach to *hypothesis testing.
1963	*Matheron publishes *Traité de Géostatistique Appliquée*, setting out the fundamentals of *geostatistics.
1965	*Tukey and Cooley introduce the *fast Fourier transform.
1969	*Akaike introduces his criterion for model comparison.
1970	*Box and *Jenkins publish *Time Series Analysis: Forecasting and Control*.
1970	*Tukey publishes *Exploratory Data Analysis*, introducing the *boxplot and the *stem and leaf diagram.
1972	*Nelder and *Wedderburn introduce the framework for *generalized linear models.
1977	*Cook introduces new *regression diagnostics.
1977	*Dempster, Laird, and Rubin introduce the *EM algorithm for handling incomplete data.
1979	*Efron introduces the *bootstrap and other *resampling methods.
1988	*Daubechies introduces her family of *wavelets.

Appendix XVI: Further Reference

Introductory

Upton, G. J. G., and Cook, I. T. (1996) *Understanding Statistics*. Oxford University Press, Oxford.

Historical

Johnson, N. L., and Kotz, S. (eds) (1997) *Leading Personalities in Statistical Sciences*. Wiley, New York.

Stigler, S. (1986) *The History of Statistics*. Belknap Press, Cambridge, MA.

The MacTutor Website. *http://www-history.mcs.st-andrews.ac.uk/history/index.html*

The York Website *http://www.york.ac.uk/depts/maths/histstat/*

Special topics

Agresti, A. (2002) *Categorical Data Analysis*, 2nd edn. Wiley, New York.

Armitage, P., Berry, G., and Matthews J. N. S. (2000) *Statistical Methods in Medical Research*, 4th edn. Blackwell, Oxford.

Bertsekas, D. P. (1999) *Nonlinear Programming*, 2nd edn. Athena Scientific, Belmont, MA.

Bertsekas, D. P., and Tsitsiklis, J. (1997) *Introduction to Linear Optimization*. Athena Scientific: Belmont, MA.

 Chatfield, C. (2003) *The Analysis of Time Series: An Introduction*, 6th edn. Chapman & Hall/CRC Press, London.

Chatfield, C., and Collins, A. J. (2000) *Introduction to Multivariate Analysis*, rev. edn. Chapman & Hall/CRC Press, London.

Chetwynd, A., and Diggle, P. (1995) *Discrete Mathematics*. Arnold, London.

Cochran, W. G. (1977) *Sampling Techniques*, 3rd edn. Wiley, New York.

Cochran, W. G., and Cox, G. M. (1957) *Experimental Designs*, 2nd edn. Wiley, New York.

Collett, D. (2003) *Modelling Binary Data*, 2nd edn. Chapman & Hall/CRC Press, Florida.

Cox, D. R. (1992) *Planning of Experiments*. Wiley, New York.

Cox, D. R., and Isham, V. (1980) *Point Processes*. Chapman & Hall/CRC Press, London.

Cressie, N. A. C. (1993) *Statistics for Spatial Data*, revd edn. Wiley, New York.

Diggle, P. J., Heagerty, P., Liang, K.-Y., and Zeger, S. L. (2002) *Analysis of Longitudinal Data*, 2nd edn. Oxford Science Publications, Oxford.

Draper, N. R., and Smith, H. (1998) *Applied Regression Analysis*, 3rd edn. Wiley, New York.

Everitt, B. S., Landau, S., and Leese, M. (2001) *Cluster Analysis*, 4th edn. Edward Arnold, London.

Everitt, B. S. (2001) *A Handbook of Statistical Analyses using S-Plus*, 2nd edn. Chapman & Hall/CRC Press, London.

Feller, W. (1968) *An Introduction to Probability Theory and its Applications, Volume 1*, 3rd edn. Wiley, New York.

Fisher, N. I. (1993) *Statistical Analysis of Circular Data*. Cambridge University Press, Cambridge.

Gilks, W. R., Richardson, S., and Spiegelhaler, D. J., (Eds) (1995) *Markov Chain Monte Carlo in Practice*. Chapman & Hall/CRC Press, London.

Gujarati, D. N. (2003) *Basic Econometrics*, 4th edn. McGraw-Hill, New York.

Hogg, R. V., and Craig, A. J. (1995) *Introduction to Mathematical Statistics*, 5th edn. Macmillan, New York.

Lindley, D. V. (1991) *Making Decisions*, 2nd edn. Wiley, New York.

Mardia, K. V., Kent, J. T., and Bibby, J. M. (1979) *Multivariate Analysis*. Academic Press, London.

McCullagh, P., and Nelder, J. A. (1989) *Generalized Linear Models*, 2nd edn. Chapman & Hall/CRC Press, London.

Miller, M., Miller, I., and Freund, J. E. (1999) *John E. Freund's Mathematical Statistics*, 6th edn. Prentice Hall, New Jersey.

Morgan, B. J. T. (1984) *The Elements of Simulation*. Chapman & Hall/CRC Press, London.

Robert, C. P., and Casella, G. (1999) *Monte Carlo Statistical Methods*. Springer, Berlin.

Ross, S. M. (2002) *Introduction to Probability Models*, 8th edn. Academic Press, San Diego.

Taha, H. A. (2002) *Operations Research: An Introduction*, 7th edn. Macmillan, New York.

Tufte, E. R. (2001) *The Visual Display of Quantitative Information*, 2nd edn. Graphics Press, Cheshire, CT.

Venables, W. N., and Ripley, B. D. (2002) *Modern Applied Statistics with S-Plus*, 4th edn. Springer-Verlag, Berlin.

Advanced

Johnson, N. L., Kotz, S., and Balakrishnan, N. (1994) *Continuous Univariate Distributions, Volume 1*, 2nd edn. Wiley, New York.

Johnson, N. L., Kotz, S. and Balakrishnan, N. (1995) *Continuous Univariate Distributions, Volume 2*, 2nd edn. Wiley, New York.

Johnson, N. L., Kotz, S., and Kemp, A. W. (1992) *Univariate Discrete Distributions*, 2nd edn. Wiley, New York.

Kotz, S., Johnson, N. L., and Read, C. B. (eds) (1982 onwards) *Encyclopaedia of Statistical Sciences*, 9 volumes + supplements. Wiley, New York.

O'Hagan, A. (1994) *Kendall's Advanced Theory of Statistics, Volume 2B*. Edward Arnold, London.

Stuart, A., and Ord, K. (1994) *Kendall's Advanced Theory of Statistics, Volume 1*, 6th edn. Edward Arnold, London.

Stuart, A., and Ord, K. (1991) *Kendall's Advanced Theory of Statistics, Volume 2*, 5th edn. Edward Arnold, London.

Useful Addresses

American Statistical Association, 1429 Duke Street, Alexandria, VA 22314–3415, USA.

Belgian Statistical Society, National Statistical Institute, Rue de Louvain, 44 1000 Bruxelles, Belgium.

Deutsche Statistische Gesellschaft Geschäftsstelle in DIW Berlin, Königin-Luise-Str. 5, 14195 Berlin, Germany.

Indian Statistical Institute, 203 Barrackpore Trunk Road, Kolkata, 700016, India.

Institute of Mathematical Statistics, PO Box 22718, Beachwood, OH 44122, USA.

International Biometric Society, 1444 I Street NW Suite 700, Washington, DC 20005–6542, USA.

International Statistical Institute, PO Box 950, 2270 AZ Voorburg, The Netherlands.

The Japan Statistical Society. c/o The Institute of Statistical Mathematics, 4-6-7 Minami-Azabu Minato-Ku, Tokyo 106-8569, Japan.

Netherlands Society for Statistics and Operations Research, PO BOX 2095, 2990 DB Barendrecht, The Netherlands.

New Zealand Statistical Association, PO Box 1731, Wellington, NZ.

Royal Statistical Society, 12 Errol Street, London, EC1Y 8LX, UK.

Società Italiana di Statistica, Salita de Crescenzi 26, 00186 Roma, Italy.

Société Française de Statistique, Institut Henri Poincaré, 11 rue Pierre et Marie Curie, 75231 Paris cedex 05, France.

South African Statistical Association, PO Box 27231, Sunnyside, Pretoria 0132, South Africa.

Statistical Society of Australia, Inc., PO Box 85, Ainslie, ACT 2602, Australia.

Statistical Society of Canada, 1485 Laperrière Street, Ottawa, Ontario K1Z 7S8, Canada.

Oxford Paperback Reference

A Dictionary of Chemistry

Over 4,200 entries covering all aspects of chemistry, including physical chemistry and biochemistry.

'It should be in every classroom and library ... the reader is drawn inevitably from one entry to the next merely to satisfy curiosity.'
School Science Review

A Dictionary of Physics

Ranging from crystal defects to the solar system, 3,500 clear and concise entries cover all commonly encountered terms and concepts of physics.

A Dictionary of Biology

The perfect guide for those studying biology – with over 4,700 entries on key terms from biology, biochemistry, medicine, and palaeontology.

'lives up to its expectations; the entries are concise, but explanatory'
Biologist

'ideally suited to students of biology, at either secondary or university level, or as a general reference source for anyone with an interest in the life sciences'
Journal of Anatomy

Oxford Paperback Reference

Concise Medical Dictionary

Over 10,000 clear entries covering all the major medical and surgical specialities make this one of our best-selling dictionaries.

'"No home should be without one" certainly applies to this splendid medical dictionary'

Journal of the Institute of Health Education

'An extraordinary bargain'

New Scientist

'Excellent layout and jargon-free style'

Nursing Times

A Dictionary of Nursing

Comprehensive coverage of the ever-expanding vocabulary of the nursing professions. Features over 10,000 entries written by medical and nursing specialists.

An A-Z of Medicinal Drugs

Over 4,000 entries cover the full range of over-the-counter and prescription medicines available today. An ideal reference source for both the patient and the medical professional.

Great value ebooks from Oxford!

An ever-increasing number of Oxford subject reference dictionaries, English and bilingual dictionaries, and English language reference titles are available as ebooks.

All Oxford ebooks are available in the award-winning Mobipocket Reader format, compatible with most current handheld systems, including Palm, Pocket PC/Windows CE, Psion, Nokia, SymbianOS, Franklin eBookMan, and Windows. Some are also available in MS Reader and Palm Reader formats.

Priced on a par with the print editions, Oxford ebooks offer dictionary-specific search options making information retrieval quick and easy.

For further information and a full list of Oxford ebooks please visit: www.askoxford.com/shoponline/ebooks/

Oxford Paperback Reference

A Dictionary of Psychology
Andrew M. Colman

Over 10,500 authoritative entries make up the most wide-ranging dictionary of psychology available.

'impressive ... certainly to be recommended'
Times Higher Educational Supplement

'Comprehensive, sound, readable, and up-to-date, this is probably the best single-volume dictionary of its kind.'
Library Journal

A Dictionary of Economics
John Black

Fully up-to-date and jargon-free coverage of economics. Over 2,500 terms on all aspects of economic theory and practice.

A Dictionary of Law

An ideal source of legal terminology for systems based on English law. Over 4,000 clear and concise entries.

'The entries are clearly drafted and succinctly written ... Precision for the professional is combined with a layman's enlightenment.'
Times Literary Supplement